TETRAHEDRON ORGANIC CHEMISTRY SERIES
Series Editors: **J E Baldwin, FRS & P D Magnus, FRS**

VOLUME 11

Organic Syntheses Based on Name Reactions and Unnamed Reactions

Related Pergamon Titles of Interest

BOOKS

Tetrahedron Organic Chemistry Series:
CARRUTHERS: Cycloaddition Reactions in Organic Synthesis
DEROME: Modern NMR Techniques for Chemistry Research
GAWLEY & AUBÉ: Principles of Asymmetric Synthesis
PAULMIER: Selenium Reagents & Intermediates in Organic Synthesis
PERLMUTTER: Conjugate Addition Reactions in Organic Synthesis
SESSLER & WEGHORN: Expanded, Contracted & Isomeric Porphyrins
SIMPKINS: Sulphones in Organic Synthesis
TANG & LEVY: Chemistry of *C*-Glycosides
WILLIAMS: Synthesis of Optically Active Alpha-Amino Acids*
WONG & WHITESIDES: Enzymes in Synthetic Organic Chemistry

JOURNALS

BIOORGANIC & MEDICINAL CHEMISTRY
BIOORGANIC & MEDICINAL CHEMISTRY LETTERS
JOURNAL OF PHARMACEUTICAL AND BIOMEDICAL ANALYSIS
TETRAHEDRON
TETRAHEDRON: ASYMMETRY
TETRAHEDRON LETTERS

Full details of all Elsevier Science publications/free specimen copy of any Elsevier Science journal are available on request from your nearest Elsevier Science office

* In Preparation

Organic Syntheses Based on Name Reactions and Unnamed Reactions

A. HASSNER

Department of Chemistry,
Bar-Ilan University, Israel

and

C. STUMER

Teva Pharmaceutical Industries,
Netanya, Israel

PERGAMON

U.K.	Elsevier Science Ltd, The Boulevard, Langford Lane, Kidlington, Oxford OX5 1GB, U.K.
U.S.A.	Elsevier Science Inc., 655 Avenue of the Americas, New York, NY 10010, U.S.A.
JAPAN	Elsevier Science Japan, Higashi Azabu 1-chome Building 4F, 1-9-15, Higashi Azabu, Minato-ku, Tokyo 106, Japan.

First Edition 1994

Reprinted 1995, 1998

Library of Congress Cataloging in Publication Data

Hassner, Alfred, 1930

Organic syntheses based on name reactions and unnamed reactions / A. Hassner & C. Stumer

p. cm. -- (Tetrahedron organic chemistry series ; v. 11)

Includes index.

1. Organic compounds--Synthesis. I. Stumer, C..

II. Title. III. Series.

QD262. H324 1994 547'. 2--dc20 93-21414

British Library Cataloguing in Publication Data

A catalogue record for this book is available from the British Library

ISBN 0 08 040280 1 Hardcover
ISBN 0 08 040279 8 Flexicover

Printed in Great Britain by BPC Wheatons Ltd, Exeter

CONTENTS

Non nova sed nove

FOREWORD

And these are the names...

The above are the opening words of Exodus, the second book of the Petateuch. Already in ancient times, names were important in association with events. As organic chemistry developed during the 20th century, researchers started associating synthetically useful reactions with the names of discoverers or developers of these reactions. In many cases such names serve merely as a mnemonic, to remember a reaction more easily; there are few chemistry undergraduates who do not know what the Friedel-Crafts reaction is.

In recent years there has been a proliferation of new reactions and reagents that have been so useful in organic synthesis that often people refer to them by name. Many of these are stereoselective or regioselective methods. While the expert may know exactly what the Makosza vicarious nucleophilic substitution, or the Meyers asymmetric synthesis refers to, many students as well as researchers would appreciate guidance regarding such "Name Reactions".

It is in this context that we perceived the necessity to incorporate the older name reactions with some newer name reactions or "unnamed reactions", that are often associated with a name but for which details, references and experimental details are not at everyone's fingertips. This was our inspiration for the current monograph "*Organic Syntheses based on Name Reactions and Unnamed Reactions*".

In particular we thought it would be useful to include cross-references of functional group transformations and an experimental procedure, so that the reader will be able to evaluate the reaction conditions at a glance; for instance is this reaction carried out at room temperature or at 200°C? For 1 h or 5 days? Are special catalysts required? How is the reaction worked up, what yield can be expected?

The choice of which reactions to include is not an easy one. First there are the well known "Name Reactions", that have appeared in various monographs or in the old Merck index. Some of these are so obvious mechanistically to the modern organic chemistry practitioner that we have in fact omitted them; for instance esterification of alcohols with acid chlorides – the Schotten-Baumann procedure. Others are so important and so well entrenched by name, like the Baeyer-Villiger ketone oxidation, that it is impossible to ignore them. In general we have kept older name reactions that are not obvious at first glance.

In some cases we have combined similar reactions under one heading, for instance the Hunsdiecker-Borodin-Cristol-Firth decarboxylative bromination. It is not a simple task to decide whether credit is due to the first discoverer of a reaction or to its developer. Often an improvement on a method is more useful than the original discovery and usually one reaction owes its inception to some previous discovery; *non nova sed nove.*

Except in the case of reactions that have been known for a long time under shared names, we often took the liberty to include in the title, as well as in the references (here to save space), only the name of the major author; for this we apologize to the co-authors, whose contributions are often seminal. For reactions named after contemporary authors, we have tried to consult the authors about choice of examples, etc. This led for instance to the Mannich-Eschenmoser methylination.

Among the newer reactions we have chosen those that are not only synthetically useful but at first glance not immediately obvious transformations. Another criterion was the stereochemical implications of the process. Yet, we admit our own bias in choosing from the plethora of novel transformations that have appeared in the literature over the past 30 years or so. Space limitation was by necessity a criterion. Nevertheless we have included approximately 450 name reactions and 2100 references. We sincerely apologize if we have inadvertedly omitted important reactions.

In all cases we have tried to include the first reported reference, a reference to an experimental procedure and whenever possible a review reference (journal or *Organic Reactions*). In general, we did not include references to books, series of monographs or to *Organic Syntheses*; chemists will of course consult these where available.

Furthermore, we have complied four indices, which should be helpful to the reader:

1. **A names index** with cross references to multiple names;
2. **A reagents index**;
3. **An index to types of reactions**, e.g. alkylations, stereoselective reductions, cyclizations etc.; and
4. Most important for the synthetic chemist is **an index to the synthesis of functional groups**, e.g. *synthesis* of alkenes *from* ketones, as well as *conversion* of ketones *to* alkenes.

We thank our families for their support and understanding during the travail on this book. Special thanks are due to my son Lawrence Hassner for constructive suggestions and valuable help.

We are grateful to the TEVA Pharmaceutical Co. for support of this project.

Alfred Hassner
Carol Stumer

A L D E R (Ene) Reaction

Sigmatropic rearrangement with H-transfer and C-C bond formation (inter or intramolecular) and chiral induction.

1 + 2 (CO_2Me, CO_2Me) → 140°, 16 h → 3 (62%) (CO_2Me, CO_2Me, OH)

1	Alder, K.	*Chem. Ber.*	**1943**	*76*	27
2	Usieli, V.	*J. Org. Chem.*	**1973**	*38*	1703
3	Achamatowicz, O.	*J. Org. Chem.*	**1980**	*45*	1228
4	Snider, B.B.	*J. Org. Chem.*	**1982**	*47*	745
5	Hill, R.	*J. Am. Chem. Soc.*	**1964**	*86*	965
6	Oppolzer, W.	*Angew. Chem. Int. Ed.*	**1978**	*17*	476

Methyl 2-hydroxy-2-carbomethoxy-4-heptenoate (3).[3] A solution of dimethyl mesoxalate **2** (1.46 g, 10 mmol) and 1-pentene **2** (0.70 g, 10 mmol) in CH_2Cl_2 was heated at 140°C for 16 h. The solvent was removed and the residue distilled under reduced pressure. The fraction collected between 90 and 105°C (0.5 torr) was diluted with Et_2O (20 mL), washed with water and dried. The residue after evaporation of the solvent, gave on distillation 1.55 g of **3** (62%), bp 89-90°C (0.2 torr).

ALPER Carbonylation

Carbonylation, hydroformylation ($CO-H_2$) of olefins catalyzed by metal (Pd or Rh) complexes.

N-Bu
1
CO
Pd $(Ph_3P)_4$
N-Bu
O
2 (79%)

NH
3
CO, $HRh(CO)(Ph_3P)_3$
$NaBH_4$ 100° 34 atm
N
O
4 (78%)

1	Alper, H.	*J. Chem.Soc. Chem. Commun.*	**1983**	*102*	1270
2	Alper, H.	*Tetrahedron Lett.*	**1987**	*28*	3237
3	Alper, H.	*J. Org. Chem.*	**1992**	*57*	3328
4	Alper, H.	*J. Am. Chem. Soc.*	**1990**	*112*	2803
5	Alper, H.	*Aldrichimica Acta*	**1991**	*24*	3

N-(n-Butyl)-α-methylene-β-lactam (2).[2] CO was bubbled through $Pd(OAc)_2$ or $Pd(Ph_3P)_4$ (0.136 mmol) in CH_2Cl_2 (4 mL). After 2 min Ph_3P (0.54 mmol) in CH_2Cl_2 (2 mL) is added followed by aziridine **1** in CH_2Cl_2. After 40 h, the solvent was removed in vacuum and the residue purified by prep tlc (silica gel, hexane:~EtOAc 8:1) to yield **2** (79%).

N-Cyclohexyl-2-pyrrolidone (4).[3] N-Cyclohexylallylamine **3** (278 mg, 2 mmol), $NaBH_4$ (75 mg, 2.25 mmol) and $HRh(CO)(Ph_3P)_3$ (18.36 mg, 0.02 mmol) in i-PrOH (0.5 mL) and CH_2Cl_2 (5 mL) was treated with CO at 34.5 atm and heated with stirring for 30 h at 100°C. Quenching with water at 20°C, extraction, evaporation and chromatography (silica, hexane:EtOAc 19:1) afforded 260 mg of **4** (78%).

AMADORI Glucosamine Rearrangement

Conversion of N-glucosides of aldoses to N-glucosides of ketoses.

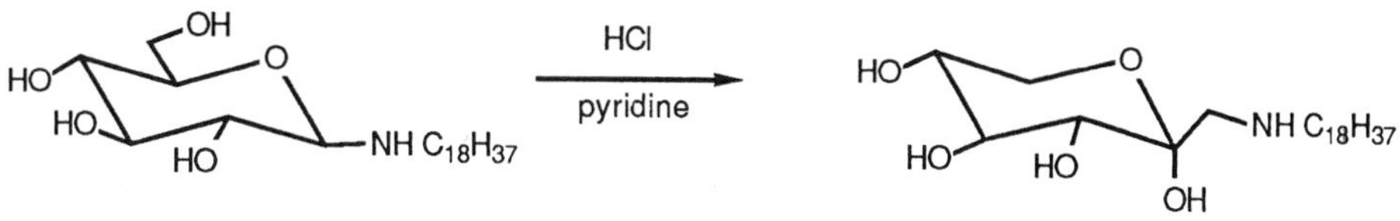

1	Amadori, M.	*Atti. accad. Lincei*	**1925**	*2*	337(6)
2	Cameron, E.J.	*J. Am. Chem. Soc.*	**1926**	*48*	2737
3	Carson, M.	*J. Am. Chem. Soc.*	**1956**	*78*	3728
4	Ames, G.R.	*J. Org. Chem.*	**1062**	*27*	390

1-Deoxy-1-N-octadecylamino-D-fructose (2).[4] A solution of N-octadecyl-D-glucosylamine **1** (2.5 g, 5.8 mmol) in pyridine (25 mL) was mixed with 0.28 N HCl (21 mL) in pyridine (5.8 mmol). Next day, water (200 mL) was added and the product was extracted with a mixture of 4:1 EtOAc:BuOH (200 mL). The organic layer was washed with water and dried (Na_2SO_4). Evaporation of the solvent afforded 0.6 g of **2** (25%), mp 104-107°C.

ANGELI - RIMINI Hydroxamic Acid Synthesis

Synthesis of hydroxamic acids from aldehydes and N-sulfonylhydroxylamine; also used as a color test for aldehydes.

CHO ... Cl (1) + SO_2NHOH (2) —MeONa→ O=C−NHOH ... Cl (3)

1 2 3

1	Angeli, A.	*Gazz. Chim. Ital.*	**1896**	*26*	17 (II)
2	Rimini, E.	*Gazz. Chim. Ital.*	**1901**	*31*	84 (II)
3	Hassner, A.	*J. Org. Chem.*	**1970**	*35*	1952
4	Lwowsky, W.	*Angew. Chem. Int. Ed.*	**1967**	*6*	897

p-Chlorobenzohydroxamic acid (3).[3] To an ice cooled solution of N-hydroxybenzene-sulfonamide **2** (365 mg, 2.1 mmol) in MeOH, a 1.93 M NaOMe solution (2.18 mL, 4.2 mmol) was added dropwise under stirring. A solution of p-chlorobenzaldehyde **1** (281 mg, 2 mmol) in MeOH (2 mL) was added and the reaction mixture was warmed to 20°C and stirred for an additional 2 h. The solution was concentrated in vacuum, diluted with Et_2O (100 mL) and extracted twice with 2 M NaOH. The organic phase yielded impure starting materials (45 mg). The aqueous phase was acidified with 36% HCl to pH 7-8 and extracted twice with EtOAc. The dried solution ($MgSO_4$), after evaporation of the solvent, gave 225 mg of **3** (68%), mp 193-195°C. The product gives a red color with $FeCl_3$.

A R B U Z O V - M I C H A E L I S Phosphonate Synthesis

Ni catalyzed phosphonate synthesis from phosphites and aryl halides. Reaction of alkyl halides with phosphites proceeds without nickel salts.

$$\underset{\mathbf{1}}{NiCl_2 + (EtO)_3P} \xrightarrow{150°} \underset{\mathbf{2}}{[(EtO)_3P]_4Ni} \xrightarrow[\mathbf{1},\ 150°]{Ph\text{-}I} \underset{\mathbf{4}\ (94\%)}{Ph\text{-}P(O)(OEt)_2}$$

$$(iPro)_3P + Me\text{-}I \xrightarrow{reflux} MeP(O)(OiPr)_2 \quad (85\%)$$

$$\text{pyridine-}N^+\text{-}O^- \longrightarrow \text{pyridine-}N^+\text{-}OMe\ MeSO_4^- \xrightarrow{LiP(O)(OEt)_2} \text{2-pyridyl-}P(O)(OEt)_2$$

1	Michaelis, A.	*Chem. Ber.*	**1898**	*31*	1048
2	Arbuzov, A.	*J. Russ. Phys. Chem. Soc.*	**1906**	*38*	687
3	Balthazar, T.M.	*J. Org. Chem.*	**1980**	*45*	5425
4	Montero, J.L.	*Tetrahedron Lett.*	**1987**	*28*	1163
5	Brill, Th.B.	*Chem. Rev.*	**1984**	*84*	577
6	Kosolapov, G.M.	*Org. React.*	**1951**	*6*	276
7	Kem, M.K.	*J. Org. Chem.*	**1970**	*36*	5118
8	Redmore, D.	*J. Org. Chem.*	**1981**	*46*	4114

Tetrakis(triethylphosphite)nickel (0) (2).[3] Stirred $NiCl_2$ (5 g) and triethylphosphite **1** (60 mL) was heated and maintained for 1 h at 150°. The solid was filtered, triturated with MeCN and washed with MeOH to give 4.6 g of **2**, mp 106-109°C.

Diethyl phenylphosphonate (4). To **2** (10 mg) in iodobenzene (10.0 g, 49 mmol) at 160° was added slowly **1** (9.37 g, 56.4 mmol). The solution (red upon each addition of **1**), faded to yellow and EtI was distilled. Vacuum distillation afforded 9.88 g of **4** (94%), bp 94-101°C (0.1 mm).

ARENS - VAN DORP Cinnamaldehyde Synthesis

Synthesis of cinnamaldehydes from aryl ketones.

$HC \equiv C-OC_2H_5$ (**2**) $\xrightarrow{EtMgBr}$ $\xrightarrow{Ph-CO-Me}$ (**1**) $Ph(Me)C(OH)-C \equiv C-OC_2H_5$ (**3**) (48%)

3 $\xrightarrow[BaSO_4]{H_2/Pd}$ $Ph(Me)C(OH)-CH=CH-OC_2H_5$ (**4**) (98%) $\xrightarrow{HCl}$ $Ph-C(Me)=CH-CHO$ (**5**)

1	Arens, J.F., v. Dorp, A.D.	*Nature*	**1947**	*160*	189
2	Isler, O.	*Helv. Chim. Acta*	**1956**	*39*	259
3	Kell, P.S.	*J. Am. Chem. Soc.*	**1959**	*81*	4117

3-Phenyl-3-hydroxy-1-ethoxybutyne (3).[3] Ethoxyacetylene **2** (12.5 g, 0.18 mol) was added to EtMgBr (from 4.34 g of Mg and 19.5 g EtBr in Et_2O) over 10 min stirring and ice cooling. After 15 min acetophenone **1** (21.5 g, 0.18 mol) in Et_2O (50 mL) was added over 1 h under ice cooling. Quenching (100 mL of 10% NH_4Cl) and distillation gave 15 g of **3** (48%), bp 98-100°C (0.1 mm).

3-Phenyl-3-hydroxy-1-ethoxybutene (4). Hydrogenation of **3** (13.1 g, 0.075 mol) in EtOAc (100 mL) with Pd/$BaSO_4$ (200 mg) gave after vacuum distillation 13 g of **4** (98.5%), bp 74.5-78°C (0.1 mm).

β-Methylcinnamaldehyde(semicarbazone) (5). **4** (1.0 g, 5.7 mmol) in 0.1 N HCl was refluxed for 15 min. The cooled solution was treated with 0.1 N NaOH, semicarbazide HCl (500mg) and NaOAc (500mg). After addition of EtOH (6 mL) and heating at 30°C the semicarbazone of **5**, mp 205-5-206.5°C separated.

ARNDT - EISTERT Homologation

Homologation of carboxylic acids or ketones.

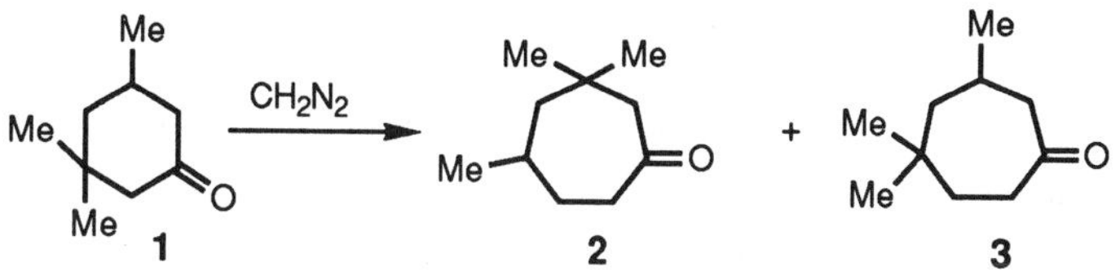

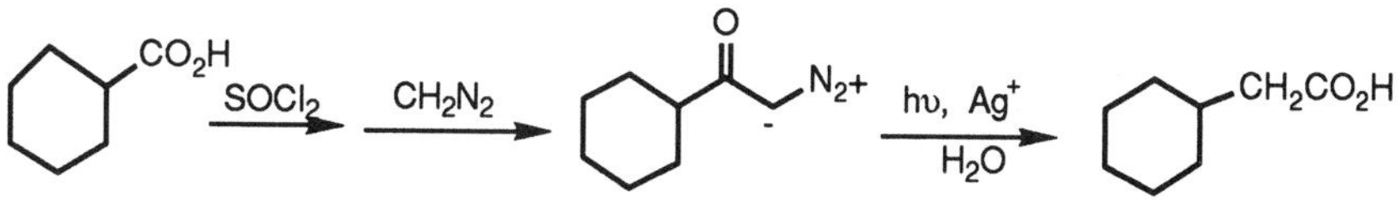

1	Eistert, B., Arndt, F.	*Chem. Ber.*	**1927**	*60* 1364
2	Barbier, F.	*Helv. Chim. Acta.*	**1940**	*23* 523
3	Gokel, G.	*Synthesis*	**1976**	181
4	Aryama, T.	*Chem. Pharm. Bull.*	**1981**	*29* 3249
5	Smith, A.B.	*J. Am. Chem. Soc.*	**1986**	*108* 3110
6	Weigand, F.	*Angew. Chem.*	**1960**	*72* 535
7	Bachmann, W.E.	*Org. React.*	**1942**	*1* 38

3,3,5 and **3,5,5-Trimethylcycloheptanone (2)** and **(3)**.[2] To a cooled solution of 3,5,5-trimethylcyclohexanone **1** (100 g, 0.71 mol) in MeOH (225 mL) and 50% KOH was added in small portions nitrosomethylurea (74 g, 0.68 mol) under stirring and cooling (0°C). After each addition the solution was allowed to become colorless before the next portion was added. The reaction mixture was allowed to stand for several hours and was filtered and neutralized with AcOH. The solvent was removed and the residue distilled to afford a mixture of **2** and **3**, bp 70-95°C/11 mm.

For an alternate procedure for generation of diazomethane from hydrazine, chloroform and KOH see ref. 4

ATHERTON - TODD Phosphoramidate Synthesis

Synthesis of phosphoramidates from formamides and dialkyl phosphite.

$$C_6H_5NH\text{-}CHO + HP(O)(OC_2H_5)_2 \xrightarrow{R_4N^{+}\,{}^{-}OH} [C_6H_5\text{-}N(CHO)\text{-}P(O)(OC_2H_5)_2] \rightarrow C_6H_5\text{-}NH\text{-}P(O)(OC_2H_5)_2$$

1 + 2 → [intermediate] → 3

1	Atherton, F.R., Todd, A.R.	*J. Chem. Soc.*	**1945**		660
2	Wadsworth, W.S.	*J. Am. Chem. Soc.*	**1962**	*84*	1316
3	Zwierzak, A.	*Synthesis*	**1982**		922
4	Lukanow, L.K.	*Synthesis*	**1985**		971

Diethyl N-Phenylphosphoramidate (3).[4] To an ice cooled stirred suspension of formylanilide **1** (0.605 g, 5 mmol) in CCl_4 (25 mL) was added a solution of 30% NaOH (10 mL) and benzyltriethylammonium bromide (0.2 g). Diethyl phosphite **2** (0.828 g, 6 mmol) in CCl_4 (5 mL) was added dropwise. Stirring continued for 1 h at ice bath temperature and 4 h at room temperature. The separated organic layer was dried (Na_2SO_4) and the solvent removed in vacuum to give crude **3**, which was purified by recrystallization to yield 0.687 g (60%), mp 96-97°C.

A U W E R S Flavone synthesis

Synthesis of benzopyran-4-ones (flavones) from o-hydroxychalcones or from benzofuran-3-ones.

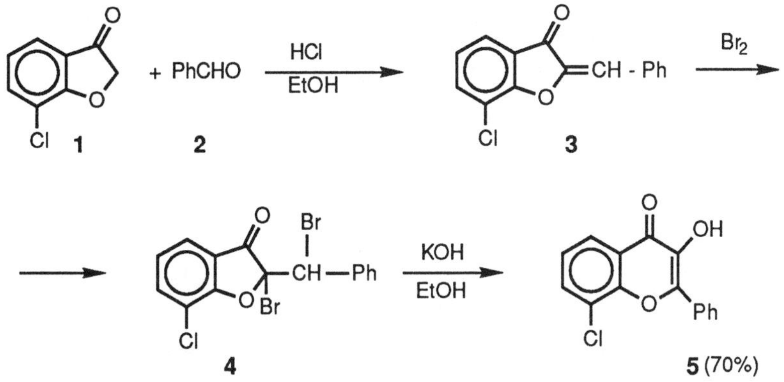

1 Auwers, K. *Chem. Ber.* **1908** *41* 4233

2 Minton, T.H. *J. Chem. Soc.* **1922** *121* 1598

3 Ingham, B.H. *J. Chem. Soc.* **1931** 895

4 Acharya, B.C. *J. Chem. Soc.* **1940** 817

7-Chloro-2-benzylidenecoumaran-3-one (3)[2] A solution of coumaranone **1** (2.0 g, 12 mmol) and **2** (3.2 g, 30 mmol) in EtOH was heated at 60°C and 36% HCl (1 mL) was added slowly. On cooling, **3** crystallized. The filtered and dried product melted at 143°C.

7-Chloro-2-benzylidenecoumaran-3-one dibromide (4). To a solution of **3** (5.0 g, 20 mmol) in $CHCl_3$ (10 mL) was added a solution of bromine (3.2 g, 20 mmol) in $CHCl_3$ (10 mL). After 24 h the solvent was removed at 20-25°C and the residue recrystallized from HOAc, mp 147°C.

8-Chloroflavonol (5). A solution of **4** (2.0 g, 5 mmol) in EtOH (150 mL) was treated with 0.1 N KOH (100 mL). The mixture was boiled for 10 min and the product was precipitated with water. Recrystallization from HOAc yielded 1 g of **5** (70%), mp 187°C.

B A E R - F I S C H E R Amino sugar synthesis

Synthesis of 3-nitro and 3-amino sugars by aldol condensation of sugar-derived dialdehydes with nitroalkanes.

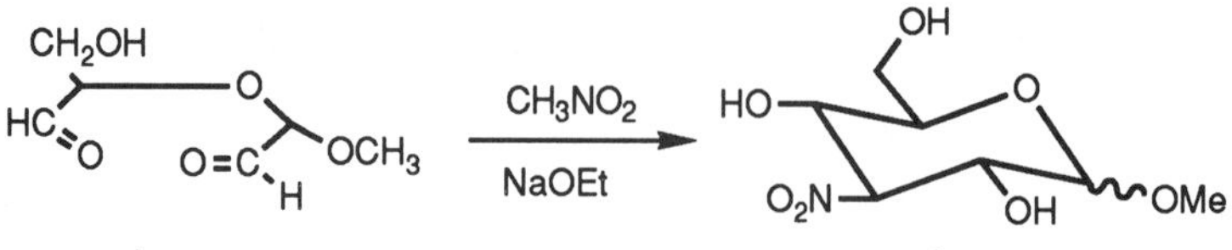

CH_3NO_2, NaOEt: **1** → **2**

1. $NaIO_4$; 2. Me CH_2NO_2: 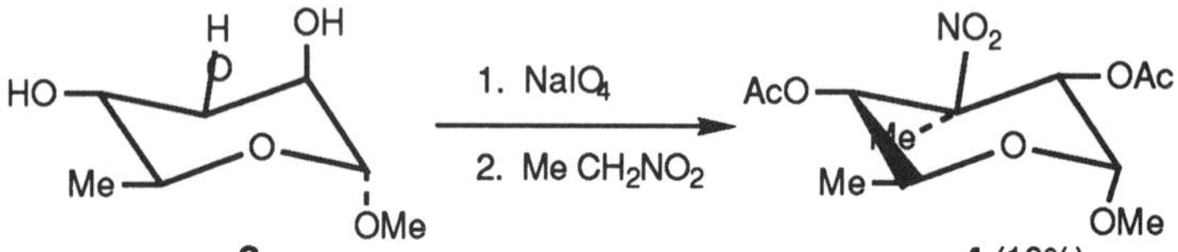 **3** → **4** (19%)

1	Baer, H.H., Fischer, H.O.L.	*Proc. Nat. Acad. Sci. USA*	**1958**	*44*	991
2	Baer, H.H.	*Adv. Carbohydr. Chem.*	**1969**	*24*	67
3	Brimacombe, J.S.	*J. Chem. Soc. Perkin I*	**1974**		62
4	Santoyo-Gonzales, F.	*Synlett*	**1990**		715

Methyl 2,4-Di-O-acetyl-3,6-dideoxy-3-C-methyl-3-nitro-α-L-glucopyrannoside (4).[3] Methyl-α-L-rhamnopyrannoside **3** (100 g, 0.55 mol) in water (1000 mL) was treated with $NaIO_4$ (200 g, 0.83 mol) at 20°C. After 3 h $NaHCO_3$ was added, the mixture poured into EtOH (4000 mL) and filtered. The filtrate was concentrated and extracted with hot EtOH. The extract was cooled, filtered and treated with nitroethane (104.5 g, 1.4 mol) followed by a solution of Na (12 g, 0.52 at. g) in EtOH (750 mL). After 4 h at 20°C the solution was treated with CO_2, filtered and concentrated. The mixture was treated with pyridine (400 mL) and Ac_2O (300 mL) at 20°C for 12 h. Workup left a residue which dissolved in Et_2O:petroleum ether (1:1) (500 mL) and cooled afforded 36 g of **4** (19%), mp 137-138°C, $[\alpha]_D$=-130°C (c 1).

B A E Y E R Oxindole Synthesis

Synthesis of oxindole from o-nitrophenylacetates.

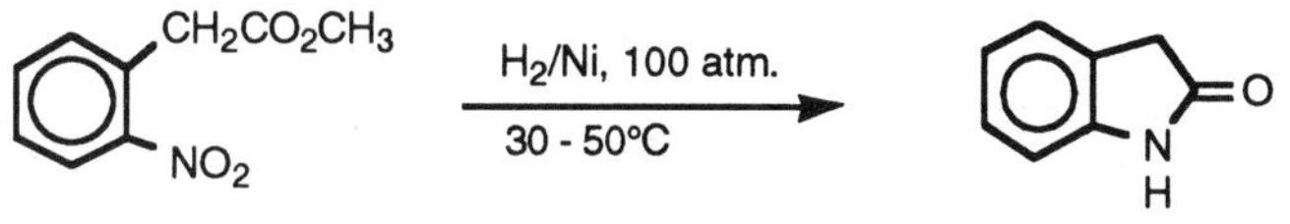

1 2 (100%)

1	Baeyer, A.	*Chem. Ber.*	**1878**	*11*	582

B A E Y E R Diarylmethane Synthesis

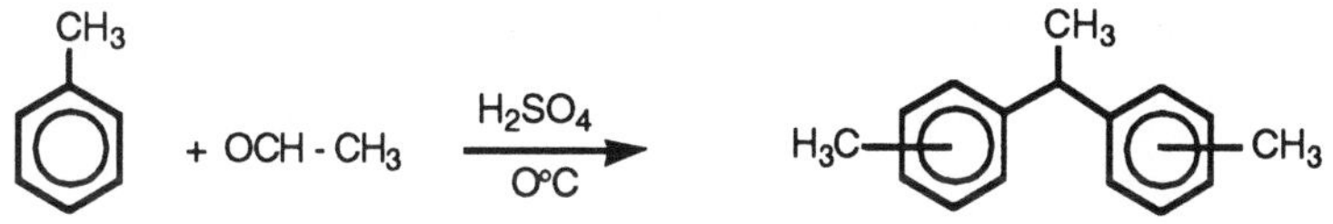

1 2 3 (88%)

1	Baeyer, A.	*Chem. Ber.*	**1872**	*5*	280
2	Matsuda, K.	*J. Org. Chem.*	**1963**	*27*	3256
3	Schnell, W.	*Angew. Chem. Int. Ed.*	**1962**	*75*	622

Ditolylethane (isomers mixtures) (3).[2] To ice cooled toluene **1** (557 g, 642 mL) was added simultaneously acetaldehyde **2** (77 g, 1.67 mol) in 420 mL of **1** and 95% H_2SO_4 (655 g) in 2 h. After stirring for 1 h, water (1000 mL) was added cautiously and the organic layer washed (5% NaOH, water). Vacuum distillation gave 101 g of **3** (88%), bp 150°C (10 mm).

BAEYER Pyridine Synthesis

Synthesis of pyridines from pyrones

1 → (1. Me_2SO_4, 2. $HClO_4$) → 2 → ($(NH_4)_2CO_3$) → 3

1	Baeyer, A.	*Chem. Ber.*	**1910**	*43*	2337
2	Nenitzescu, C.D.	*Liebigs Ann.*	**1959**	*625*	74
3	Cavallieri, L.F.	*Chem. Rev.*	**1947**	*41*	525
4	Dimroth, K.	*Angew. Chem.*	**1960**	*72*	331

2-Methyl-4,5-dimethoxypyridine (3).[2] **1** (70 g, 0.2 mol) and dimethyl sulfate (31 g, 0.25 mol) were heated for 2 h at 50°C and poured into 2x excess of ice cold 20% $HClO_4$. After 2 h the product was filtered and added to 10% $(NH_4)_2CO_3$ (175 mL), saturated with $(NH_4)_2SO_4$ and extracted with EtOAc. Evaporation and distillation afforded **3**, bp 75-80°C (1 mm); hydrochloride, mp 164°C.

BAEYER - DREWSON Indoxyl Synthesis

Conversion of o-nitrobenzaldehyde and ketones to indoxyls.

1 → (1N NaOH, acetone) → → 2 (46%)

1	Baeyer, A., Drewson, W.	*Chem. Ber.*	**1882**	*15*	2856
2	Hinkel, L.E.	*J. Chem. Soc.*	**1932**		985
3	Hassner, A.	*Tetrahedron Lett.*	**1962**		975

BAEYER - VILLIGER Ketone Oxidation

Synthesis of esters or lactones from ketones with retention of configuration.

1 → (H_2O_2, HOAc) → 2

1	Baeyer, A., Villiger, V.	*Chem. Ber.*	**1899**	*32*	3625
2	Hassner, A.	*J. Org. Chem.*	**1978**	*43*	1774
3	Sarapanami, C.R.	*J. Org. Chem.*	**1986**	*51*	2322
4	Hassal, C.H.	*Org. React.*	**1957**	*9*	73

Bicyclic lactone (2).[2] To **1** (0.79 g, 5 mmol) in 90% AcOH (5 mL) at 0°C was added 30% H_2O_2 (2.5 mL) in 90% AcOH (3 mL), kept at 0°C for 24 h poured into H_2O and extracted with petroleum ether (bp 40-60°C). The organic extract was washed ($NaHSO_3$, H_2O) and evaporated to give 0.57 g of **2** (65%).

BAEYER - VILLIGER Tritylation

Introduction of a triphenylmethyl group into an aromatic ring.

PhOH (**1**) + Ph_3COH (**2**) → (H_2SO_4, HOAc, rt) → **3** (90%) (4-CPh_3-phenol)

1 + **2** → (H_2SO_4 cat., 100°) → **4** (2-CPh_3-phenol)

1	Baeyer, A., Villiger, V.	*Chem. Ber.*	**1902**	*35*	3013
2	Hardy, D.V.M.	*J. Chem. Soc.*	**1929**		1000
3	Shulgin, A.T.	*J. Org. Chem.*	**1962**	*27*	3868

BAKER - VENKATARAMAN Flavone synthesis

Rearrangement of aromatic keto esters of phenols to flavones.

HO, OH, COMe **1** — PhCOCl **2** → PhCOO, OCOPh, COMe **3** (73%) — K_2CO_3 / PhMe, 100° →

PhCOO, OH O, Ph, O **4** (82%) — HOAc / Na OAc → HO, O, Ph, O **5** (90%)

1	Baker, W.	*J. Chem. Soc.*	**1933**		1381
2	Venkataraman, K.	*J. Chem. Soc.*	**1934**		1767
3	Kramm, E.	*J. Org. Chem.*	**1984**	*49*	3212
4	Levine, E.	*Chem. Rev.*	**1954**	*54*	493

Resacetophenone dibenzoate (3).[1] Resacetophenone **1** (30.4 g, 0.2 mol) benzoyl chloride **2** (5.2 g, 0.4 mol) and pyridine (60 mL), were heated on a steam bath for 15 min. The mixture was treated with water, dil HCl and then EtOH to give 53 g of **3** (73.6%), mp 80°C (MeOH).

1-Phenyl-3(2-hydroxy-4-benzoyphenyl)-propandione-1,3 (4). A mixture of **3** (20 g, 55 mmol), PhMe (200 mL) and K_2CO_3 (60 g) was stirred on a steam bath for 2 h. After usual work up 16.5 g of **4** (82%) was obtained.

7-Hydroxyflavone (5). **4** (5 g, 13.8 mmol) in AcOH (50 mL) and NaOAc (10 g) was boiled for 6 h, then poured into water to give 2.6 g of **5** (90%), mp 240°C.

BALLY - SCHOLL Benzanthrene Synthesis

Synthesis of polycondensed aromatics from antraquinone and triols.

O HO OH HO + O

H_2SO_4

Cu, 80°

O

1 **2** **3** (5%)

1	Bally, O.	*Chem. Ber.*	**1905**	*38*	194
2	Scholl, R.	*Chem. Ber.*	**1911**	*44*	1656
3	Warren, F.L.	*J. Chem. Soc.*	**1938**		401
4	Allen, C.F.H.	*Org. Synth.*	Coll. Vol.	*II*	62

BALSON Alkylation

Alkylation of aromatics with HF (see Friedel-Crafts).

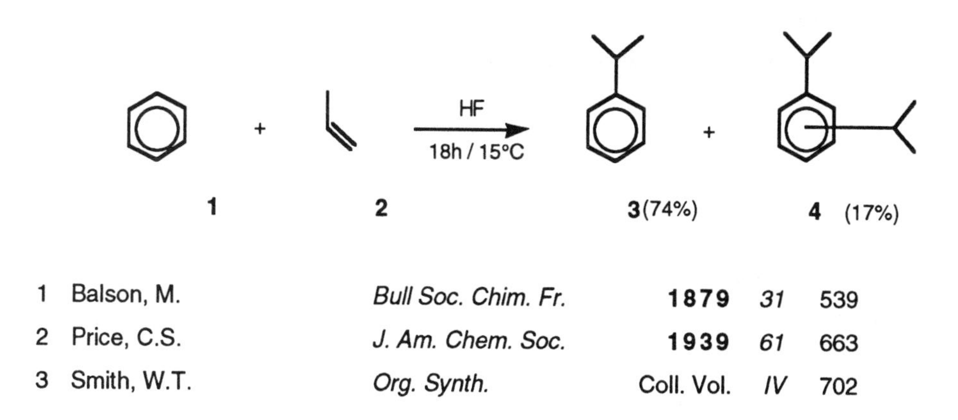

1	Balson, M.	*Bull Soc. Chim. Fr.*	**1879**	*31*	539
2	Price, C.S.	*J. Am. Chem. Soc.*	**1939**	*61*	663
3	Smith, W.T.	*Org. Synth.*	Coll. Vol.	*IV*	702

B A M B E R G E R Imidazole cleavage

Cleavage of imidazoles to enediamides, useful in synthesis of 2substituted imidazoles.

HN N
CH$_2$
CH$_2$
COOC$_2$H$_5$
1
O
PhCCl
NaHCO$_3$
PhCONH NHCOPh
C=CH
CH$_2$
CH$_2$
COOC$_2$H$_5$
2
Δ
Ph O
N N-C-Ph
CH$_2$
CH$_2$
COOC$_2$H$_5$

1	Bamberger, E.	*Liebigs Ann.*	**1893**	*273*	342
2	Babad, E.	*J. Heterocycl. Chem.*	**1969**	*6*	235
3	Grace, M.E.	*J. Am. Chem. Soc.*	**1980**	*102*	6784
4	Kimoto, H.	*J. Org. Chem.*	**1978**	*43*	3403
5	Altman, J.	*J. Chem. Soc. Perkin I*	**1984**		59

Ethyl 4,5-Dibenzamidopent-4-enoate (2)[5] Ethyl 3-imidazol-4(5)-ylpropanoate **1** (9.2 g, 54 mmol) in EtOAc (140 mL) was treated with benzoyl chloride (15.7 g, 112 mmol) in EtOAc (40 mL) and 1M $NaHCO_3$ (380 mL) added simultaneously in 1 h under ice-cooling. The reaction mixture was stirred for 1 h, then a further portion of benzoyl chloride (15.7 g, 112 mmol) in EtOAc) and 1M $NaHCO_3$ (280 mL) was added in the same manner, followed by an additional portion of 1M $NaHCO_3$ (200 mL). The reaction mixture was stirred for 24 h, then the organic layer was separated, concentrated, and the residue dissolved in THF (300 mL). The THF solution was stirred with 10% $NaHCO_3$ (600 mL) for 24 h to decompose any N-formyl intermediate and to remove benzoic acid. Extraction with EtOAc, drying (Na_2SO_4), solvent evaporation and recrystallization of the residue from EtOAc:hexane afforded 16.24 g of **2** (84%), mp 128-129°C.

BAMBERGER Phenylhydroxylamine Rearrangement

Rearrangement of N-arylhydroxylamine to aminophenol.

H-N-OH (1) —10% H_2SO_4, 95°, 45 min→ [H^+ H-N-OH → NH, +] —H_2O→ NH_2, OH 2 (70%)

1 Bamberger, E.	*Chem. Ber.*	**1894**	*27*	1348
2 Hughes, E.D.	*Quart. Rev. (London)*	**1925**	*6*	45

BAMBERGER Benzotriazine Synthesis

From pyruvic acid hydrazone **2** and aryldiazoamine salts **1**.

NH_2, OMe —HNO_2, 0°→ N_2^+, OMe (1) —N-NH_2, CO_2H (2), KOH→ —HOAc, H_2SO_4, 95°→ N, Me, N, N, OMe (3)

1 Bamberger, E.	*Chem. Ber.*	**1892**	*25*	3201
2 Abramovitch, R.A.	*J. Chem. Soc.*	**1955**		2326

BAMFORD - STEVENS - CAGLIOTI - SHAPIRO Olefination

Conversion of ketones to olefins via tosyl hydrazones with NaOR, LAH, LDA or BuLi.

NaOMe, Δ

N-N-Ts
H

1 → **2** (91%)

LiAlH$_4$, THF
24 h, Δ

HO, =N-NH-Ts

3 → **4** (81%)

1	Bamford, W., Stevens, T.,	*J. Chem. Soc.*	**1952**		4735
2	Farnum, D.G.	*J. Org. Chem.*	**1973**	*28*	870
3	Nikon, A.	*J. Org. Chem.*	**1981**	*46*	4692
4	Stadler, H.	*Helv. Chim. Acta*	**1984**	*67*	1379
5	Shapiro, R.H.	*Org. React.*	**1976**	*23*	405
6	Caglioti, R.	*Tetrahedron Lett.*	**1962**		1261
7	Caglioti, R.	*Tetrahedron*	**1963**	*19*	1127
8	Nikon, A.	*J. Org. Chem.*	**1970**	*35*	1509

(E)-Spiro[12,11]tetracos-13-ene (2).[3] Spiro tosylhydrazone **1** (0.275 g, 0.53 mmol) in diglyme (8 mL) under N_2 was treated with MeONa (0.11 g, 2 mmol). After1 h reflux the mixture was extracted with Et_2O, the extract washed (brine) and evaporated to afford 0.16 g of **2** (91%), mp 64-66°C, (pentane).

Coriolane derivative (4).[8] To a suspension of $LiAlH_4$, (0.5 g, 13.2 mmol) in THF (12 mL) under N_2 was added **3** (0.5 g, 1.14 mmol). After 24 h reflux the mixture was washed (water, 15% NaOH and water) extracted with Et_2O and the organic extract concentrated. The oily residue, 0.22 g of crude **4** (81%), was purified by sublimation and recrystallization ($EtOH/H_2O$) to give 0.11 g of **4** (40%), mp 101-101.5°C.

B A R B I E R Reaction

In situ Grignard generation in the presence of an electrophile.

1 —Mg→ 2 (70%)

Me_2SiCl_2 + Mg + allyl bromide → $Me_2Si(CH_2CH{=}CH_2)_2$ (94%)

—Mg→

1	Barbier, P.	*C.R.*	**1899**	*128*	110
2	Grignard, V.	*C.R.*	**1900**	*130*	1322
3	Ashby, R.	*Pure & Appl. Chem.*	**1980**	*52*	545
4	Huang, X.Z.	*Tetrahedron Lett.*	**1988**	*29*	1395
5	Blomberg, C.	*Synthesis*	**1977**		18
6	Hassner, A.	*J. Organomet. Chem.*	**1978**	*156*	227

Cyclopentanone (2).[5] Magnesium (1.2 g, 0.05 at g) was heated under Ar. After cooling Et_2O (10 mL) was added followed by 5-cyano-1-iodobutane **1** (10 g, 0.05 mol), so that the mixture refluxed gently. A slight coloration disappeared on addition of a few drops of 1,2-dibromoethane. The reaction mixture was stirred for another 12-15 h at 20-25°C and hydrolyzed (ice-NH_4Cl). The Et_2O layer was washed with $NaHSO_3$ solution. The solvent was evaporated and the residue refluxed 24 h in a solution of oxalic acid (10 g) in water (60 mL). After extraction with Et_2O, evaporation of the solvent and distillation there was obtained 2.56-3.4 g of **2** (61-79%), bp 130-131°C.

BARBIER - WIELAND Degradation

A multi-step procedure for chain degradation of esters.

$$\mathbf{1} \xrightarrow[Ac_2O]{Ph\,MgBr\ \mathbf{2}} \mathbf{3} \xrightarrow[NaIO_4]{RuO_4} \mathbf{4}\ (80\%)$$

1	Barbier, P.	*C.R.*	**1913**	*156*	1443
2	Wieland, E.	*Chem. Ber.*	**1912**	*45*	484
3	Sarel, S.	*J. Org. Chem.*	**1959**	*24*	2081
4	Fetisson, M.	*C.R.*	**1961**	*252*	139
5	Djerassi, C.	*Chem. Rev.*	**1946**	*38*	526

3α-Acetoxy-24,24-diphenylchol-23-ene (3).[3] Methyl 3α-hydroxycholanate **1** (78 g, 0.2 mol) was refluxed with PhMgBr **2** (3 mol) in benzene for 24 h. The product was acetylated with Ac_2O (60 mL) in pyridine (100 mL) and after evaporation of the solvents, the residue was refluxed in AcOH (200 mL) for 20 h. Cooling gave **3**, recrystallizated from Me_2CO, mp 160°C, α_D^{20} = +67°.

3-α-Acetoxynorcholanic acid (4).[3] **3** (2 g, 3.7 mmol) in acetone (200 mL) was treated with RuO_4 (120 mg) and 5% $NaIO_4$ at 20°. $NaIO_4$ (4.5 g, 0.021 mol) was added in portions over 4 h. A few mL of iPrOH was added, the catalyst was filtered, the solvent evaporated and the residue treated with water. Extraction with Et_2O and reextraction with aq. Na_2CO_3 gave after chromatography (SiO_2, PhH) 1.5 g of **4**, from acetone 1.20-1.25 g (80-83%), mp 177-178°C, α_D^{20} = 51°.

B A R T - S C H E L L E R Aromatic arsonylation

Aromatic arsonylation substitution of an aromatic amine by As via diazonium salts.

NH_2 / NO_2 **1** —(HNO_2, $NaBF_4$)→ BF_4^- $\overset{+}{N}\equiv N$ / NO_2 **2** —(AsO_3Na_3, $CuCl_2$ 60°)→ AsO_3Na_3 / NO_2 **3** (75%)

1	Bart, H.	*Ger. Pat.*	**1910**		250.264
		Frd.	**1910**	*10*	1254
2	Scheller	*Brit. Pat.*	**1927**		261.026
		C.A.	**1927**	*21*	3371
3	Ruddy, A.V.	*J. Am. Chem. Soc.*	**1942**	*64*	823
4	Cowdry, W.A.	*Quart. Rev.*	**1952**	*6*	363

S C H I E M A N N AromaticFluorination

Substitution of an aromatic amino group by fluorine via a diazonium salt using fluoroborates.

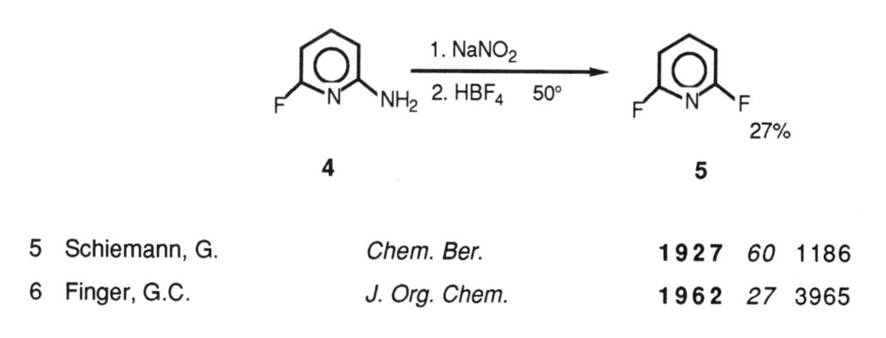

5	Schiemann, G.	*Chem. Ber.*	**1927**	*60*	1186
6	Finger, G.C.	*J. Org. Chem.*	**1962**	*27*	3965

B A R T O N Nitrite photolysis

Long range functionalization of alcohols via nitrites leading to γ-hydroxy oximes.

.NO

ONO H hν O. H H NO O NOH OH

O OAc HO -H NOCl ONO O OAc hν HON O OAc HO -H

1 2 3 (21%)

1	Barton, D.H.R.	*J. Am. Chem. Soc.*	**1960**	*82*	2640
2	Barton, D.H.R.	*J. M. Chem. Soc.*	**1961**	*83*	4076
3	Barton, D.H.R.	*Pure Appl. Chem.*	**1968**	*16*	1
4	Baldwin, S.W.	*J. Am. Chem. Soc.*	**1982**	*104*	4990
5	Barton, D.H.R.	*Aldrichimica Acta*	**1990**	*23*	3

Aldosterone acetate oxime (3).[1] Corticosterone acetate **1** in dry pyridine was treated with excess of NOCl at 20°C to give corticosterone acetate-11-nitrite **2**, mp 176°C α_D = 316° ($CHCl_3$). The nitrite (4 g) in PhMe (200 mL) was irradiated at 32°C under pressure of N_2 for 2-5 h. The IR band of nitrite disappeared. A crystalline product was obtained, 885 mg of **3** (21.2%), mp 175-194°C, α_D = +198°C.

BARTON Deamination

Free radical deamination of primary amines via isocyanides.

$$H_3C(CH_2)_{16}\text{-}CH_2\text{-}NH_2 \xrightarrow{HC(=O)\text{-}O\text{-}C(=O)CH_3} H_3C(CH_2)_{16}CH_2NH\text{-}CHO \xrightarrow{pTsCl}$$

1 — **2** (91%)

$$H_3C(CH_2)_{16}CH_2\text{-}N\equiv C \xrightarrow[AIBN]{n\text{-}Bu_3SnH} H_3C(CH_2)_{16}\text{-}CH_3$$

3 (95%) — **4** (81%)

1	Barton, D.H.R.	*J. Chem. Soc. Perkin I*	**1980**		2657
2	Swindell, C.S.	*J. Org. Chem.*	**1990**	*55*	3
3	Barton, D.H.R.	*Aldrichimica Acta*	**1990**	*23*	3

1-Formamidooctadecane (2).[1] To a saturated solution of octadecylamine **1** (2 g, 7.43 mmol) in Et_2O-pentane was added acetic formic anhydride (0.8 g, 10 mmol) dropwise at 0°C. After 1 h pentane was added to turbidity and the mixture was cooled to -20°C. Formamide **2** was filtered and recrystallized from MeOH to give 2 g of **2** (91%), mp 66°C.

1-Isocyanooctadecane (3). A solution of **2** (0.712 g, 2.39 mmol) in pyridine (30 mL) was treated with pTsCl (0.77 g, 4.05 mmol) and after 2 h was poured onto ice water. The precipitate was filtered, washed and recrystallized ($CHCl_3$-MeOH) to yield 0.635 g of **3** (95%), mp 35°C.

Octadecane (4). A solution of **3** (0.279 g, 1 mmol) and azoisobutyronitrile (AIBN) (0.1 g) in dry xylene (50 mL) was added dropwise to a solution of tri-n-butyl stannane (0.64 g, 2.2 mol equiv). A solution of AIBN (0.1 g) in xylene (50 mL) was slowly added at 80°C over 5 h. The solvent was removed in vacuum, the residue dissolved in pentane and iodine in pentane was added until the iodine colour persisted. The solvent was evaporated and **4** was isolated by preparative TLC (silica gel, pentane). Sublimation in vacuum gave 0.205 g of **4** (81%), mp 29°C.

B A R T O N Decarboxylation

Decarboxylation of a mixed anhydride (thiohydroxamic-carboxylic) and interception of radicals as a sulfide, selenide or bromo derivative.

$CH_3(CH_2)_{14}$-COCl + NaO-N(thiazolethione) → R-C(O)-O-N(thiazolethione) —PhS-SPh, Δ [4]→ $CH_3(CH_2)_{14}SPh$

1 2 3 5 (74%)

2-Naphthoyl chloride + N-hydroxypyridine-2-thione → mixed ester —$BrCCl_3$, AIBN, 130°→ 2-bromonaphthalene

6 7 8 9 (75%)

1	Barton, D.H.R.	*J. Chem. Soc. Chem. Commun.*	**1983**		939
2	Barton, D.H.R.	*Tetrahedron Lett.*	**1984**	*25*	5777
3	Barton, D.H.R.	*Tetrahedron Lett.*	**1985**	*26*	5939
4	Barton, D.H.R.	*Aldrichimica Acta*	**1990**	*23*	3

Pentadecyl phenyl thioether (5).[2] From palmitoyl chloride and the hydroxamic sodium salt **2** there was obtained the ester **3** which by heating in neat PhSSPh (30 equivalents) afforded **5** in 74% yield.

2-Bromonaphthalene (9).[3] In the same manner the acyl chloride **6** was reacted with hydroxamic derivative **7** to give the mixed ester **8**, which heated in $BrCCl_3$/PhCl in the presence of azoisobutyronitrile (AIBN) at 130°C afforded **9** in 85% yield.

B A R T O N - K E L L O G G Olefination

Olefin synthesis (tetrasubstituted) from hydrazones and thioketones via Δ^3-1,3,4-thiadiazolines.

$Ph_2C=N-NH_2$ (**1**) $\xrightarrow{Pb(OAc)_4}$ Ph_2CN_2 (**2**) → (with thioketone) thiadiazoline $\xrightarrow[\Delta]{Ph_3P}$ olefin

3 $\xrightarrow[H_2S\ HCl]{HC(OEt)_3}$ **4** (88%) $\xrightarrow[PPh_3]{2}$ **5** (90%)

1	Barton, D.H.R.	*J. Chem. Soc. Perkin I*	**1972**		305
2	Barton, D.H.R.	*Chem. Soc.*	**1970**		1225
3	Kellog, R.M.	*Tetrahedron Lett.*	**1970**		1987
4	Kellog, R.M.	*J. Org. Chem.*	**1972**	*37*	4045
5	Barton, D.H.R.	*J. Chem. Soc. Perkin I*	**1974**		1794

Diphenyldiazomethane (2).[5] From lead tetraacetate (889 mg, 2 mmol) in CH_2Cl_2 (5 mL), **1** ((295 mg, 1.5 mmol) and TEA (5 mL) at -20°C was obtained 291 mg of **2** (100%).

(-) Thiocamphor (4). (-) Camphor **3** (15.2 g, 0.1 mol) and triethyl orthoformate (15.9 g, 0.15 mol) in MeOH (50 mL) were saturated simultaneously with H_2S and HCl gas over 1.5 h at 0°C. The mixture poured onto ice afforded **4**. Chromatography and sublimation at 85°C (20 mm) gave 7.12 g (88%) of **4**, mp 116°C, α_D^{24} = -21.4°.

(-)-2-Diphenylethylenecamphane (5). **2** (585 mg, 3 mmol) (from **1**, lead tetraacetate and TEA in CH_2Cl_2 at -20°)[5] and **4** (505 mg, 3 mmol) in THF (5 mL) were heated to reflux under N_2 for 3 h. After chromatography, the product was refluxed with Ph_3P (870 mg) in THF (5 mL) for 16 h and evaporated. The residue in petroleum ether was treated with 1 mL of MeI (exothermic) and stirred 2 h. Chromatography (silica) afforded 545 mg of **6** (90%), mp 69.5 - 72.5°C (EtOH).

BARTON - MC COMBIE Deoxygenation

Deoxygenation of secondary alcohols to hydrocarbons (via xanthates)

1 → NaH, CS_2, MeI → R-O-C(=S)-S-Me **2**

2 + Bu_3SnH **3** → 110° → **4** (94%)

(imidaz)$_2$ C=S → Bu_3SnH, AIBN 120°

1	Barton, D.H.R., McCombie, S.W.	*J. Chem. Soc. Perkin I*	**1975**		1574
2	Cristol, S.J.	*J. Org. Chem.*	**1982**	*47*	132
3	Barton, D.H.R.	*Tetrahedron*	**1986**	*42*	2329
4	Crich, D.	*Aldrichimica Acta*	**1987**	*20*	36
5	Hartwig, W.	*Tetrahedron*	**1983**	*39*	2609
6	McClure, C.K.	*J. Org. Chem.*	**1991**	*56*	2326

1,6-Anhydro-2-deoxy-3,4-0-isopropylidene-D-galatose (4).[1] A mixture of 1,6-anhydro-3,4-0-isopropylidene-β-D-galactose **1** (900 mg, 4.45 mmol), NaH dispersion (80%, 270 mg), imidazole (5 mg) and THF (12 mL) was stirred for 0.5 h at 20°C. CS_2 (2 mL) was added and the stirring was continued for 1 h. Methylation (MeI 0.5 mL) and usual work up gave a yellow oil (xanthate **2**) which was heated under reflux in PhMe (40 mL) under Ar during addition over 1 h of a solution of tributylstannane **3** (1.6 g, 4.5 mmol). Refluxing was continued for 16 h, then the mixture was evaporated and chromatographed on silica gel. Evaporation gave 780 mg of **4** (94%), bp 75°C (2 mm), α^{22}_D= -141° (c 3.5).

B A U D I S C H Nitrosophenol Synthesis

Synthesis of o-nitrosophenols

$(NH_4)_3\ (Fe(CN)_5NH_3)$ / $NH_2OH,\ H_2O_2$ → OH, NO **2**

OH, Cl **3** — $NaNO_2,CuSO_4$ → O, $Cu/_2$, NO, Cl **4**

1	Baudisch, O.	*Naturwissenschaften*	**1939**	*27*	769
2	Baudisch, O.	*Science*	**1940**	*92*	336
3	Baudisch, O.	*J. Am. Chem. Soc.*	**1941**	*63*	622
4	Cronheim, G.	*J. Org. Chem.*	**1947**	*12*	1,7,20

o-Nitrosophenol (2).[3] An aqueous solution of pentacyanoamine ferroate, ammonium salt (2.0 g) cooled in iced water, was treated with PhH (50 mL), ligroine (50 mL) and hydroxylamine (2.0 g). Under shaking hydrogen peroxide (4.0 mL) was added and after 1 h the organic layer changed color to deep green due to the formation of **2.**

2-Nitroso-4-chlorophenol copper chelate (4).[4] A solution of p-chlorophenol **3** (12.85 g, 0.1 mol) in AcOH (30 mL) diluted with water (50 mL) was treated with NaOAc at pH = 4.2. To this mixture was added a solution of $NaNO_2$ (17.3 g, 0.25 mol) and $CuSO_4$ (12.5 g, 0.05 mol) in water (500 mL). After 24 days at 25°C, filtration of the precipitate afforded 80-95% of copper chelate **4** in 90% purity. The product was crystallized from a mixture EtOH : $CHCl_3$ (2:3).

B A Y L I S - H I L L M A N Vinyl alkylation

Amine catalyzed conversion of acrylates to α-(hydroxyalkyl) acrylates or of vinyl ketones to α-(hydroxyalkyl) vinyl ketones.

C_9H_{11}-CH=O + (methyl vinyl ketone) $\xrightarrow[THF]{DABCO}$ C_9H_{11}-CH(OH)-C(=CH_2)-C(=O)-CH_3

1 **2** **3**

Me-CHO + CH_2=CH-CO_2Et $\xrightarrow{DABCO}$ CH_3-CH(OH)-CH(CO_2Et)-CH_2-N^+R_3 ⟶ CH_3-CH(OH)-C(=CH_2)-CO_2Et + DABCO

1 Baylis, A.B., Hillman, M.E.D.	*Ger. Pat. 2155113 C.A.*	**1972**	*77*	3417
2 Basavaiah, D.	*Tetrahedron Lett.*	**1986**	*27*	2031
3 Basavaiah, D.	*Tetrahedron Lett.*	**1987**	*28*	4591, 4351
4 Basavaiah, D.	*Tetrahedron Lett.*	**1990**	*31*	1621

4-Hydroxy-3-methylenetridecan-2-one (2).[2] A solution of decanal **1** (3.12 g, 20 mmol), methyl vinyl ketone **2** (1.4 g, 20 mmol) and 1,4-diazabicyclo-octane (DABCO) (0.33 g, 3 mmol) in THF (5 mL) was allowed to stand at 20°C for 10 days. The reaction mixture was taken up in Et_2O (25 mL), washed with 2N HCl, $NaHCO_3$ solution and dried ($MgSO_4$). Purification by column chromatography (5% EtOAc in hexane) and distillation gave 2.95 g of **2** (65), bp 117-120°C/0.5 mm.

B E C H A M P Arsonilation

Arsonilation of phenols and anilines.

NH_2 (1) + AsO_4H_3 —165°→ NH_2, AsO_3H_2 **2** (30%)

OH + AsO_4H_3 —heat→ OH, AsO_3H_2 (33%)

1	Bechamp, A.J.	*C.R.*	**1863**	*56*	1172
2	Ehrlich, P.	*Chem. Ber.*	**1907**	*40*	3292
3	Brown, H.P.	*J. Am. Chem. Soc.*	**1934**	*56*	151
4	Hamilton, C.S.	*Org. React.*	**1944**	*2*	428

1-Amino-2-naphthalenearsonic acid (2).[3] α-Naphthylamine **1** (50 g, 0.34 mol) was melted by heating to 150°C and arsenic acid (10 g, 0.07 mol) was added under stirring. The mixture was maintained for 10 min and cooled to 75-90°C. It was then heated to 165-170°C under constant stirring and was maintained for 10-15 min at this temperature. After cooling to 100°C, the mixture was poured into 0.5 N NaOH (250 mL), the sodium salt solution was treated with Norite, filtered and the product precipitated with 6N HCl at pH 3. Purification via the sodium salt afforded 5-6 g of **2** (16-32%), mp 175-176°C.

BECKMANN Rearrangement or fragmentation

Acid catalyzed rearrangement of oximes to amides or cleavage of oximes to nitriles.

1 → 2 (96%): P_2O_5 - $MeSO_3H$, 100°C, 1 h

3 → 4 (73%): $SOCl_2$, pyr, 0°

Ph - CH(Tol) - C(=NOH) - Me (5) → Ph - CH(Tol) - Cl + MeCN (6[6]): PCl_5, 0°

1	Beckmann, E.	*Chem. Ber.*	**1886**	*19*	988
2	Conley, R.T.	*J. Org. Chem.*	**1963**	*28*	210
3	Eaton, P.E.	*J. Org. Chem.*	**1973**	*38*	4071
4	Nishiyama, H.	*Tetrahedron*	**1988**	*44*	2413
5	Johnson, C.R.	*J. Am. Chem. Soc.*	**1990**	*112*	6729
6	Hassner, A.	*Tetrahedron Lett.*	**1965**		525
7	Popp, I.	*Chem. Rev.*	**1958**	*58*	370
8	Heldt, W.Z.	*Org. React.*	**1960**	*11*	1

ε-Caprolactam (2).[3] To $MeSO_3H$ (360 g) was added under dry conditions and stirring P_2O_5 (36 g) until all dissolved (1-2 h). The product is stable if stored dry. **1** (2 g, 20 mmol) was added to the $MeSO_3H$:P_2O_5 (50 g) under efficient stirring. After 1 h at 100°C, the cooled mixture was quenched with saturated $NaHCO_3$ (200 mL). Extraction with $CHCl_3$, evaporation of $CHCl_3$ and recrystallization from petroleum ether gave 1.92 g of **2** (96%), mp 65-68°C.

ω-Hexenenitrile (4).[5] To oxime **3** (99 mg, 0.5 mmol) in CH_2Cl_2 (2 mL) was added P_2O_5 (70 mg, 0.5 mmol) at 0°C. After 1 day stirring at 20°C, Et_2O (2 mL) and NEt_3 (0l2 mL) were added. Washing with brine and chromatography on silica gel afforded 43 mg, of **4** (73%).

B E N A R Y Conjugated Aldehyde Synthesis

Formation of polyunsaturated aldehydes from vinyl halides and enaminoaldehydes.

Bu Br Li O N Me Ph Bu CHO

1 **2** **3** (57%)

1	Benary, E.	*Chem. Ber.*	**1930**	*63*	1573
2	Normant, H.	*C..R.*	**1958**	*247*	1744
3	Schiess, P.	*Helv. Chim. Acta.*	**1972**	*55*	2363
4	Näff, F.	*Helv. Chim. Acta.*	**1974**	*57*	1317

(E,E) and (E,Z)-2,4-Decadienal (3).[4] 1-Bromo-1-heptene **1** (8.85 g, 50 mmol) in Et_2O (20 mL) at -10°C was treated with Li (0.7 g, 0.1 at) in Et_2O (20 mL). After 2 h stirring at -8 to -12°C, (E)-3-(N-methyl-N-phenylamino) acrolein **2** (8.05 g, 50 mmol) in Et_2O (50 mL) was added over 15 min. The reaction mixture became doughy. Stirring was continued for 2 h while the temperature rose to 20°C. Excess Li was filtered and the mixture was poured onto iced 20% aqueous H_2SO_4. Extraction with Et_2O, washing and evaporation gave 4.32 g of **3** (57%), bp 57-61°C (0.01 mm), 14% (E,Z) and 86% (E,E) by GC (20% Carbowax 20M, 200°C).

3 (via Grignard reagent). **1** (4.42 g, 25 mmol) and Mg (0.6 g, 25 mat) in THF followed by **2** (4.02 g, 25 mmol) and usual work up gave 1.32 g of **3** (33%), bp 95-103°C, as a mixture of 12% (E,Z) and 88% (E,E).

B E R N T H S E N Acridine synthesis

Acridine synthesis from diphenylamine and carboxylic acids.

1 + 2 (CO_2H) — $ZnCl_2$, 260° → 3 (C_6H_5) (25%)

4 (NH_2) + HCOOH 5 — 155-175°, HCl → 6 (NH_2) (59%)

1	Bernthsen, A.	*Liebigs Ann.*	**1878**	*192*	1
2	Popp, F.D.	*J. Org. Chem.*	**1962**	*27*	2658
3	Albert, F.	*J. Chem. Soc.*	**1948**		1225
3	Buu-Hoi, M.P.	*J. Chem. Soc.*	**1955**		1082

5-Phenylacridine (3).[1] Diphenylamine **1** (16.9 g, 0.1 mol), **2** (12.2 g, 0.1 mol) and anh. $ZnCl_2$ (41 g, 0.3 mol) were heated for 10 h at 260°C. The brown solid dissolved in warm alcohol was poured into conc. NH_4 OH (50 mL) and a fter 24 h diluted with 500 ml of water. Filtration, washing and extraction with benzene, decolorization with charcoal and evaporation in vacuum afforded 8-9 g of **3**, (25%).

2-Aminoacridine (6).[3] **4** (37 g, 0.2 mol), glycerol (150 g), formic acid (9.2 g, 0.2 mol) and conc. HCl (17 ml) were heated to 155°C, maintained at this temp. for 30 min and heated to 175°C for 30 min. The product was treated with sodium acetate (10 g) in water (400 mL). After filtration of N,N'-diphenylproflavine, the filtrate was treated with NaOH solution and boiled to precipitate the crude product. After recrystallization from alcohol there was obtained **6**, 23 g (59%), mp 216°C.

B I G I N E L L I Pyrimidone synthesis

Pyrimidone synthesis from urea, an aldehyde and a β-keto ester.

$O{=}C(NH_2)_2$ + $O{=}C(CH_3)H$ + $C_6H_5{-}C({=}O){-}CH_2{-}COOC_2H_5$ —heat→ **4** (30%)

1 + **2** + **3**

1	Biginelli, P.	*Chem. Ber.*	**1891**	*24*	2962
2	Folkers, K.	*J. Am. Chem. Soc.*	**1933**	*55*	3361
3	Swett, I.	*J. Am. Chem. Soc.*	**1973**	*95*	8741
4	Zaugg, H.,E.	*Org. React.*	**1965**	*14*	88

2-Keto-4-methyl-5-carbetoxy-6-phenyl-1,2,3,4-tetrahydropyrimidine (4).[2] To cooled AcOH (10 mL) at 10°C was added under stirring, urea **1** (3 g, 68 mmol), Me-CHO **2** (3.3 g, 75 mmol) and ethyl benzoylacetate **3** (12 g, 83 mmol). The reaction mixture was heated on a steam bath for 20 h, conc. HCl (1 mL) was added and heating was continued for another 24 h. The cooled mixture was poured into water (200 mL) and after a few days the product was filtered and washed with 50% EtOH to give 5.85 g (30%) of **4**, mp 165-165.5°C (EtOH).

BIRCH - HÜCKEL - BENKESER Reduction

Reduction of aromatics, unsaturated ketones, conjugated dienes by alkali metals in liquid ammonia or amines.

Na / liq. NH_3: 1 → 2 (85%)

Li / $EtNH_2$: 3 → 4 + 5 (80:20)

1	Huckel, W.	*Liebigs Ann.*	**1939**	*540*	156
2	Birch, A.I.	*J. Chem. Soc.*	**1944**		430
3	Benkeser, R.A.	*J. Am. Chem. Soc.*	**1961**	*77*	3230
4	Benkeser, R.A.	*J. Org. Chem.*	**1964**	*29*	955
5	Moody, C.J.	*Tetrahedron Lett.*	**1986**	*27*	5253
6	Webster, F.X.	*Synthesis*	**1987**		923

5,8-Dihydro-1-naphthol (2).[2] To 1-naphthol **1** (10.0 g, 0.069 mol) was added powdered $NaNH_2$, (2.7 g, 0.069 mol), liquid NH_3 (100 mL), tert BuOH (12.5 g) and then Na (3.2 g, 0.139 at) in small pieces. After evaporation of the NH_3, the residue was extracted with Et_2O. The oil that separated on acidification solidified and after recrystallization from petroleum ether (80-100°C) gave 89.5 g of **2** (85%), mp 71-74°C.

$\Delta^{9,10}$-Octalin and $\Delta^{1,9}$-octalin (4) and **(5)**[4] A mixture of **3** (25.6 g, 0.2 mol), and. Et_2NH (250 mL) was treated with small pieces of Li (11.55 g, 1.65 atoms). After 14 h and solvent evaporation the residue was treated with water (caution). The mixture was extracted with Et_2O, the extract washed, dried and distilled to give 19-20 g of **4** and **5** (75-80%), bp 72-77°C (14 mm), 80:20 by GC (Apiezon L, 148°).

BISCHLER Benzotriazine synthesis

Benzotriazine synthesis from o-nitrophenylhydrazines.

$NH\text{-}NH_2$, O_2N, NO_2 **1** —reflux, HCO_2H→ $NH\text{-}NHCHO$, O_2N, NO_2 **2** (83%) —5% Pd/C, H_2→ H_2N, N, N, N **3**

1	Bischler, A.	*Chem. Ber.*	**1899** *22* 2801
2	Abramovitz, A.	*J. Chem. Soc.*	**1955** 2326
3	Ramesh, D.	*Synth. Commun.*	**1986** *16* 1525

BISCHLER - MOHLAU Indole synthesis

Indole synthesis from α-aminoacetophenones.

HN, O **1** —Ph - NH_2.HBr, EtOH, Δ, 2h→ N, H **2** (11%)

1	Mohlau, E.	*Chem. Ber.*	**1881** *14* 171
2	Bischler, A.	*Chem. Ber.*	**1892** *25* 2860
3	Nelson, L. R.	*J. Am. Chem. Soc.*	**1958** *80* 5957
4	Buu Hoi, N.P.	*J. Chem. Soc. (C)*	**1971** 2606
5	Bigot, P.	*J. Chem. Soc. Perkin Tr.*	**1972** 2573

BISCHLER - NAPIERALSKI Isoquinoline synthesis

Isoquinoline synthesis from amides of phenethylamines.

P_2O_5, Xylene; **1** → **2** (31%)

1	Bischler, A., Napieralski, B.	*Chem. Ber.*	**1893**	*26*	1903
2	Morrison, C.G.	*J. Org. Chem.*	**1964**	*29*	2771
3	Ramesh, D.	*Synth. Commun.*	**1986**	*16*	1523
4	Thygarayan, B.S.	*Chem. Rev.*	**1954**	*54*	1033
5	Fodor, G.	*Angew. Chem. Int.Ed.*	**1972**	*11*	919
6	Govindachari, T.R.	*Org. React.*	**1951**	*6*	74

2,3,4,6,7,12-Hexahydroindolo[2,3a]quinolizine-4-one (2).[2] To stirred **1** (3.0 g, 11.8 mmol) in boiling xylene (410 mL), was added P_2O_5 (3.15 g) at 45 min intervals. The cooled, filtered solid, was added to ice water (1,000 mL), made basic with 40% KOH and extracted with CH_2Cl_2. Evaporation and crystallization from EtOH gave 0.94 g of **2** (31%), mp 234-235°C.

BLANC Cyclization

Formation of cyclopentanones by cyclization-decarboxylation of adipic acids.

CO_2H, CO_2H, CH_2 **1** → Ac_2O, 300° (30 min) → CH_2, O **2** (69%)

1	Blanc, G.	*C.R.*	**1907**	*144*	1356
2	Bachmann, N.	*J. Am. Chem. Soc.*	**1949**	*71*	3540

BLANC - QUELLET Chloroalkylation

Lewis acid catalyzed aromatic chloromethylation (Blanc), chloroalkylation (Quellet).

$ClCH_2OCH_3$ **2**; $TiCl_4$, HCl; **1** → **3** (76%) CH_2Cl CH_2Cl

MeO-Ph **4**; Me CHO **5**; $ZnCl_2$, HCl → **6** (Cl, CH-Me, MeO) → **7** (40%) MeO

1. Grassi, G., Masselli, C.	*Gazz. Chim. Ital.*	**1898**	*28*	477
2. Blanc, G.	*Bull. Soc. Chim. Fr.*	**1923**	*33*	313
3 Tashiro Masashi	*J. Org. Chem.*	**1978**	*43*	1413
4 Fuson, R.	*Org. React.*	**1942**	*1*	63
5 Quellet, R.	*C.R.*	**1932**	*195*	155
6 Quellet, R.	*Bull. Soc. Chim. Fr.*	**1940**	*7*	196
7 Neda, V.	*J. Soc. Chem. Ind. Jpn.*	**1944**	*47*	565

2,2'-Dichloromethyl-4,4'-ditert-butyldiphenylmethane (3).[3] To cooled (-5°C) **1** (35 g, 125 mmol) and chloromethyl methyl ether **2** (80.5 g, 100 mmol) in CS_2 (150 ML) was added $TiCl_4$ (20 mL). The mixture was stirred for 1 h, poured into ice water (300 mL) and the organic layer extracted with PhH. Evaporation gave 36 g of **3** (76%), mp 90-91°C (EtOH).

p-Methoxyvinylbenzene (7).[6] **4** (125 g, 1.25 mol), **5** (44 g, 1 mol), 16% HCl (50 mL), and $ZnCl_2$ (25 g) in petroleum ether (150 mL) was saturated with HCl gas at 5°C. After 1 h the mixture was poured on ice, the organic layer concentrated and the residue (**6**), was heated with two volu-mes of pyridine for 10 h at 115°C. Distillation gave 56-70 g of **7** (35-40%), bp 95-96°C (16 mm).

B L U M Aziridine synthesis

Synthesis of aziridines from epoxides via amino alcohols or azido alcohols and reaction with phosphines or phosphites.

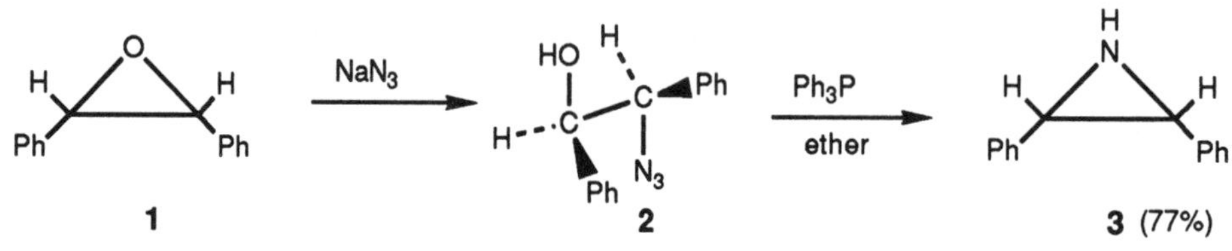

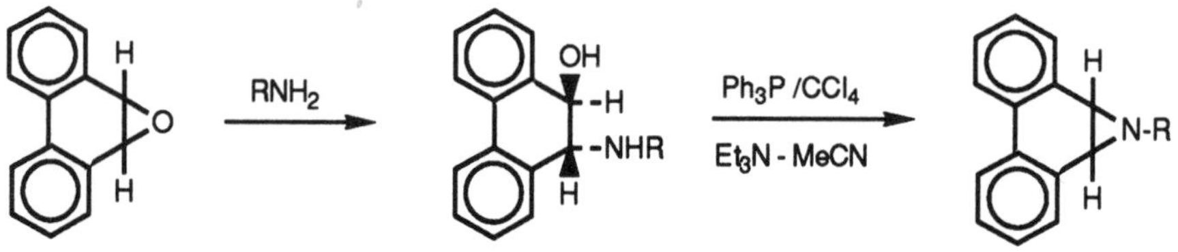

1 Blum, J. *J. Org. Chem.* **1978** *43* 397, 4273

2 Shudo, K. *Chem. Pharm Bull.* **1976** *24* 1013

Threo-2-Azido-1,2-diphenylethanol (2).[1] A mixture of cis-stilbene oxide **1** (3.92 g, 20 mmol) and NaN_3 (4.48 g, 70 mmol) in 50% aqueous acetone (60 mL) was refluxed for 3h. The solvent was removed in vacuum and the residue extracted with $CHCl_3$. The organic solution was washed with water, dried ($MgSO_4$) and concentrated. Distillation of the residue afforded 3.70 g of **2** (77%) as a pale yellow oil, bp 122°C/0.15 mm.

Cis-2,3-Diphenylaziridine (3). A solution of **2** (0.84 g, 3.5 mmol) and triphenylphosphine (0.92 g, 3.5 mmol) in dry Et_2O (25 mL) was refluxed for 1 h. Et_2O (50 mL) was added and the mixture was allowed to stand overnight at 5°C to allow complete precipitation of triphenyphosphine oxide. Column chromatography on silica gel yielded 0.53 g of **3** (77%).

BODROUX - CHICHIBABIN Aldehyde synthesis

Aldehyde synthesis from Grignard reagents and trialkyl orthoformate.

$$\underset{\mathbf{1}}{C_6H_5\text{-}Br} \xrightarrow[\text{ether}]{Mg} C_6H_5\text{-}MgBr \xrightarrow[\text{reflux, 6h}]{HC(OEt)_3\ \mathbf{2}} \underset{\mathbf{3}\ (60\%)}{C_6H_5\text{-}CH{=}O}$$

1	Chicibabin, A.E.	*J. Russ. Phys. Chem. Soc.*	**1903**	*35*	1284
2	Bodroux, F.	*C.R.*	**1904**	*138*	92
3	Smith, L.I.	*J. Org. Chem.*	**1941**	*6*	437

BOUVEAULT Aldehyde synthesis

Aldehyde synthesis from Grignard or Li derivatives with a formamide.

$$\underset{\mathbf{4}}{\text{2-}CH_3O\text{-thiophene}} \xrightarrow{Li} \xrightarrow[25^\circ]{HC(=O)\text{-}NMe_2\ \mathbf{5}} \underset{\mathbf{6}\ (65\%)}{CH_3O\text{-thiophene-}CHO}$$

1	Bouveault, L.	*C.R.*	**1903**	*137*	987
2	Bouveault, L.	*Bull. Soc. Chim. Fr.*	**1904**	*31*	1306(3)
3	Sice, J.	*J. Am. Chem. Soc.*	**1953**	*75*	3697
4	Einchorn, J.	*Tetrahedron Lett.*	**1986**	*27*	1791

5-Methoxy-2-thienaldehyde (6).[3] 5-Methoxy-2-thienyllithium prepared from **4** (11.4 g, 0.1 mol) and Li in Et_2O (125 mL) was added slowly to ice cooled DMF **5** (8.0 mL, 0.11 mol) in Et_2O (75 mL) with efficient stirring and let stand at 20° overnight. It was poured into ice, extracted with Et_2O and distillation gave 9.27 g of **6** (65%), bp 79-81°C, (0.9 mm); mp 24-26°C from petroleum ether .

BOGER Heterocycle Synthesis

Hetero Diels-Alder reaction of active olefins (enamines) with triazenes, tetrazenes with loss of N_2 and formation of new N-heterocycles.

1 + 2 —45°→ 3 (71%)

1 + 4 → 5

1	Boger, D.L.	*J. Org. Chem.*	**1981**	*46*	2179
2	Boger, D.L.	*J. Org. Chem.*	**1982**	*47*	3763
3	Boger, D.L.	*J. Org. Chem.*	**1983**	*48*	621
4	Boger, D.L.	*J. Am. Chem. Soc.*	**1985**	*107*	5745

3-Ethyl-4-n-propylpyridine (3).[1] A solution of 4-pyrrolidinohept-3-ene **1** (132 mg, 0.8 mmol) in $CHCl_3$ (0.5 mL) was added to a stirred solution of 1,2,4-triazine **2** (95 mg, 1.2 mmol) in $CHCl_3$ (0.5 mL) under N_2 at 25°C. The resulting dark orange solution was warmed at 45°C for 20 h. Chromatography (SiO_2, 50% Et_2O-hexane) afforded 92 mg of pure **3** (71%).

B O O R D Enol ether synthesis

Synthesis of chloroethers and enol ethers from aliphatic aldehydes.

$$\underset{\mathbf{1}}{CH_3\text{-}CHO} + \underset{\mathbf{2}}{HO\text{-}Et} \xrightarrow{HCl} \underset{\mathbf{3}}{CH_3\text{-}CH(Cl)\text{-}O\text{-}Et} \xrightarrow{Br_2} \underset{\mathbf{4}\ (70\%)}{CH_2(Br)\text{-}CH(Br)\text{-}O\text{-}Et}$$

$$\mathbf{4} \xrightarrow{BrMg\text{-}CH_2\text{-}CH{=}CH_2} \underset{\mathbf{6}\ (77\%)}{CH_2{=}CH\text{-}CH_2\text{-}CH_2\text{-}CH(Br)\text{-}O\text{-}Et} \xrightarrow[ZnCl_2]{Zn,\ \Delta} \underset{\mathbf{7}\ (72\%)}{H_2C = CH\text{-}CH_2\text{-}CH{=}CH\text{-}O\text{-}Et}$$

1	Boord, C.E.	*J. Am. Chem. Soc.*	**1930**	*52*	651
2	Boord, C.E.	*J. Am. Chem. Soc.*	**1931**	*53*	1505
3	Boord, C.E.	*J. Am. Chem. Soc.*	**1933**	*55*	3293
4	Crombie, L.	*Quart. Rev. (London)*	**1952**	*6*	131

1-Ethoxypenta-1,4-diene (7).[3] A mixture of paraldehyde **1** (20.0 g, 0.45 mol) and anh. EtOH **1** (20 g, 0.43 mol) cooled to -5°C, was saturated with dry HCl gas. After absorption of 20 g HCl a flow of dry N_2 was passed through. The organic layer was separated and dried ($CaCl_2$)to give 41-42 g of **3** (87-92%). Ice-cooled **3** (42.5 g, 0.4 mol) was treated with bromine (62.5 g, 0.4 mol) under stirring. After 6 h dry N_2 was bubbled through the solution, then the mixture was dried ($CaSO_4$). There was obtained 60-66 g of **4** (66-73%), bp 73-75°C (2 mm), n_D^{25} = 1.5017. To a Grignard solution obtained from allyl bromide (40 g, 0.33 mol) and Mg (19.5 g) in Et_2O (250 mL), was added **4** (58 g, 0.25 mol) in Et_2O (60 mL). The mixture was poured into AcOH and after distillation there was obtained, 37-39 g of **6** (77-82%), bp 72-75°C (21 mm).

A mixture of **6** (38 g, 0.2 mol), Zn (55 g, 0.85 at) and $ZnCl_2$ (0.2 g) in BuOH (50 mL) was heated until **7** distilled out. There was obtained 9.5-10 g of **7** (72-76%), n_D^{25} = 1.3887.

B O R C H Reduction

Reductive amination of aldehydes or ketones by cyanoborohydride (or triacetoxyborohydride)[6] anion. Selective reduction of carbonyls to alcohol, oximes to N-alkylhydroxylamines, enamines to amines.

$$OCH\text{-}(CH_2)_3\text{-}CHO \; (\mathbf{1}) + CH_3\text{-}NH_2 \; (\mathbf{2}) \xrightarrow{NaBH_3CN} \text{N-methylpiperidine } \mathbf{3} \; (43\%)$$

$$Ph\text{-}CH_2\text{-}C(=O)\text{-}COONa \; (\mathbf{4}) + NH_4Br \xrightarrow{LiBH_3CN} Ph\text{-}CH_2\text{-}CH(NH_2)\text{-}COOH \; (\mathbf{5})$$

1	Borch, R.F.	*J. Am. Chem. Soc.*	**1969**	*91*	3996
2	Borch, R.F.	*J. Am. Chem. Soc.*	**1971**	*93*	2897
3	Borch, R.F.	*J. Chem. Soc. Perkin I*	**1984**		717
4	Lane, C.F.	*Synthesis*	**1975**		135
5	Hutchins, R.O.	*Org. Prep. Proc. Int.*	**1979**	*11*	20
6	Abdel-Magid	*Tetrahedron Lett.*	**1990**	*31*	5595

M-Methylpiperidine (3).[2] To the gluteraldehyde **1** bisulfite addition product (3.08 g, 10 mmol) and **2** .HCl (1.3 g, 20 mmol) in MeOH (50 mL) was added $NaBH_3CN$ (500 mg, 9 mmol). After 48 h at 25°C, it was acidified to pH=2, evaporated and the residue dissolved in water and basified with 6 N KOH. After extraction with Et_2O (8x25 mL), drying ($MgSO_4$) and evaporation, the residue in EtOH (3 mL) was treated with picric acid (10 mmol) in EtOH to give 1.38 g of **3** picrate (43%), mp 147-149°C.

Phenylalanine (5). **4** (186 mg, 1 mmol), NH_4Br (500 mg, 5 mmol) and $LiBH_3CN$ (100 mg, 2 mmol) in MeOH (20 mL) was stirred for 48 h at 25°C, treated with 36% HCl (5 mL) and after 1 h evaporated in vacuum. The residue in 3 mL of water was passed through Dowex 50 (H^+ form), 100 mequiv capacity. Washing (water) and elution with 1 N NH_4OH afforded 81 mg of **5** (49%).

BORSCHE - BEECH Aromatic Aldehyde Synthesis

Synthesis of aromatic aldehydes and of alkyl aryl ketones from aldoximes or semicarbazones and aromatic diazonium salts.

NH_2 $\xrightarrow{HNO_2}$ $N_2]^+$ + $H_2C=NOH$ ⟶ CHO (14%)

2 1 3

NH_2 Cl $\xrightarrow{HNO_2}$ $\xrightarrow{Me-CH=NOH}$ O Cl (36%)

1	Borsche, C.	*Chem. Ber.*	**1907**	*40*	737
2	Beech, W.F.	*J. Chem. Soc.*	**1954**		1297
3	Woodward, R.B.	*Tetrahedron*	**1958**	*2*	1

Pyridine-3-aldehyde (3).[2] A solution of 3-aminopyridine **2** (23.5 g, 0.24 mol), 36% HCl (68 mL), $NaNO_2$ (17.5 g, 0.25 mol) and water (75 mL) was made neutral (NaOAc) and treated with formaldoxime **1**. The mixture was made acid (pH-3) and after $FeCl_3$ (150 g) was added, it was boiled for 1 h. Usual work up gave 3.6 g of **3** (14%), bp 95-100°C/16 mm. p-Chloroaceto-phenone was prepared in the same manner from p-chloroaniline and acetaldoxime in 36%.

BOUVEAULT - BLANC Reduction

Reduction of esters to alcohols by means of sodium in alcohol.

COOCH$_3$ COOH → (Na / EtOH) → OH COOH → O O

1 2 (20%)[2] 3

1	Bouveault, L., Blanc, G.	*C.R.*	**1903**	*136*	1676
2	Paquette, L.A.	*J. Org. Chem.*	**1962**	*27*	2274
3	Ruhlmann, K.	*Synthesis*	**1972**		236
4	Chaussar, J.	*Tetrahedron Lett.*	**1987**	*28*	1173

BOUVEAULT - HANSLEY - PRELOG - STOLL Acyloin synthesis

Condensation of two esters to an α-hydroxyketone by means of rapidly stirred (8000 rpm) Na suspension in boiling toluene or xylene.

COOCH$_3$ COOCH$_3$ → (Na, Δ / xylene (3.5L)) → O OH

1 (0.2 mol) 2 (75%)

1	Bouveault L.	*C.R.*	**1905**	*140*	1593
2	Hansley, V.L.	*U.S. Pat. 2.228.268.C.A.*	**1941**	*35*	2354
3	Prelog, V.	*Helv. Chim. Acta.*	**1947**	*30*	1741
4	Stoll, M.	*Helv. Chim. Acta.*	**1947**	*30*	1815
5	Cramm, D.J.	*J. Am. Chem. Soc.*	**1954**	76	2743
6	Finley, K.T.	*Chem. Rev.*	**1964**	*64*	573

B O U V E A U L T - L O C Q U I N Aminoacid synthesis

Synthesis of aminoacids from malonate esters by nitrosation.

$$C_6H_5-CH_2-CH(C\text{-}OOEt)_2 + EtO\text{-}N=O \xrightarrow[2.\ H^+]{1.\ EtONa} C_6H_5\text{-}CH_2\text{-}C(=N\text{-}OH)\text{-}COOEt$$

1 **2** **3** (97%)

$$C_6H_5-CH_2-C(=N\text{-}OH)-COOEt \xrightarrow[H_2]{\text{Raney Ni}} C_6H_5-CH_2-CH(NH_2)-COOEt$$

3 **4** (53%)

1 Bouveault, L., Locquin, R. *C.R* **1902** *135* 179

2 Hauser, C.R. *J. Am. Chem. Soc.* **1947** *69* 1387

Phenylalanine ethyl ester (4).[2] a) **Ethyl α-Oximino-β-phenylpropionate (3).** A solution of diethyl benzylmalonate **1** (50.0 g, 0.2 mol) in EtoH (92 mL) was treated with EtONa (13.6 g, 0.2 mol) and ethyl nitrite **2** (22.5 g, 0.3 mol) at 0°C. The reaction mixture was kept at -20°C for 20 h. After evaporation of the solvent under vacuum (no heating), cold conc. HCl was added to bring the pH to 5. Extraction with Et_2O and evaporation of the solvent gave 40 g of **3** (97%), mp 54-57°C (mp 56-57°C from ligroin, bp.90-120°C).

A solution of **3** (16.8 g, 80 mmol) in EtOH (850 mL) and Raney nickel was hydrogenated at 70-75°C. Removal of the catalyst by filtration and evaporation of the solvent gave 8.3 g of **4** (53%), bp 141-142°C (10 mm).

BOYLAND - SIMS o-Hydroxyaniline Synthesis

Oxidation of anilines with persulfates to o-aminophenols.

Me N Me — $K_2S_2O_8$ → Me N Me, $OSO_3K(H)$ — conc. HCl → Me N Me, OH

1 2 (40%) 3 (83%)

1 Boyland, E, Sims, P. *J. Chem. Soc.* **1953** 3623

2 Behrman, E.J. *Org. React.* **1988** *35* 432

o-Dimethylaminophenol (3).[1] To dimethylaniline **1** (5.0 g, 41 mmol) in water (250 mL), acetone (400 mL) and 2N KOH (30 mL) was added $K_2S_2O_8$ (11.2 g, 41 mmol) as a saturated aqueous solution at 20°C over 8 h with stirring. The solution was evaporated to 250 mL, washed with Et_2O (3x150 mL), evaporated to dryness under reduced pressure, and the residue extracted with hot 95% EtOH (3x50 mL). The combined extracts were diluted with Et_2O (1500 mL) to give o-dimethylaminophenyl potassium sulfate **2** (4.2 g) (40%).

The potassium salt **2** (0.46 g) in water (2 mL) was treated with HCl (2 mL) to afford o-dimethylaminophenyl hydrogen sulfate **2** (310 mg, 82%), mp 217-218°C (EtOH).

Heating of **2** (0.4 g, 2 mmol) in conc. HCl (5 mL) for 1 h, and partial neutralization of the ice cooled solution with 2N NaOH, gave 210 mg of **3** (83%), mp 44-45°C (from aq. EtOH).

von BRAUN Amine degradation

Degradation of tertiary amines to secondary amines with cyanogen bromide (BrCN).

BrCN
55°
48% HBr
reflux
N
CN
N^{+}
H_2Br^{-}
1 2 3

Ph - N + Br CN ⟶ Ph - N CN + Ph - N^{+} Br^{-}

Me N Br CN ⟶ H N

1	v. Braun, J.	*Chem. Ber.*	**1907**	*40*	3914
2	Elderfield, R.C.	*J. Am. Chem. Soc.*	**1950**	*72*	1334
3	Wolf, V.	*Liebigs Ann.*	**1957**	*610*	67
4	Boekelheide, V.	*J. Am. Chem. Soc.*	**1955**	*77*	4079
5	Rapoport, H.	*J. Am. Chem. Soc.*	**1967**	*89*	1942
6	Fodor, G.	*Can. J. Chem.*	**1969**	*47*	4393
7	Hageman, H.A.	*Org. React.*	**1953**	*7*	198

4-Pipecoline (3).[2] To a solution of BrCN (48 g, 0.46 mol) in PhH (100 mL) was added 1-isopropyl-4-pipecoline **1** (58 g, 0.41 mol) in PhH (275 mL) over 1 h at 40°C. The mixture was heated for 45 min at 55-60° C and was maintained at 20° for 36 h. The basic material was extracted with HCl (100 mL) and the solvent distilled to give 44 g of residue. The neutral product was refluxed with 48% HBr (300 mL) for 10 h. After distillation of HBr, the residue was leached in a mixture of EtOAc:EtOH (80:20). Filtration of insoluble NH_4Br and concentration gave **3**, mp 171-173°C.

B R E D E R E C K Imidazole synthesis

Synthesis of imidazoles from formamide (acetamide) and α-diketones, α-ketols, α-aminoketones, α-oximinoketones.

$CH_3-CO-CO-CH_3$ **1** + $HCONH_2$ **2** —(180°C)→ 4,5-dimethylimidazole **3** (50%)

$(CH_2)_7$ cyclic α-diketone + $HCONH_2$ —(Δ)→ $(CH_2)_7$-fused imidazole

$Ph-CO-CH_3$ —(1) Br_2; 2) $EtCO_2K$)→ $Ph-CO-CH_2-O-CO-Et$ —(HCO_2H, Δ; $(NH_4)_2CO_3$)→ 4(5)-phenyl-2-ethylimidazole (70%)

1	Bredereck, H.	*Chem. Ber.*	**1953**	*86*	88
2	Grimmett, V.	*Adv. Heteroc. Chem.*	**1970**	*12*	113
3	Bredereck, H.	*Angew. Chem.*	**1959**	*71*	753
4	Schubert, H.	*Z. Chem.*	**1967**	*7*	461
5	Novelli, A.	*Tetrahedron Lett.*	**1967**		265

4,5-Dimethylimidazole (3).[1] A mixture of acetoin **1** (10 g, 0.116 mol) and formamide **2** (50 mL, 66.7 g, 1.5 mol) was heated for 4 h, distilled in vacuum and the fraction with bp 165-175°C/11 mm was collected. There was obtained 6 g of **3** (50%) as an oil which crystallized, mp 117°C; **3** hydrochloride, mp 305°C.

B R O O K Silaketone rearrangement

Rearrangement of sila ketones to silyl ethers (with chirality transfer).

Ph
Ph2 - Si - C - Ph ⟶ (⁻OEt) Ph2 Si - C - Ph2 (O⁻, OEt) ⟶ Ph2 Si - O - C̄Ph2 (OEt) ⟶ Ph2 - Si - O - CHPh2 (OEt)
1 O
2 (74%)

Me, Np, Ph - Si - C(=O) - Me ⟶ (tBuOK) tBuO - Si(Me)(Np) - O - C(H)(Ph) - - - Me

1	Brook, A.G.	*J. Org. Chem.*	**1962**	*27* 2311
2	Brook, A.G.	*Acc. Chem. Res.*	**1974**	*7* 77
3	Wilson, S.R.	*J. Org. Chem.*	**1981**	*47* 747
4	Kuwajima, J.	*Tetrahedron Lett.*	**1980**	*21* 623

Benzhydryloxy ethoxy diphenyl silane (2).[2] To a solution of benzoyl triphenylsilane **1** (2.5 g, 6.9 mmol) in benzene (25 mL) was added a solution of sodium ethoxide in ethanol (2 mL, 0.8 mmol). The solution changed colour and after 11 min faded almost completely. The solution was washed with water and the solvent removed in vacuum. The oily residue was dissolved in hot ethanol (15 mL) and cooled to give **2** (2.1 g, 74%), mp 67-75°C. Recrystallization from ethanol gave 1.8 g (64%), mp 77-78,C.

B R O W N Stereoselective reduction

Stereoselective reduction of ketones to alcohols by means of borohydride reagents (Li s-Bu_3BH or t-BuClBR* for formation of chiral alcohols).

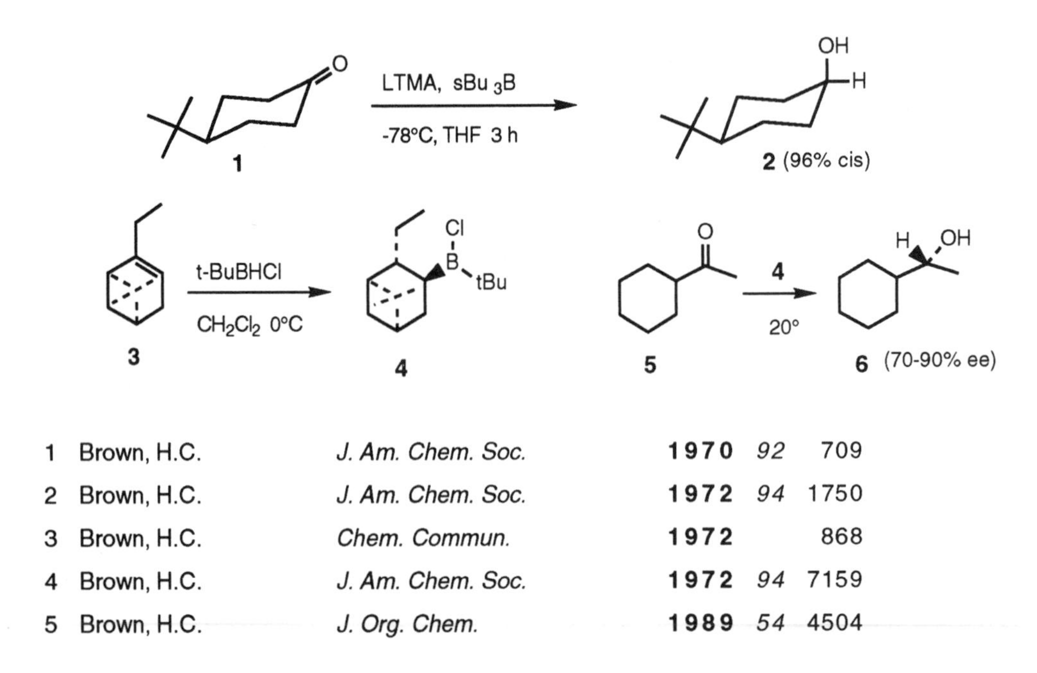

1	Brown, H.C.	*J. Am. Chem. Soc.*	**1970**	*92*	709
2	Brown, H.C.	*J. Am. Chem. Soc.*	**1972**	*94*	1750
3	Brown, H.C.	*Chem. Commun.*	**1972**		868
4	Brown, H.C.	*J. Am. Chem. Soc.*	**1972**	*94*	7159
5	Brown, H.C.	*J. Org. Chem.*	**1989**	*54*	4504

Cis-4-tert-butylcyclohexanol (2).[4] To 1.0 M lithium trimethoxyaluminium hydride (LTMA) (5.0 mL) in THF under N_2, was added sec-butylborane (from 2-butene and diborane) 1.25 mL, 5 mmol). After 30 min the mixture was cooled to -78°C and **1** (390 mg, 2.5 mmol) was added. After 3 h, hydrolysis and oxidation (H_2O_2) gave **2** (96.5% cis and 3.5% trans).

(S)-Cyclohexylethanol (6).[5] To 5.5 mmol of **4** in THF (from Li-t$BuBH_3$, HCl followed by (-)-2-ethylapopinene **3** α_D=-42.78°) was added **5** (0.64 g, 5 mmol) under N_2. After 2 days the solvent was removed, the residue dissolved in Et_2O (20 mL), diethanolamine (2.2 equiv) was added and stirred for 2 h. After filtration and washing with pentane, the filtrates were concentrated and chromatography gave 0.42 g of **6** (65%), 90% ee.

BROWN Hydroboration

Hydroboration - regioselective and stereoselective (syn) addition of BH_3 (RBH_2, R_2BH) to olefins. Synthesis of alcohol including optically active alcohols from olefins. Also useful in synthesis of ketones by "stitching" of olefins and CO.

$Me_2S.BH_3$ + 2 → 3 (100%) —TMEDA→ 4 (BH_2:TMEDA:H_2B) —$BF_3.Et_2O$→ 5 (BH_2) —H_2O_2 / NaOH→ 6 (OH)

7 + 3 → —H_2O_2 / NaOH→ 8 (92%), 100% ee

cyclohexene + BH_2 → —(propene)→ —CO→ ketone

1	Brown, H.C.	*J. Am. Chem. Soc.*	**1956**	*78*	2583
2	Brown, H.C.	*J. Org. Chem.*	**1978**	*43*	4395
3	Masamune, S.	*J. Am. Chem. Soc.*	**1986**	*108*	7401
4	Hoffmann, R.W.	*Angew. Chem. Int. Ed.*	**1982**	*21*	555
5	Brown, H.C.	*J. Am. Chem. Soc.*	**1986**	*108*	2049
6	Srebnik, M.	*Aldrichimica Acta*	**1987**	*20*	9
7	Brown, H.C.	*J. Org. Chem.*	**1989**	*54*	4504

cont. on next page

Isopinocampheol (6).[2] To a hot solution of borane-methyl sulfide **1** (2.0 mL, 20.0 mmol) in Et_2O (11.3 mL) was added (+)-α-pinene **2** (7.36 mL, 46.0 mmol), which led to quantitative formation of **3** After addition of TMEDA 1.51 mL, 10 mmol), reflux was continued for 30 min. The adduct was filtered and washed with pentane to give 3.32 g of **4** (80%), mp 140-141°C (Et_2O). A solution of **4** (3.32 g, 8 mmol) in THF (16 mL) was treated with $BF_3.Et_2O$ (1.97 mL, 16 mmol). After 1 h, the solid TMEDA.$2BF_3$ was removed and the solution of **5** was oxidized with alkaline H_2O_2 to give **6** (100%).

(-) 3-Hydroxytetrahydrofuran (8)[5] To a suspension of (-) Ipc_2BH (diisopinocamphenyl borane **3** (7.1 g, 25 mmol), in THF, see above, at -25°C was added 2,3 -dihydrofuran **7** (1.9 mL, 25 mmol). The reaction mixture was stirred at the same temperature for 6 h. The solid **3** disappeared, and formation of trialkyl borane was complete. The mixture was brought to 0°C and acetaldehyde (5.6 mL, 100 mmol) was added dropwise and stirring was continued for another 6 h at 25°C. Excess acetaldehyde was removed under vacuum (25°C, 12 mm Hg), and 20 mL of THF was added. The boronate thus obtained was oxidized with 25 mL of 3N NaOH and 3.75 mL of 30% H_2O_2, and maintained for 5 h at 25°C. The aqueous layer was saturated with K_2CO_3, extracted with 3.25 mL Et_2O and the organic layer dried ($MgSO_4$). The solvent was evaporated, the residue filtered through silica; pentane eluent removed α-pinene whereas the Et_2O eluent afforded the alcohol **8** which on distillation yielded 1.87 g, bp 80°C/15 mm (92%), GC purity 99%, α_D = -17.3° (c 2.4 MeOH 100% ee).

BRUYLANTS Amination

Amination - alkylation of aldehydes via α-cyanoamines.

C=O + CN^- +

1 2 3 (82%)

3 + C_2H_5MgBr

4 5 (78%)

1 Bruylant, P. *Bull. Soc. Chim. Belge* **1924** *33* 467

2 Bruylant, P. *Bull. Soc. Chim. Belge* **1926** *35* 139

3 Bersch, H.W. *Arch. Pharm.* **1978** *311* 1029

4 Ahlbrecht, H. *Synthesis* **1985** 743

N-(4-hex-2-ene)-pyrrolidine (5).[3] Pyrrolidine **2** (0.71 g, 0.1 mol) is neutralized by 4N HCl and mixed with a solution of KCN (6.5 g, 0.1 mol) in H_2O (50 mL). The mixture is added slowly to a stirred solution of crotonaldehyde **1** (7.0 g, 0.1 mol) in Et_2O (100 mL) at 0°C. The mixture is stirred for 3 h at 20°. The water layer is separated, washed with Et_2O (4.x 30 mL), the combined Et_2O layers washed (H_2O), dried (Na_2SO_4) and evaporated. The residue is distilled to yield 12.38 g of **3** (82%), bp 51°C (1.5 10^{-2} torr).

To a stirred solution of **3** (10.57 g, 70 mmol) in THF (20 mL) under Ar, a solution of EtMgBr (1 molar, 22 mmol) in THF is added slowly at 0°C. The resulting mixture is stirred for 3 h at 20°, then diluted with Et_2O (50 mL). The mixture is washed with water, the solvent removed and the residue distilled to give 8.35 g of **5** (78%), bp 83°C (19 mm).

BUCHERER - LE PETIT Naphthol (Naphthylamine) Synthesis

Synthesis of naphthylamines from naphthols and naphthols from naphthylamines.

CH_3 OH 1 $\xrightarrow[200°C]{NaHSO_3,NH_4OH}$ CH_3 NH_2 SO_3^- → CH_3 NH_2 **2** (93%)

NH_2 $\xrightarrow{NaHSO_3}$ OH

1	Le Petit, R.	*Bull. Soc. Ind. Mulhouse*	**1903**	*73*	326
2	Bucherer, H.T.	*J. Prakt. Chem.*	**1904**	*69*	49(2)
3	Johnson, W.S.	*J. Am. Chem. Soc.*	**1944**	*66*	210
4	Seeboth, H.	*Angew. Chem.*	**1958**	*70*	312
5	Drake, N.L.	*Org. React.*	**1942**	*1*	105
6	Weidon, B. C. L.	*Quart. Rev.*	**1952**	*6*	380

1-Methyl-2-naphthylamine (2).[3] A mixture of 1-methyl-2-naphthol **1** (7.8 g, 50 mmol), $NaHSO_3$ (10 g, 0.1 mol), 28% NH_4OH (40 mL) and water (10 mL) was heated to 200-205°C for 48 h. The cooled solution was made alkaline with NaOH solution and extracted with Et_2O. The dried ether solution was treated with HCl gas to afford **2** hydrochloride which was filtered, treated with alkali and extracted with Et_2O. Evaporation of the solvent gave 7.2 g of **2** (93%), mp 46-50°C. Recrystallized from petroleum ether (bp 60-70°C), mp 49-50°C.

BUCHNER - CURTIUS - SCHLOTTERBECK Homologation

Ring enlargement of benzene derivatives by carbenes generated from diazo compounds (better in the presence of a Rh catalyst). Conversion of aldehydes to ketones by diazo compounds (Schlotterbeck); see also Pfau-Platter.

$N_2CH\text{-}CO_2Me$ **1**, $(Cl_2[NO_2]_2C_6HCOO)_4Rh_2$ **2** → **3** (85%) CO_2Me

4 CHO, CH_2N_2, 12 d → **6** (70%)

1	Buchner, E., Curtius, T.	*Chem. Ber.*	**1885**	*18*	2371
2	Buchner, E.	*Chem. Ber.*	**1896**	*29*	106
3	Schotterbeck, F.	*Chem. Ber.*	**1907**	*40*	479
4	Ramonczay, J.	*J. Am. Chem. Soc.*	**1950**	*72*	2337
5	Doering, W.v.	*J. Am. Chem. Soc.*	**1957**	*79*	352
6	Anciaux, A.J.	*J. Org. Chem.*	**1981**	*46*	873

Tetrakis(2,4-dichloro-3,5-dinitrobenzoato)dirhodium (II) (2).[6] To a solution of $RhCl_3.3H_2O$ (250 mg, 10 mmol) and 2,4-dichloro-3,5,-dinitrobenzoic acid (3 g, 10mmol) in EtOH (100 mL), $NaHCO_3$ (220 mg) was added and the mixture was refluxed under N_2. Cold filtration afforded **2** as a green solid, after washing with Me_2CO and Et_2O, in 69% yield.

1-Carboxymethyl-2,4,6-cycloheptatriene (3).[6] Methyl diazoacetate **1** (0.5 g, 5 mmol) was added to PhH (8.4 g, 100 mmol) and catalyst **2** (0.02 mmol) over 2 h while the mixture was stirred at 20°C till the 2175 cm^{-1} band disappeared. The mixture was distilled under vacuum to afford 0.58 g of **3** (85%).

B U R G E S S Dehydration reagent

Thermolysis of tertiary and secondary alcohols with (carbomethoxysulfamoyl) triethylammonium inner salt **5** to give olefins; also conversion of amides to nitriles.

$$OCN\text{-}SO_2Cl + MeOH \longrightarrow MeO_2C\text{-}N\text{-}SO_2Cl + NEt_3 \longrightarrow MeO_2C\text{-}\overset{-}{N}\text{-}SO_2\overset{+}{N}Et_3$$

1 **2** **3** **4** **5**

6 —(**5**, 55°)→ **7** (69%)

OH —(**5**)→

1	Burgess, E.M.	*J. Org. Chem.*	**1973**	*38*	26
2	O'Grodnick, J.S.	*J. Org. Chem.*	**1974**	*39*	2124
3	Goldsmith, D.J.	*Tetrahedron Lett.*	**1980**	*21*	3543
4	Claremon, D.A.	*Tetrahedron Lett.*	**1988**	*29*	2155
5	Burgess, E.M.	*Org. Synth.*	**1977**	*56*	40

Inner Salt (5).[1] Anhydrous MeOH **2** (18.8 g, 0.55 mol) in PhH (25 mL) was added over 30 min to **1** (65.72 g, 0.5 mol) in PhH (200 mL) with cooling. The solvent was removed in vacuum and the residue recrystallized from PhMe to give 61.0 g of **3** (91.9%), mp 70-71°C. **3** (3.47 g, 20 mmol) in PhH (50 mL) was added dropwise to Et_3N (4.6 g, 45 mmol) in PhH (25 mL). Et_3N .HCl was filtered, the filtrate evaporated and the residue crystallized from PhMe to give 3.87 g of **5** (81%), mp 71-72°C.

3,4-Epoxy-2-methylbut-1-ene (7).[1] 3,4-Epoxy-2-methylbutan-2-ol **6** (3 g, 29 mmol) was added neat to **5** (9.5 g, 40 mmol) at 20°C under N_2. The mixture was warmed to 55°C and 1.41 g of **7** (69%) distilled.

B U R T O N Trifluoromethylation

Trifluoromethylation of aryl iodides with Cd (Cu) reagents.

$$\underset{\mathbf{1}}{CF_2\text{-}CClBr} + Cd \xrightarrow{DMF} \underset{\mathbf{2}}{CF_3\text{-}CdCl} + CdBr_2$$

$$\underset{\mathbf{3}}{p\text{-}I\text{-}C_6H_4\text{-}NO_2} \xrightarrow[\text{Cu Br, 60°}]{F_3C\ Cd\ Cl\ \mathbf{2}} \underset{\mathbf{4}\ (75\%)}{p\text{-}CF_3\text{-}C_6H_4\text{-}NO_2}$$

1	Burton, D.J.	*J. Am. Chem. Soc.*	**1985**	*107*	5014
2	Burton, D.J.	*J. Am. Chem. Soc.*	**1986**	*108*	832
3	Clark, J.H.	*J. Chem. Soc. Chem. Commun.*	**1988**		638

Trifluoromethyl Cd chloride (2).[1] To activated Cd (22.4 g, 0.2 at) in DMF (50 mL) was condensed F_2C ClBr **1** (8.6 mL, 0.1 mol). After an exothermic reaction a precipitate formed. After 2 h stirring at 25°, the brown solution was filtered under nitrogen and the precipitate was washed with 10-15 mL of DMF. The solution of **2** (1 M) thus obtained was utilized in the next reaction.

p-Trifluoromethylnitrobenzene (4) To a solution of **2** (0.1 mol) and CuBr (7.17 g, 0.05 mol), was added at 0°C a solution of p-iodonitrobenzene **3** (8.71 g, 0.035 mol). The mixture was heated for 4-6 h at 60-70°C and steam distilled. Extraction of the organic layer and redistillation gave 3.93 g of **4** (75%).

CANNIZZARO Oxidation - Reduction

A redox reaction between two aromatic aldehydes (or an aromatic aldehyde and formaldehyde) to a mixture of alcohol and acid.

CHO, OMe (benzene) —KOH→ CO_2H, OMe (benzene) + CH_2OH, OMe (benzene)

$$PhCH{=}O + CH_2{=}O \xrightarrow{30\%\ NaOH} PhCH_2\text{-}OH + HCO_2H$$

1	Wöhler, F.	*Ann.*	**1832**	*3*	252
2	Cannizzaro, S.	*Ann.*	**1853**	*88*	129
3	Bruce, R.A.	*Org. Prep. Proced Int.*	**1987**		19
4	Geissmann, T.A.	*Org. React.*	**1944**	*2*	95

o-Methoxybenzyl alcohol (2) and **o-Methyloxybenzoic acid (3).**[3] To a solution of KOH (120 g, 2 mol) in water are added o-methyoxybenzaldehyde **1** (136 g, 1 mol) under efficient stirring and external cooling with water. Stirring was maintained until a stable emulsion was obtained. After 24 h at 30°C, the mixture was diluted with water and extracted with Et_2O. Evaporation of the solvent and vacuum distillation of the residue afforded 55 g of **2** (79%), bp 245-255°C. Acidification of the aqueous solution, extraction with Et_2O and distillation of the solvent gave **3**, mp 98-100°C.

C A R R O L L Allylation

Carboxylation and subsequent allylation of alkyl heterocycles (via Claisen rearrangement of allyl esters).

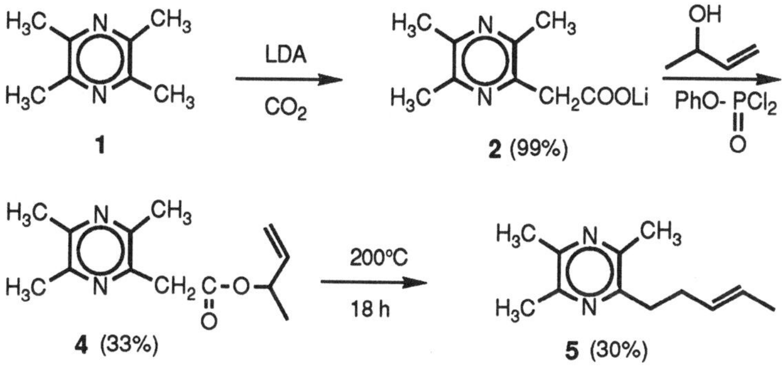

1 Carroll, M.F.	*J. Chem. Soc.*	**1940**		704
2 Kimel, W.	*J. Org. Chem.*	**1958**	*23*	153
3 Stephen, W.	*J. Org. Chem.*	**1984**	*49*	722
4 Podraza, K.F.	*J. Heterocycl. Chem.*	**1986**	*23*	581

2-(3-Pentenyl)-3,5,6-trimethylpyrazine (5).[4] To LDA (from iPr_2NH (7.8 g, 77.7 mmol) in Et_2O and 1.6 M nBuLi (43.6 mL, 77.2 mmol) in hexane was added 2,3,5,6-tetramethylpyrazine **1** (10 g, 73.5 mmol) In Et_2O (100 mL). After 30 min at 0°C, dry CO_2 was passed through for 30 min. Filtration, washing (Et_2O) and drying gave 13.5 g of **2** (99%).
To **2** (1 g, 5.4 mmol) in dimethoxyethane (25 mL) was added sequentially: pyridine (0.85 g, 10 mmol), phenyl dichlorophosphate (1.71 g, 8.1 mmol) and 3-butene-2-ol **3** (9.78 g, 10.8 mmol). After 18 h at rt the mixture was poured into ice water and extracted with $CHCl_3$. Purification by HPLC (50% EtOAc in hexane) yielded 420 mg of **4** (33%).
Heating of **4** (100 mg, 0.42 mmol) for 18 h at 200°C yielded 20-28 mg of **5** (25-35%).

CHAN Reduction of Acetylenes

Stereospecific reduction of acetylenic alcohols to E- allylic alcohols by means of sodium bis(2-methoxyethoxy)aluminium hydride (SMEAH).

$NaAlH_2(OCH_2CH_2OCH_3)_2$ (SMEAH), heat

1 → **2** (83%)

1 Chan, Ka-Kong *J. Org. Chem.* **1976** *41* 62

2 Chan,Ka-Kong *J. Org. Chem.* **1976** *41* 3497

3 Chan, Ka-Kong *J. Org. Chem.* **1978** *43* 3435

(4S,6R)-(-)-(E)-6,10-Dimethylundec-2-en-4-ol (2).[2] To a solution of (4S,6R)-(-)-6,10-dimethylundec-2-yn-4-ol **1** (1.9 g, 9.66 mmol) in dry Et_2O (75 mL) was added dropwise with stirring to a 70% benzene solution of sodium bis(2-methoxyethoxy)aluminium hydride (3 mL) in Et_2O (60 mL). The resulting mixture was stirred and refluxed under Ar for 20 h then cooled in an ice bath and cautiously decomposed by the dropwise addition of dilute H_2SO_4 (20 mL). Extraction with Et_2O, washing with $NaHCO_3$, evaporation of the solvent and distillation gave 1.61 g of **2** (83.8%) as a colorless oil, bp 85°C (0.11 mm) (bath temperature), $[\alpha]_D^{25}$ = -9.0°C (c 3.79 $CHCl_3$); no Z isomer was detectable.

CHAPMAN Rearrangement

O to N aryl migration in O-aryliminoethers.

CH3O, N, Ph, Cl, 1 + HO, Cl, COOCH3, 2 —NaOEt→ CH3O, N, Ph, O, Cl, 3, COOCH3 —1. 210° 2. NaOH→ CH3O, N, Cl, COOH, O, 4 (27%)

1	Chapman, A.W.	*J. Chem. Soc.*	**1925**	*127*	1992
2	Dauben, W.G.	*J. Am. Chem. Soc.*	**1950**	*72*	3479
3	Crammer, F.	*Angew. Chem.*	**1956**	*68*	649
4	Roger, R.	*Chem. Rev.*	**1969**	*69*	503
5	Schulenberg, J.W.	*Org. React.*	**1965**	*14*	1

N-(4-Methoxyphenyl)-N-benzoyl-3-chloro-3-chloroantranilic acid (4).[2] To ice cooled NaOEt (from 0.11 g Na and 10 mL EtOH) was added in rapid succession methyl 2-chloro-6-hydroxybenzoate **2** (0.88 g, 4.7 mmol) and N-(4-methoxyphenyl)benzimidyl chloride **1** (1.14 g, 4.7 mmol) in Et_2O (30 mL). The mixture was stirred vigorously and allowed to stand at rt for 48 h. The oily solid obtained after evaporation was heated to 210-215°C for 70 min under N_2. The product was diluted with water (5.4 mL), EtOH (10.8 mL) and treated with 1 M NaOEt (5.4 mL). The mixture was refluxed for 90 min, the solvent evaporated and the aqueous solution acidified. The dark oil was heated with NaOH (7.5 g) in EtOH (22 mL). After removal of benzoic acid by exhaustive extraction with water, the product was recrystallized from EtOH to give 0.36 g of **4** (27.7%), mp 139.5-140.5°C.

CHICHIBABIN Amination

α-Amination of pyridines, quinolines and other N-heterocycles.

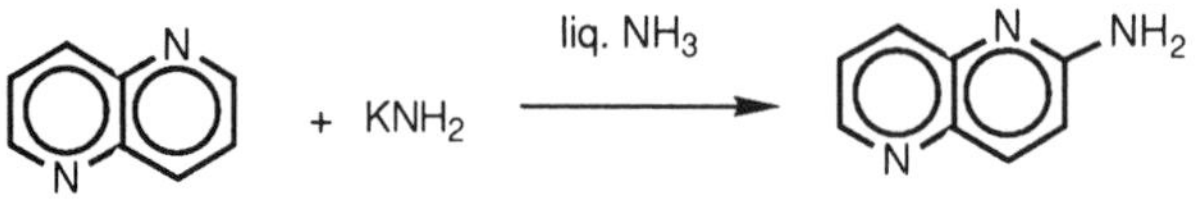

1 + KNH_2 —(liq. NH_3)→ 2 (36%)

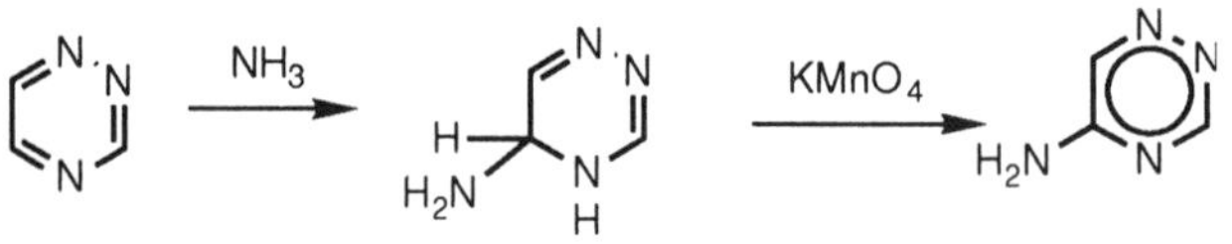

1	Chichibabin, A.	*J. Russ. Phys. Chem. Soc.*	**1914**	*46*	1216
2	van der Plas, H.C.	*J. Org. Chem.*	**1981**	*46*	2134
3	Bunnette, J.F.	*Chem. Rev.*	**1951**	*49*	375
4	Rykowscy, A.	*Synthesis*	**1985**		884
5	Leffler, M.T.	*Org. React.*	**1942**	*1*	19

2-Amino-1,5-Naphthyridine (2).[2] To a solution of potassium amide (from 0.15 g of K) in liquid ammonia (17 mL) was added 1,5-naphthyridine **1** (0.2 g, 1.5 mmol). After 15 min stirring, $KMnO_4$(0.80 g) was added and after another 10 min $(NH_4)_2SO_4$ was added to destroy the KNH_2. The NH_3 was evaporated, conc NH_4OH was added and the solution was continuously extracted with $CHCl_3$ for 48 h. The residue obtained after distillation of the $CHCl_3$ was purified by preparative TLC ($CHCl_3$:EtOH 10:1) to give 80 mg of **2** (36%), mp 238-240°C.

CHICHIBABIN Pyridine synthesis

Pyridine synthesis from aromatic acetaldehydes and ammonia.

CH_2 - CHO + NH_3 ; H_3CO ; OCH_3 ; 1 ; 2 ; EtOH, Δ → [Ar, Ar, $ArCH_2$, N, H] → 1N HCl, EtOH Δ →

H_3CO ; H_3CO ; N ; OCH_3 ; OCH_3 + CH_3 ; OCH_3 ; OCH_3

3 (13%)

1	Chichibabin, A.	*J. Russ. Phys. Chem. Soc.*	**1906**	*37*	1229
2	Eliel, E.L.	*J. Am. Chem. Soc.*	**1953**	*75*	4291
3	Sprung, M.M.	*Chem. Rev.*	**1940**	*26*	301
4	Frank, R.L.	*Org. Synth. Coll.*		*IV*	451
5	Mc Gill, C. K.	*Adv. Heterocycl. Chem.*	**1988**	*44*	1

3,5-Diveratrylpyridine (3).[2] A solution of homoveratric aldehyde **1** (45.5 g, 0.25 mol) in anh. EtOH (250 mL) was refluxed for 1 h under a flow of dry NH_3 **2**. The solvent was removed and the residue was heated with 1N HCl in EtOH (100 mL). The HCl and EtOH were removed in vacuum and the residue was slurried in acetone. The hydrochloride of **3** was filtered, dissolved in 98% EtOH (350 mL) at reflux and retreated with 25% NH_4OH (100 mL). On cooling the base separated. It was filtered and dried to afford 5.66 g of **3** (12.9%). Recrystallization gave 4.41 g, mp 170-173°C.

CIAMICIAN Photocoupling

Reductive coupling of ketones to diols.

hv, iPrOH

1	Ciamician, G.	*Chem. Ber.*	**1900**	*33*	2911
2	De Mayo, P.	*Quart. Rev. (London)*	**1961**	*15*	415
3	Göth, H.	*Helv. Chim. Acta*	**1965**	*48*	1395

CIAMICIAN - DENNSTEDT Cyclopropanation

Cyclopropanation of alkenes with dichlorocarbene derived from $CHCl_3$ and sometimes subsequent ring enlargement of fused cyclopropanes.

1 → 1. Na, 2. $CHCl_3$ → → 2 (10%)

$Ph\text{-}CH{=}CH_2 + CHCl_3$ (3) → NaOH, Dibenzo(18)crown-6 → 4

1	Ciamician, G., Dennstedt, N.	*Chem. Ber.*	**1881**	*14*	1153
2	Parham, W.E.	*J. Am. Chem. Soc.*	**1955**	*77*	1177
3	Vogel, E.	*Angew. Chem.*	**1960**	*72*	8
4	Makosza, M.	*Angew. Chem. Int. Ed.*	**1974**	*13*	665

1,1-Dichloro-2-phenylcyclopropane (4).[4] To styrene **3** (10.4 g, 0.1 mol), $CHCl_3$ (11.9 g, 0.1 mol) and 50% NaOH, was added dibenzo(18)-crown-6 (0.36 g, 1 mmol). After a mild exothermic reaction and work up there was obtained 16.25 g of **4** (87%), bp 112°C/15 torr.

CLAISEN - GEUTER - DIECKMANN Ester condensation

Synthesis of open chain or cyclic β-ketoesters by aldol type condensation.

$$2\ Me_2N-\underset{\|\atop O}{C}-(CH_2)_8-COOCH_3 \ (\mathbf{1}) \xrightarrow[\Delta\ 24h]{NaOMe} Me_2N-\underset{\|\atop O}{C}-(CH_2)_7-CH(COOCH_3)-\underset{\|\atop O}{C}-(CH_2)_8-\underset{\|\atop O}{C}-NMe_2 \ \ \mathbf{2}(75\%)$$

$$\mathbf{3} \xrightarrow[heat]{NaH} \mathbf{4}(73\%)$$

3: C-OtBu, C-OtBu; 4: OtBu

1	Geuter, A.	*Arch. Pharm.*	**1863**	*106*	97
2	Claisen, L.	*Chem. Ber.*	**1887**	*20*	651
3	Dieckmann, W.	*Chem. Ber.*	**1894**	*27*	965
4	Cohen, H.	*J. Org. Chem.*	**1973**	*38*	1425
5	Taylor, R.J.K.	*Synthesis*	**1983**		996
6	Tanabe, J.	*Chem. Lett.*	**1986**		1813
7	Thyagarajan, B.S.	*Chem. Rev.*	**1954**	*54*	1029
8	Schaefer, J. P.	*Org. React.*	**1967**	*15*	1

2-t-Butoxycarbonylcyclopentanone (4).[4] To a stirred suspension of NaH (24 g, 1 mol) in PhH (400 mL) (N_2) was added **3** (5.0 g, 20 mmol) and t-BuOH (2.0 mL) in one portion and the mixture was boiled for 30 min. Another portion of **3** (120 g, 0.465 mol) in PhH (200 mL) was added dropwise for 45 min and reflux was continued for 4.5 h. The mixture was neutralized with AcOH, water (750 mL) was added followed by extraction Et_2O (2x500 mL). Evaporation of the solvent and vacuum distillation afforded 65.5 g of **4** (73%), bp 80-85°C (2 torr), Rf:0.25 silica gel Et2O:hexane 1:2.

CLAISEN - IRELAND Rearrangment

Rearrangement of allyl phenyl ethers to o- (or p-)allylphenols or of allyl vinyl ethers to γ,δ–unsaturated aldehydes or ketones (Claisen). Rearrangement of allyl esters as enolate anions to γ,δ–unsaturated acids (Ireland).

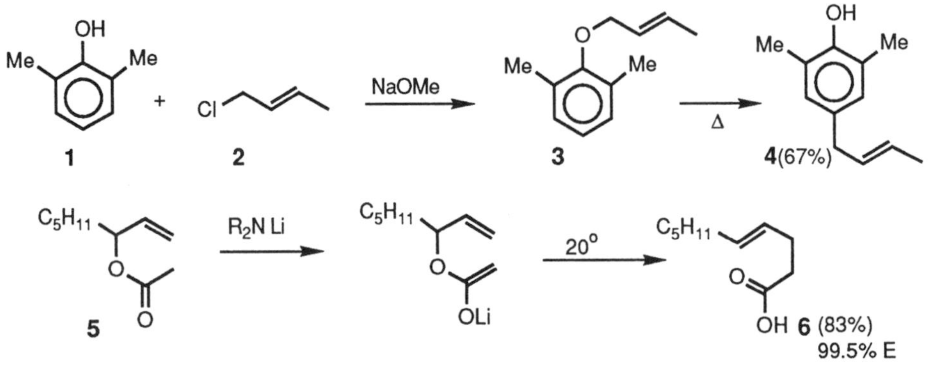

1	Claisen, L.	*Chem. Ber.*	**1912**	*45*	3157
2	Rhoades, S.J.	*J. Am. Chem. Soc.*	**1955**	*73*	5060
3	Daub, D.W.	*J. Org. Chem.*	**1986**	*51*	3404
4	Tarbell, D. S.	*Org. React.*	**1944**	*2*	1
5	Ireland, E.	*J. Am. Chem. Soc.*	**1972**	*94*	5897

2,6-Dimethyl-4-(α-methylallyl)phenol (4).[2] To refluxing **1** (122 g, 1 mol) in MeOH containing an equivalent of NaOMe was added **2** (90.5 g, 1 mol) over 20 min. After 2 h reflux, workup afforded 75.7 g of **3** (43%), bp 55-56°C (0.3 mm). **3** (17.6 g, 0.1 mol) in dimethylaniline was refluxed for 3 h to give 11.8 g of **4** (67%), bp 89-90°C (0.5 mm).

4-Decenoic acid (6).[5] N-Isopropylcyclohexyl amine (1.70 g, 12.1 mmol) in THF (20 mL) at 0°C was treated with BuLi (5 mL, 11.1 mmol) in hexane. After 10 min **5** (1.64 g, 10 mmol) was added dropwise at -78°C. After 5 min stirring the mixture was warmed to 20°C, poured into 5% NaOH (20 mL), extracted with Et_2O, the aqueous solution treated with HCl, extracted with CH_2Cl_2 and the solvent evaporated. This yielded 1.36 g of **6** (83%), 99.5% E.

C L A Y - K I N N E A R - P E R R E N Phosphonyl Chloride Synthesis

Synthesis of alkyl phosphonyl chlorides from alkyl chlorides or from ethers with PCl_3 - $AlCl_3$.

$$\underset{\mathbf{3}}{C_2H_5Cl} + \underset{\mathbf{1}}{AlCl_3} + \underset{\mathbf{2}}{PCl_3} \xrightarrow{4^\circ} \underset{\mathbf{4}}{C_2H_5ClAlCl_3PCl_3} \xrightarrow[0^\circ]{HCl} \underset{\mathbf{5}\ (43\%)}{C_2H_5\text{-}PCl_2}$$

$$\underset{\mathbf{6}}{(C_2H_5)_2O} + \mathbf{1} + \mathbf{2} \xrightarrow{0^\circ} \underset{43\%}{\mathbf{5}}$$

1	Clay, J.P.	*J. Org. Chem.*	**1951**	*16*	892
2	Kinnear, M.M., Perren, E.A.	*J. Chem. Soc.*	**1952**		3434
3	Hamilton, C.S.	*Org. Synth. Coll. vol*	*IV*		950

Ethylphosphonyl dichloride (5).[1] **A. From ethyl chloride:** To cooled stirred $AlCl_3$ **1** (13.3 g, 0.1 mol) and PCl_3 **2** (13.7 g, 0.1 mol) was added **3** (19.3 g, 0.3 mol). After 2 h stirring, the mixture was kept for 24 h at 4°C. The crystalline complex **4** (18 g) was filtered, dissolved in CH_2Cl_2 (200 mL), cooled to 0°C and treated with 32% HCl (25.3 mL). The mixture was cooled for another 2 h and filtered. The filtrate was concentrated and the residue distilled to give 7.6 g of **5** (43%), bp 174.5°C.

B. From diethyl ether: Et_2O **6** (18.5 g, 0.25 mol) was added to a mixture of **1** (66.5 g, 0.5 mol) and **2** (68.5 g, 0.5 mol) at 0°C. The mixture was heated for 7 h at 100°C (sealed tube). The crystalline product was dissolved in CH_2Cl_2 and hydrolyzed with water. After filtration and distillation 28 g of **5** (43%) was isolated.

CLEMMENSEN Reduction

Reduction of ketones or aldehydes to hydrocarbons by means of zinc amalgam and acid.

CH_3 Zn(Hg) HCl reflux CH_3

1 **2** (90%)

Zn(Hg) HCl HO OH

1	Clemmensen, E.	*Chem. Ber.*	**1913**	*46*	1838
2	Dauben, W.G.	*J. Am. Chem. Soc.*	**1954**	*76*	3864
3	Sanda, G.	*Tetrahedron Lett.*	**1983**	*24*	4425
4	Starschewsky, W.	*Angew. Chem.*	**1959**	*71*	726
5	Vedejs, E.	*Org. React.*	**1975**	*22*	401

cis-9-Methyldecalin (2).[2] cis-10-Methyl-2-decalone **1** (8.0 g, 48.2 mmol) was heated under reflux with amalgamated zinc (40 g) in AcOH (35 mL) and 32% HCl (17.5 mL). Reflux was maintained for 17 h and every 2 h there was added HCl (2 mL). Water (60 mL) was added and the mixture was steam distilled. Neutralization of the distillate with Na_2CO_3, extraction with pentane and evaporation of the solvent followed by distillation from potassium afforded 6.57 g of **2** (90%), bp 91.5 - 92.0°C/20 mm, n_D^{25} = 1.4791.

C L O K E - W I L S O N Cyclopropylketone Rearrangement

Rearrangement of cyclopropyl ketones or imines to dihydrofurans or dihydropyrroles, thermally, photochemically, or by Lewis acids.

MeO 1 CO_2Et O

Al_2O_3 24h

+ MeO 2 CO_2Et O

MeO CO_2Et O 3 (100%)

1	Cloke, J.B.	*J. Am. Chem. Soc.*	**1929**	*51*	1174
2	Wilson, C.L.	*J. Am. Chem. Soc.*	**1947**	*69*	3002
3	Alonso, M.E.	*J. Org. Chem.*	**1980**	*45*	4532
4	Hudlicky, T.	*Org. React.*	**1986**	*33*	247

2,4-Dimethyl-3-carboxyethyl-5(p-methoxyphenyl)-4,5-dihydrofuran (3).[3] A mixture of syn and anti cyclopropyl-β-ketoesters **1** and **2** was left in contact with neutral alumina activity I in $CHCl_3$ for 24 h. The starting materials **1** and **2** disappeared and **3**, homogeneous by HPLC and NMR in quantitative yield, was isolated.

COLLMAN Carbonylation Reagent

Dipotassium iron tetracarbonyl in the synthesis of aldehydes and ketones from alkyl halides.

$$K(sec\text{-}C_4H_9)_3BH + Fe(CO)_5 \xrightarrow{\Delta} K_2[Fe(CO)_4]$$

1 2 3 (98%)

4 (Br) —3→ 5 (100%) (CHO)

OTs, OTs —3→ =O

1	Collman, J.P.	*Acc. Chem. Res.*	**1986**	*1*	136
2	Collman, J.P.	*J. Am. Chem.Soc.*	**1973**	*95*	4089
3	Collman, J.P.	*Acc. Chem. Res.*	**1975**	*8*	342
4	Collman, J.P.	*J. Am. Chem. Soc.*	**1977**	*99*	2515
5	Glaisy, J.A.	*J. Am. Chem. Soc.*	**1978**	*100*	2545
6	Glaisy, J.A.	*J. Org. Chem.*	**1978**	*43*	2280

Dipotassium iron tetracarbonyl catalyst (3).[6] A THF solution of 0.5 M of K-selectride (potassium sec butylborohydride) **1** (70 mL, 35 mmol) was refluxed with $Fe(CO)_5$ **2** (3.22 g, 15.6 mmol) for 4 h. After cooling, the white solid was filtered (Schlenk technique) and washed with hexane (50 mL). Vacuum drying afforded 4.0 g of **3** (98%), mp 270-273°C (dec.) - Caution: The by-product $(sec\text{-}C_4H_9)_3B$ is spontaneously inflammable, the waste solvents must be treated with a mild oxidizing agent (NaOCl).

Nonanal (5). Octyl bromide **4** (89.44 mg, 0.46 mmol), **3** (94.5 mg, 0.0384 mmol) and Et_3P (132.5 mg, 0.508 mmol) were stirred for 12 h. Glacial AcOH (200 mL) and tridecane (100 mL) (as reference standard) was added. GC analysis indicated 100% yield of **5**.

COMBES Quinoline Synthesis

Quinoline synthesis from anilines and β-diketones.

NH_2 + O= O= $\xrightarrow[\Delta]{CaSO_4}$ N O $\xrightarrow[20°]{HF}$ N

1 **2** **3** (83%) **4** (96%)

1	Combes, A.	*Bull. Soc. Chim. Fr.*	**1882**	*49*	89(2)
2	Johnson, W.S.	*J. Am. Chem. Soc.*	**1944**	*66*	210
3	Born, J.L.	*J. Org. Chem.*	**1972**	*37*	3952
4	Bergstrom, F.W.	*Chem. Rev.*	**1944**	*35*	156
5	Siefert, W.	*Angew. Chem. Int. Ed.*	**1962**	*1*	215

4-(2-Naphthylimino)pentanone-2 (3).[2] A mixture of β-naphthylamine **1** (5.0 g, 0.0346 mol), acetylacetone **2** (5.85 g, 0.585 mol) and drierite (10 g) was heated on a water bath for 3-4 h. The crude product 7.55 g (96%), mp 89-96°C, after recrystallization from petroleum ether gave 6.5 g, of **3** (83%), mp 98-99°C.

2,4-Dimethylbenzo(g)quinoline (4). A mixture of **3** (13.4 g, 0.059 mol) in HF (300 mL) was maintained for 24 h at 20°C. The residue obtained after removing the HF was neutralized with 10% K_2CO_3 solution, extracted with Et_2O and the solvent was evaporated to yield 11.75 g of **4** (96%), mp 91-92.5°C.

CONIA Cyclization

Thermal cyclization of dienones, enals, ynones, diones, ketoesters, etc., to monocyclic, spirocyclic bicyclic derivatives, (ene reaction of unsaturated enol).

1 + 2 → 3 (18%) —350°→ 4 (100%)

5 —370°, 30mm→ 6 ⇌ 7 (93%)

R : H, Me, CO_2Et

1 Conia, J.M.	*Tetrahedron Lett.*	**1965**		3305, 3319
2 Conia, J.M.	*Bull. Soc. Chim. Fr.*	**1966**		278, 281
3. Krapcho, A.P.	*Synthesis*	**1974**		416
4 Conia, J.M.	*Angew. Chem. Int. Ed.*	**1975**	*14*	473

2-(Penten-4'-yl) cyclohexanone (3). To **1** (10.9 g, 0.1 mol) and **2** (14.8 g, 0.1 mol) in PhH (50 mL) was added dropwise 2N sodium t-amylate. After 12 h reflux and neutralization with HCl the organic layer was washed and dried. The residue obtained after evaporation of the solvent was distilled in vacuum to give 3.0 g of **3** (18%), bp 114-116°C (10 min).

Methyl-1-spiro-(4,5)-decanone-6 (4).[4] Heating of **3** at 350°C for 1 h gave **4** as a colorless oil in quantitative yield; **4**-DNPH (from MeOH), mp 146°C.

COOPER - FINKBEINER Hydromagnesiation

Ti catalyzed reaction of Grignard reagents with olefins or acetylenes.

PrMgBr (1) + 2 —1) $TiCl_4$, 2) O_2→ CH_2-CH_2OH (3)

$Me_3Si-C\equiv C\ CH_2OH$ (4) —i BuMgBr, Cp_2TiCl_2→ $Me_3Si(BrMg)C=C(H)CH_2OMgBr$ (5) —CH_3CH_2CN→ 6 (82%)

$C_4H_9-C\equiv C-SiMe_3$ —i BuMgBr/Cp_2TiCl_2→ $C_4H_9(H)C=C(SiMe_3)H$ (>98%Z)

1	Cooper, G.D., Finkbeiner, H.L.	*J. Org. Chem.*	**1962**	*27*	3395
2	Sato, F.	*J. Chem. Soc. Chem. Comm,.*	**1981**		718
3	Sato, F.	*Tetrahedron Lett.*	**1983**	*24*	1804
4	Sato, F.	*J. Chem. Soc. Chem. Comm.*	**1983**		162
5	Sato, F.	*Tetrahedron Lett.*	**1984**	*25*	5063

β-(Δ^3-Cyclohexenyl) ethanol (3).[1] To Grignard reagent **1**, prepared from Mg (13.2 g 0.55 at g) and propyl bromide (61.3 g, 0.5 mol) in Et_2O (150 mL), was added cooled **2** (54 g, 0.5 mol), followed by $TiCl_4$ (1 mL). After 2 h reflux a second portion of $TiCl_4$ (0.45 mL) was added and reflux was continued for another 14 h. The mixture was oxidized with dry air. Distillation yielded 25 g (40%) of **3**, bp 92-94°C.

3-Trimethylsilyl-2-ethylfuran (6).[5] A catalytic amount of Cp_2TiCl_2 (0.12 g, 0.48 mol) was added to iBuMgBr in Et_2O (43 mL, 0.4 M sol 17 mmol) under Ar at 0°C. After 5 min stirring, **4** (0.18 g, 6.8 mmol) was added and the mixture was stirred at 25°C for 6 h. To this solution was added propionitrile (0.48 g, 8.8 mmol). The mixture was stirred for 2 h at 25°C, quenched with 2 N HCl and extracted with Et_2O. The extract was dried, evaporated and the residue chromatographed (silica gel) to afford 0.94 g of **6** (82%).

COPE Rearrangement

3,3-Sigmatropic rearrangement of 1,5-dienes.

CN, CO_2Et, **1**, Br, NaOEt, **2**, CN, CO_2Et, **3** (34%), 150°, CN, CO_2Et, **4** (66%)

O, cis, 60°, O

1	Cope, A.C.	*J. Am. Chem. Soc.*	**1940**	*62*	441
2	McDowell, D.W.	*J. Org. Chem.*	**1986**	*51*	183
3	Baldwin, J.E.	*J. Org. Chem.*	**1987**	*52*	676
4	Vogel, E.	*Liebigs Ann.*	**1958**	*615*	1
5	Lutz, R. P.	*Chem. Rev.*	**1984**	*84*	205
6	Blechert, S.	*Synthesis*	**1989**		71

Ethyl 2-cyano-4-methyl-4-ethyl-hepta-2,6-dienoate (4).[1] Ethyl 2-cyano-4-methyl-3-hexenoate **1** (85 g, 0.5 mol) in EtOH (200 mL), was treated with NaOEt (from 12.5 g, 0.5 at. Na in 100 mL EtOH). To the sodium salt of **1** was added dropwise under stirring allyl bromide **2** (66.55 g, 0.55 mol). The mixture was refluxed until neutral (1 h), and after cooling, water was added. The mixture was extracted with benzene and after solvent evaporation the residue was distilled in vacuum. The first fraction gave 67 g of **3**, bp 90-143°C (1 mm). After shaking with 20% $NaHSO_3$ the product was redistilled through a Widmer column to give 37 g of **3** (34%), bp 94.5-96°C (1 mm). **3** (12 g, 55 mmol) was refluxed for 20 min and distilled in vacuum to yield 8 g of **4** (66%), bp 147-148°C (16 mm), n_D^{20} = 1.4780.

COPE - MAMLOC - WOLFENSTEIN Olefin synthesis

Olefin formation by elimination from tert.amine N-oxides.

NMe2 → NMe2 O →

1 2 3 (90%)

1	Mamloc, L., Wolfenstein, R.	*Chem. Ber.*	**1900**	*33*	159
2	Cope, A.C.	*Tetrahedron Lett.*	**1949**	*71*	3929
3	Bluth, M.	*Tetrahedron Lett.*	**1984**	*25*	2873
4	De Puy, C.H.	*Chem. Rev.*	**1960**	*60*	448
5	Fujita, J.	*Synthesis*	**1978**		934
6		*Org. Synth. Coll.*		*II*	381

Cis cyclooctene (3).[2] To ice cooled N,N-dimethylcyclooctylamine **1** (5.0 g, 32 mmol) in MeOH (10 mL), was added over 30 min 35% hydrogen peroxide (10.0 g, 99 mmol). Stirring was continued until the amine had been consumed (24 h). The excess of peroxide was destroyed by stirring with platinum black (0.25 g) for 5 h. The platinum was removed by filtration and the filtrate was concentrated in vacuo to a syrup (max. temp 30-40°C), to give **2**. Amine oxide **2** was heated under vacuum (10 mm) in a N_2 atmosphere and the temperature was raised 1-2°/min. Decomposition and distillation began at 100°C and ended at 130°C. The distillate was treated with HCl, and congelated with dry ice. The product which separated as an oil was distilled to give 3.22 g of **3** (90%), mp -16°, bp 65°C (59 mm) 145-146°C (760 mm), n_D^{20} = 1.4684.

COREY Oxidizing Reagents for Alcohols

Pyridinium chlorochromate **1** or CrO_3-dimethylpyrazole **4** for oxidation of alcohols to ketones or aldehydes.

2 OH — $N^{+}H$ $ClCrO^{+}_{4}$ **1** → **3** (93%)

5 OH — CrO_3 / 3,5-dimethylpyrazole (N, NH) **4** → **6** (98%)

1 Corey, E.J.	*Tetrahedron Lett.*	**1973**		2647
2 Dauben, W.G.	*J. Org. Chem.*	**1977**	*42*	682
3 Corey, E.J.	*Tetrahedron Lett.*	**1979**		399

Isophorone (3).[2] To a sturred slurry of **1** (prepared from 6 M HCl, CrO_3 and pyridine at 0°)[1] (4.30 g, 20 mmol) in CH_2Cl_2 (30 mL) was added in one portion a solution of 1,5,5-trimethylcyclohex-2-ene-1-ol **2** (1.40 g, 10 mmol) in CH_2Cl_2 (10 mL) at 20°C. After 3 h stirring the mixture was extracted with Et_2O, the extract was washed with 5% NaOH, 5% HCl and saturated $NaHCO_3$. After evaporation of the solvent the product was distilled bulb to bulb to afford 1.33 g of **3** (92%), bp 213-214°C, n_D^{20} = 1.4795.

Acetophenone (6). 3,5-Dimethylpyrazole (580 mg, 6 mmol) was added to a suspension of CrO_3 (600 mg, 6 mmol) in CH_2Cl_2 (20 mL) and the mixture was stirred at 20° under Ar for 15 min. To this red solution of **4**, 1-phenylethanol **5** (263 mg, 2.2 mmol) in CH_2Cl_2 (2 mL) was added in one portion and the mixture was stirred for 30 min at 20° (GC monitoring on Carbowax 20M). Evaporation, extraction with Et_2O and evaporation left a residue which was dissolved in pentane and filtered through silica. Evaporation gave 260 mg of **6** (98%).

C O R E Y Enatioselective borane reduction

Enantioselective reduction of ketones by borane or catecholborane catalyzed by oxazaborolidine **3**

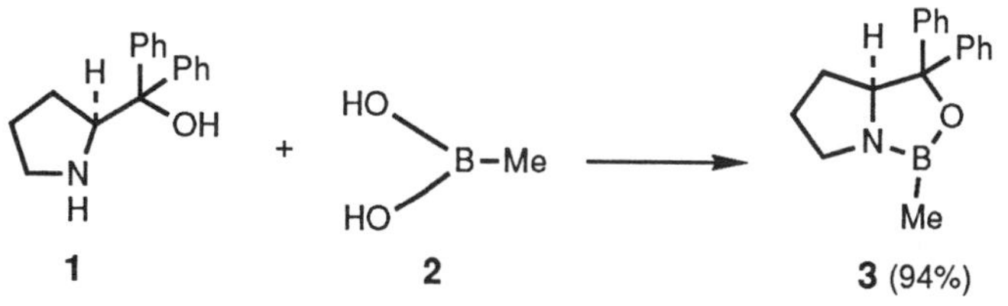

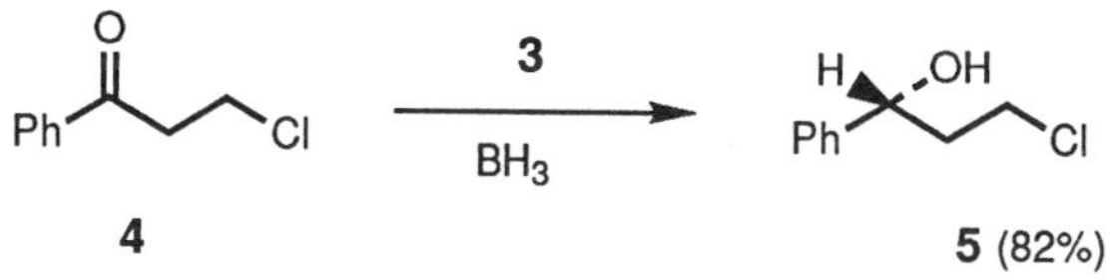

1	Corey, E.J.	*J. Am. Chem. Soc.*	**1987**	*109*	5551
2	Corey, E.J.	*J. Org. Chem.*	**1988**	*53*	2861
3	Corey, E.J.	*Tetrahedron Lett.*	**1989**	*30*	6275
4	Corey, E.J.	*Tetrahedron Lett.*	**1990**	*31*	611
5	Todd K. Jones	*J. Org. Chem.*	**1991**	*56*	763

Oxazaborolidine (3).[2] A solution of (S)-(-)2-diphenylhydroxymethylpyrrolidine **1** (96 mg, 0.38 mmol) and methyl boronic acid **2** (23 mg, 0.38 mmol) in PhH (6 mL) was stirred at 20°C for 1.5 h in the presence of 4Å molecular sieves (0.8 g). Filtration and concentration of the filtrate led to 95 mg of **3** (94.5%).

R(+)3-Chloro-1-phenyl-1-propanol (5).[3] β-Chloropropiophenone **4** (0.162 g, 1 mmol) in THF was added to 0.6 equiv. of BH_3 and 0.1 equiv. of **3** at 0°C in THF over 20 min. After 30 min, one adds MeOH and 1.2 equiv. of HCl in Et_2O, followed by removal of the volatiles. Addition of PhMe afforded the crystal-line (S)-diphenylproline (recycled as catalyst). The solution after concentration afforded 0.162 g of **5** (99%), 94% ee. Recrystallyzation (hexane) gave a first crop (82%), mp 57-58°C, $[\alpha]_D^{25}$ = +24° (c=1, $CHCl_3$).

COREY Homologative epoxidation

Reaction of ketones with S-ylides derived from $Me_3S^+ I^-$ or $Me_3SO^+ I^-$ to give epoxides.

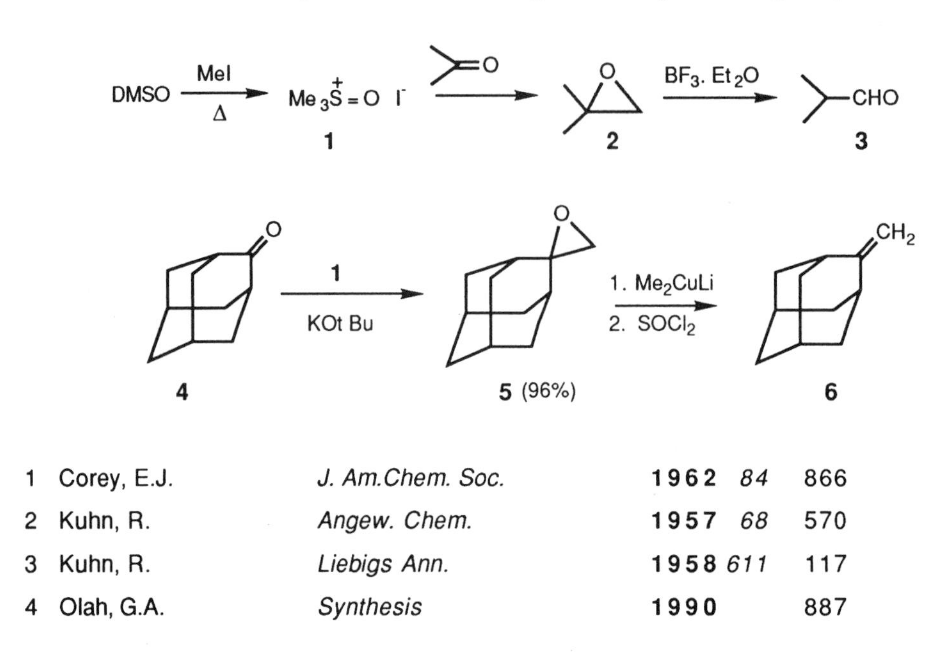

1	Corey, E.J.	*J. Am.Chem. Soc.*	**1962**	*84*	866
2	Kuhn, R.	*Angew. Chem.*	**1957**	*68*	570
3	Kuhn, R.	*Liebigs Ann.*	**1958**	*611*	117
4	Olah, G.A.	*Synthesis*	**1990**		887

2-Methyleneadamantane epoxide (5).[4] A mixture of 2-adamantanone **4** (1.5 g, 10 mmol), **1** (2.20 g, 10 mmol) and t-BuOK (97% 1.15 g, 10 mmol) in DME (50 mL) is refluxed with good stirring under N_2 for 8 h. The mixture was quenched with H_2O and extracted with Et_2O (3.50 mL). Evaporation gave 1.57 g of **5** (96%), mp 176°C.

2-Methylideneadamantane (6). To etheral Me_2CuLi under N_2 and efficient stirring was added **5** (0.4 g, 2.44 mmol) in Et_2O (5 mL) at 0°C. After 30 min freshly distilled $SOCl_2$ (2.40 g, 20 mmol) was added over 5 min. Stirring was continued for another 5 min followed by quenching with water, extraction with Et_2O and chromatography (silica gel/pentane) to give 0.32 g of **6** (81%), mp 83°C.

COREY - KIM Oxidizing Reagent

Oxidation of alcohols to ketones by means of N-chlorosuccinimide (NCS) or NBS and Me_2S.

1 Corey, E.J., Kim, C.U. *J. Am. Chem. Soc.* **1972** *94* 7586

2 Jamauki, M. *Chem. Lett.* **1989** 973

4-tert-Butylcyclohexanone (4).[1] To a stirred solution of N-chlorosuccinimide (400 mg, 3.0 mmol) in PhMe (10 mL) was added at 0°C under Ar, methyl sulfide **2** (0.3 mL, 4.1 mmol). A white precipitate appeared immediately after addition of the sulfide. The mixture was cooled to -25°C, and a solution of 4-tert-butylcyclohexanol **3** (312 mg, 2.0 mol) in PhMe (2 mL) was added dropwise. Et_2O (20 mL) was added, the organic layer was washed with 1% HCl (5 mL) and twice with water (15 mL). Removal of the solvent left 310 mg of **4** (100%), mp 44-47°C.

COREY - WINTER Alkene synthesis

Alkene synthesis from glycols via cyclic 1,2-thionocarbonates.

HO HO O OH O O **1** Δ N N C=S 2 **2** S = C O O O OH O O **3** (85%) P(OMe)3 150° O OH O O **4** (58%)

1	Corey, E.J., Winter	*J. Am. Chem. Soc.*	**1963**	*85*	2677
2	Horton, D.	*Tetrahedron Lett.*	**1964**		2531
3	Corey, E.J.	*J. Am. Chem. Soc.*	**1965**	*87*	934
4	Horton, D.	*J. Org. Chem.*	**1970**	*35*	3558
5	Dante, E.	*Tetrahedron Lett.*	**1972**		4435

5,6-Dideoxy-1,2-0-isopropylidene-α-D-xylohexofuranos-5-ene (4).[2] 1,2-0-Isopropylidene-α-D-glucofuranose **1** (11.0 g, 55 mmol) and bis(imidazol-1-yl)thione **2** (11.75 g, 66 mmol) in BuOH (100 mL), after 4 h of reflux gave 10.45 g of **3** (85%), mp 205,206°C (MeOH) $[\alpha]_D^{18}$ = -17.3°.

3 (5.0 g, 19 mmol) in trimethyl phosphite (20 mL) under N_2 was maintained for 60 h at 150°C and after cooling was poured into 1N NaOH (250 mL) under efficient stirring. Extraction with $CHCl_3$ and solvent evaporation gave 2.78 g of a crude product (TLC silica gel CH_2Cl_2;Et_2O 3:1), R_f 0.5. Crude **4** in Et_2O was treated with petroleum ether to give after 12 h at 0°C 1.92 g of **4** (58%).

CORNFORTH Rearrangement

Thermal rearrangement of 4-carbonyl substituted oxazoles via nitrilium ylids (**2**).

Ph—C≡N⁺ ... Δ PhMe

1 **2** **3** (90%)

1 Cornforth, J.W.	*The Chemistry of Penicillin*	**1949**	*698*	705
2 Dewar, M.J.S.	*J. Org. Chem.*	**1975**	*40*	1521

Ethyl 2-phenyl-5-(1-aziridinyl)oxazole-4-carboxylate (3).[2] A solution of 2-phenyl-5-ethoxyoxazole-4-carboxylic acid chloride (1.257 g, 5 mmol) in PhH (20 mL) was stirred with aziridine (0.215 g, 5 mmol) and triethylamine (0.5 g, 5 mmol) in PhH (40 mL) for 3 h at 20-°C. After washing with water, the solvent was removed and the residue crystallized from petroleum ether to give 1.032 g of **1** (80%).

A solution of **1** (1.032 g, 4 mmol) in PhMe was heated to reflux for 17 h. After removing the solvent the residue was recrystallized from petroleum ether and gave 0.928 g of **3** (90%).

CRIEGE Glycol oxidation

Oxidation of 1,2-glycols to two carbonyl moieties by lead tetraacetate (LTA).

$$CH_3-(CH_2)_7-CH(OH)-CH(OH)-(CH_2)_7-COOH + Pb(OAc)_4 \longrightarrow O{=}CH-(CH_2)_7-COOH + CH_3-(CH_2)_7-CHO$$

1   2   3 (90%)   4 (82%)

OH, OH, LTA, OHC, CHO

ref. 6

1	Criege, R.	*Liebigs Ann.*	**1930**	*481*	263
2	Criege, R.	*Chem. Ber.*	**1931**	*64*	260
3	Chi-yi Hsing	*J. Am. Chem. Soc.*	**1939**	*61*	3589
4	Criege, R.	*Angew. Chem.*	**1958**	*70*	173
5	Michailovici, M. L.	*Synthesis*	**1970**		209
6	Nakajima, N.	*Chem. Ber.*	**1956**	*89*	2274

Pelargonic aldehyde and azelaic aldehyde (2) and **(3).**[3] A suspension of 9,10-dihydroxyoctadecanoic acid **1** (0.674 g, 2.1 mmol) in AcOH (10 mL) was treated with lead tetraacetate **2** (0.95 g, 2.9 mmol) and shaken for 24 h at rt. The mixture was filtered and the filtrate was heated for 5 min at 45°C. Water (15 mL) and semicarbazine hydrochloride (0.5 g) was added followed by NaOAc (0.6 g) and a few drops of MeOH. After 24 h at 0°C the semicarbazones were filtered and separated by dissolving **3** in NaOH and reprecipitation with HCl. This gave 0.348 g of **4** (82%), mp 99-100.5°C and 0.434 g of **3** (90%), mp 163-164°C.

C U R T I U S Rearrangement

Degradation of acid hydrazides or acyl azides to amine or amine derivatives.

HO CONH-NH$_2$ HO 1 → HNO_2 / HO^- → HO NH_2 HO 2 15%

F COOH F → $(PhO)_2P\text{-}N_3$, TEA / t-BuOH, 2h reflux → F NH-C O_2-tBu F (63%)

1	Curtius, T.	*Chem. Ber.*	**1890**	*23*	3023
2	Caldwell, W.T.	*J. Am. Chem. Soc.*	**1939**	*61*	3584
3	Newcastle, G.W.	*Synthesis*	**1985**		220
4	Thor nton, T.J.	*Synthesis*	**1990**		295
5	Saunders, J.M.	*Chem. Rev.*	**1948**	*43*	205
6	Cohen, L.D.	*Angew. Chem.*	**1961**	*73*	259
7	Smith, P.A.S.	*Org. React.*	**1946**	*3*	337

Desoxycholamine (2).[2] To a solution of desoxycholic acid hydrazide **1** (22.0 g, 54 mmol) in 1 N HCl (65 mL) was added a solution of $NaNO_2$ (4.2 g, 61 mmol) in water (50 mL) under good stirring and ice cooling. After an additional 15 min, the azide was filtered and the wet cake was heated gently in AcOH (210 mL) until gas evolution ceased. After 1 h the mixture was filtered through a sinter and made alkaline to pH=9. The precipitate was dissolved in alcoholic KOH and reprecipitated with a large amount of water. Filtration, drying and recrystallization from anhydrous MeOH afforded 3g of **2** (15%).

DAKIN Oxidation

Oxidation of aldo (keto) phenols to polyphenols (see Bayer-Villiger oxidation).

Me, C=O, OH, Me, Me, 1; $Me_4N^+O^-H$, H_2O_2, 55° → Me, $^-O-C-O-OH$, OH, Me, Me, 2 → Me-C=O, O, OH, Me, Me, 3 → OH, OH, Me, Me, 4 (25%)

1	Dakin, H.D.	*Am. Chem. J.*	**1909**	*42*	477
2	Baker, J.	*J. Chem. Soc.*	**1953**		1615
3	Lee, J.B.	*Quart. Rev.*	**1969**	*21*	454

DAKIN - WEST Acylation

An acylative decarboxylation of α-amino acids or α–thio acids.

Ph-S - CH_2CO_2H 1 → (Ac_2O, Lutidine) → Ph-S - CH_2COMe 2 (21%)

1	Dakin, H., West, R.	*J. Biol. Chem.*	**1928**	*78*	91
2	Dyer, E.	*J. Org. Chem.*	**1968**	*33*	880
3	Knorr, L.	*Angew. Chem. Int. Ed.*	**1965**	*4*	705
4	Buchanan, G. L.	*Chem. Soc. Rev.*	**1988**	*17*	91

Thiophenoxyacetone (2).[2] Thiophenoxyacetic acid **1** (3.0 g, 17.9 mmol) in Ac_2O (30 mL) and 2,6-lutidine was refluxed for 12 h. After removal of the solvents the product was distilled to give 0.6 g of **2** (21%), phenylhydrazone, mp 82.5°C.

DANHEISER Annulation

Regiocontrolled synthesis of five membered rings from silylallenes and Michael acceptors in the presence of $TiCl_4$.

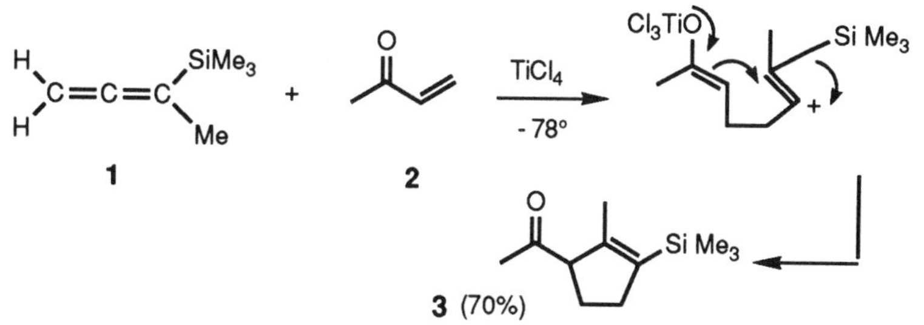

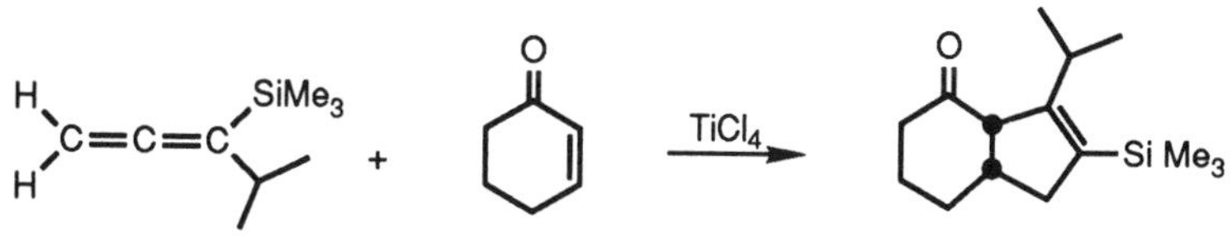

1 Danheiser, R.L. *J. Am. Chem. Soc.* **1981** *103* 1604

2 Danheiser, R.L. *Tetrahedron* **1983** *39* 935

3 Danheiser, R.L. *Org. Synth.* **1988** *66* 8

1-Methyl-2-trimethylsilyl-5-acetylcyclopentene (3).[1] Titanium tetrachloride (0.283 g, 1.5 mmol) was added to a mixture of 1-methyl-1-trimethylsilylallene **1** (0.126 g, 1 mmol) and methyl vinyl ketone **2** (0.07 g, 1 mmol) in CH_2Cl_2 at -78°C. The red mixture was stirred for 1 h at -78°C, and then was quenched by addition of water and Et_2O. Extraction with Et_2O, evaporation of the solvent and chromatography afforded 0.125 - 0.144 g of **3** (68-75%).

DANISHEFSKY Dienes

Silyloxydienes in regio and stereo controlled Diels-Alder and hetero Diels-Alder reactions.

O OMe + Me_3SiCl $\xrightarrow[Et_3N]{ZnCl_2}$ Me_3SiO OMe $\xrightarrow[\Delta]{4}$ O CHO

1 2 3 (68%) 5 (72%)

OMe Me_3SiO + O=C(H)Ph ⟶ OMe O MeSiO Ph

1 Danishefsky, S. *J. Am. Chem. Soc.* **1974** *96* 7807

2 Danishefsky, S. *J. Am. Chem. Soc.* **1982** *104* 6457

Trans-1-methoxy-3-trimethylsilyloxy-1,3-butadiene (3)[1] Anhydrous powdered $ZnCl_2$ (2.0 g, 1.5 mmol) was added to TEA (115 g, 1.1 mmol) and the mixture was stirred for 1 h until the salt was suspended in TEA. To this suspension was added a solution of trans-1-methoxybutene-3-one 1 (50.0 g, 0.5 mol) in PhH (150 mL) followed by trimethylchlorosilane **2** (108.5 g, 1 mol). After 30 min the temperature was raised to 40°C and stirring was continued overnight. The cooled mixture was added to Et_2O (100 mL) and filtered; the combined filtrate and washings were concentrated in vacuum and the residue distilled to give 58.2 g of **3** (68%), bp 54-55°C/3 mm.

4-Methyl-4-formylcyclohex-2-ene-1-one (5). A mixture of **3** and methacrolein **4** in PhH was refluxed for 24 h. After workup **5** was obtained in 72% yield.

DARAPSKI Amino Acid Synthesis (see Curtius)

Amino acid synthesis from ethyl cyanoacetates.

CO_2Et CN $\xrightarrow{NH_2NH_2}$ R ... $NHNH_2$ CN $\xrightarrow{HNO_2}$ NH_2 CO_2H

1 **2** (100%) **3** (40%)

1 Darapski	*J. Prakt. Chem.*	**1915**	*92*	297
2 Gagnon, P.E.	*Can. J. Chem.*	**1951**	*29*	70

DARZENS - NENITZESCU Acylation

Zn-Cu catalyzed Friedel-Crafts type acylation of olefins with acyl chlorides.

\+ Cl O $\xrightarrow[Cu_2Cl_2]{Zn}$ O + O

1 **2** **3** (63%) **4**

1 Darzens, G.	*C.R.*	**1910**	*150*	707
2 Nenitzescu, C.D.	*Liebigs Ann.*	**1931**	*491*	189
3 Shono, T.	*J. Org. Chem.*	**1983**	*48*	2503

DARZENS Epoxide Synthesis

Synthesis of glycidic esters (amides) from an aldehyde or ketone and an α-haloester (amide).

Ph-CHO + $ClCH_2C(O)NEt_2$ —tBuOK→ Ph-HC—CH-C(O)-NEt_2 (epoxide, O bridging)

1 **2** **3** (88%)

1	Darzens, G.	*C.R.*	**1904**	*139*	1214
2	Tung, T.T.	*J. Org. Chem.*	**1963**	*28*	1514
3	Gladiale, S.	*Synth. Commun.*	**1982**	*12*	355
4	Balester, M.	*Chem. Rev.*	**1955**	*55*	283
5	Newman, M.S.	*Org. React.*	**1949**	*5*	414

Cis and Trans Epoxypropanamide (3).[2] t-BuOK (K16 g and t BuOH 400 mL) was added to **1** (42.4 g, 0.4 mol) and **2** (59.8 g, 0.4 mol) under N_2 at 10°C over 90 min. After 1 h the solvent was removed at 50°C (40 min). Work up gave a viscous oil (87.1 g 99%) which treated with Et_2O (150 mL) and hexane (300 ml) gave 77 g of **3** (88.4%), mp 43-47°C. Crystallization gave trans **3**, mp 88.8-90°C and cis **3**, mp 52.4-53°C.

DAVIDSON Oxazole Synthesis

Synthesis of triaryl oxazoles from α-hydroxyketones (benzoins).

Ph - CH - OH / Ph - C = O → (Cl - C(=O) - Ph) → → (NH_4OAc) →

1 Davidson, D.	*J. Org. Chem.*	**1937**	*2*	328
2 Wiley, R.H.	*Chem. Rev.*	**1945**	*37*	408

DELEPINE Aldehyde Oxidation

Mild oxidation of aldehydes to carboxylic acid using silver salts.

1 → (Ag_2O) → **2** (90%)

1 Delepine, M.	*Bull. Soc. Chim. Fr.*	**1909**	*5*	879(4)
2 Stummer, C.	*Unpublished results*			
3 Harrison, R.I.	*Org. Synth. Coll.*		*IV*	493;919;972

3-Chlorosalicylic acid (2).[2] To $AgNO_3$ (8.5 g, 50 mmol) in water (50 mL) was added NaOH (2.2 g, 55 mmol) in water (5 mL). Silver oxide was filtered, washed (water), suspended in 10% NaOH (100 mL) and heated to 60-65°C. 2-Hydroxy-3-chlorobenzaldehyde **1** (7.83 g, 50 mmol) was added slowly. After 30-40 min the mixture was filtered and the filtrate acidified to pH 4. Crystallization from water gave 7.75 g of **2** (90%), mp 178°C.

DAVIS Oxidizing reagent

2-Sulfonyloxaziridines as aprotic neutral oxidizing reagents; oxidation of amines, sulfides, selenides and asymmetric oxidation.

1 + 2 → O_3 → 3 (66 %)

Na HMDS, -78° → (85%, (S) 93% ee)[6]

1	Davis, F.A.	*J. Org. Chem.*	**1982**	*47*	1174
2	Davis, F.A.	*Tetrahedron Lett.*	**1983**	*24*	1213
3	Davis, F.A.	*J. Org. Chem.*	**1986**	*51*	4083, 4240
4	Zajak, W.W.	*J. Org. Chem.*	**1988**	*53*	5856
5	Davis, F.A.	*J. Org. Chem.*	**1990**	*55*	3715
6	Davis, F.A.	*J. Am.Chem.Soc.*	**1990**	*112*	6679

cis-4-(Nitromethyl)cyclohexanecarboxylic acid (3).[4] To a solution of 2-(phenylsulfonyl)-3-phenyloxaziridine **2** (0.523 g, 2.0 mmol) in $CHCl_3$ (10 mL) was added 3-azabicyclo[3.2.2] nonane **1** (0.125 g, 1 mmol). The reaction mixture was stirred for 15 min, then the solvent was removed by rotary evaporation and replaced by CH_2Cl_2. This solution was ozonized at -78°C. The CH_2Cl_2 solution was then extracted with saturated $NaHCO_3$ solution. The aqueous layer was neutralized with HCl and then extracted with CH_2Cl_2. The CH_2Cl_2 solution was rotary evaporated, and the residue subjected to PLC. The major fraction that was isolated was recrystallized from EtOH to provide 0.123 g of **3** (66%), mp 83-85°C.

DELEPINE Amine Synthesis

Synthesis of primary amines from alkyl halides with hexamethylenetetramines.

$$\mathbf{1} \xrightarrow{CuBr_2} \mathbf{2}\ (\text{Br}) \xrightarrow{(CH_2)_6N_4\ \mathbf{3}} \mathbf{4}\ (NH_3^+\ Cl^-)\ (40\%)$$

$$PhCH_2Br + \mathbf{3} \xrightarrow{NaI} PhCH_2(N_4C_6H_{12})^+ I^+ \xrightarrow[2.\ NaOH]{1.\ HCl(g)} PhCH_2NH_2 \quad (82\%)$$

1	Delepine, M.	*Bull. Soc. Chim. Fr.*	**1885**	*13*	356
2	Galat, A.	*J. Am. Chem. Soc.*	**1939**	*61*	3585
3	Henry, A.	*J. Org. Chem.*	**1990**	*55*	1796
4	Angyal, S.T.	*Org. Synth.*	Coll. Vol.	*IV*	121

2-Fluorophenacylamine Hydrochloride (4).[3] A mixture of 2-fluoroacetophenone **1** (25 g, 0.18 mol) and $CuBr_2$ (65 g, 0.29 mol) in $CHCl_3$ (160 mL) and EtOAc (160 mL) was refluxed for 4 h. CuBr was filtered, the filtrate evaporated and the oily residue taken up in $CHCl_3$ (100 mL) and filtered into hexamethylenetetramine **3** (24 g, 0.17 mol) in $CHCl_3$ (300 mL). After 16 h the adduct was filtered and washed with $CHCl_3$. The dried adduct was suspended in MeOH (450 mL), cooled and treated with 32% HCl (63 mL). After 3 days the supernatant liquid was decanted from the ammonium salts, the solvent removed under vacuum and the residue extracted with 3 portions of anhydrous EtOH (25 mL, 200 mL and 200 mL). The cooled extract afforded 12.3 g of **4** (40% overall yield), mp 197-220°C.

DE MAYO Cycloaddition

Photochemical 2+2 cycloaddition.

1 + 2 → hυ → [cyclobutane intermediate] + regioisomer → 3 (60%) + 4

1	De Mayo, P.	*Proc. Chem. Soc. London*	**1962**		119
2	De Mayo, P.	*Can. J. Chem.*	**1962**	*41*	440
3	De Mayo, P.	*J.Org. Chem.*	**1969**	*34*	794
4	De Mayo, P.	*Acc. Chem. Res.*	**1971**	*4*	41
5	Weedon, A.C.	*The Chemistry of Enols* (Wiley)	**1990**		591

Cycloocta-1,5-dione-2 -(or **3-)carboxylic acid methyl ester (3)** + **(4).**[3] A solution of cyclohexa-1,3-dione **1** (1.00 g, 8.9 mmol) in methyl acrylate **2** (100 g, 1.16 mmol) was irradiated with a λ 450 W medium pressure lamp under N_2 through a pyrex filter for 5 h. The semicrystalline residue obtained after removal of the solvent, was crystallized from MeOH, the mother liquor, separated by preparative TLC (PhH:Et_2O) and the main fraction combined with the crystals to give 1.06 g of **3** (60%), mp 104-105°C (MeOH). The second minor fraction (from TLC) gave **4**, mp 89-91°C (MeOH).

DEMJANOV Rearrangement

Deamination of primary amines to rearranged alcohols (via diazonium compounds) with ring contraction or enlargement for alicyclic amines.

CH_3 HNO_2 CH_2NH_2 → CH_3 CH_2—N^+_2 —95°→ CH_3 OH

1 2 (78%)

HNO_2 NH_2 → CH_2OH

1	Demjanov, N.J.	*J. Russ. Phys. Chem. Soc.*	**1903**	*35*	26
2	Kottany, R.	*J. Org. Chem.*	**1965**	*30*	350
3	Smith, P.A.	*Org. React.*	**1960**	*11*	154

2-Methylcycloheptanol (2).[2] To an ice cooled solution of 85% H_3PO_4 (17.52 g, 0.159 mol) and water (120 mL) was added with stirring 1-methyl-2-(aminomethyl)cyclohexane **1** (25.4 g, 0.2 mol). To the white salt obtained, a solution of $NaNO_2$ (13.8 g, 0.2 mol) in water (30 mL) was added. After an hour of stirring a solution of $NaNO_2$ (2.76 g, 0.04 mol) in water (5 mL) and 85% H_3PO_4 (1 mL) was added. The mixture was heated for 1 h at 95°C and extracted with Et_2O. The extract was washed with HCl, evaporated and distilled to give 20 g of **2** (78%), bp 73-78°C (20 mm).

DESS - MARTIN Oxidizing Reagent

Oxidation of alcohols to aldehydes or ketones by means of periodinanes.

1 —($KBrO_3$, H_2SO_4)→ 2 —(Ac_2O)→ 3 (87%)

4 3,4,5-$(CH_3O)_3$-C_6H_2-CH_2OH + 3 → 3,4,5-$(CH_3O)_3$-C_6H_2-CHO 5

1	Dess, P.B., Martin, J.C.	*J. Am. Chem. Soc.*	**1978**	*100*	300
2	Dess, P.B., Martin, J.C.	*J. Am. Chem. Soc.*	**1979**	*101*	5294
3	Yagupolsky, L.M.	*Synthesis*	**1977**		574
4	Dess, P.B., Martin, J.C.	*J. Org. Chem.*	**1983**	48	4155
5	Robins, J.C.	*J. Org. Chem.*	**1990**	*55*	5186

Periodinane (3).[4] $KBrO_3$ (76 g, 0.45 mol) was added over 0.5 h to 2-iodobenzoic acid **1** (85.2 g, 0.34 mol) and 0.73 M H_2SO_4 (730 mL) below 55°C. The mixture was stirred for 3.6 h at 65°. Filtration (0°C), washing (H_2O 1000 mL, and EtOH 2x50 mL) gave 89.1 g of **2** (93%). A stirred slurry of **2** (25 g, 89 mmol) in Ac_2O (84 g, 0.83 mol) and AcOH (70 mL) was heated to 100°C for 40 min. Vacuum distillation of the solvent, filtration of the slurry (N_2) and washing (Et_2O) gave 35.1 g of **3** (93%), mp 124-126°C.

3,4,5-Trimethoxybenzaldehyde (5). 3,4,5-Trimethoxybenzyl alcohol **4** (0.44 g, 2.23 mmol) in CH_2Cl_2 (8 mL) was added to **3** (1.06 g, 2.47 mmol) in CH_2Cl_2 (10 mL) and after 20 min Et_2O (50 mL) was added. The suspension was added to 1.3 M NaOH (20 mL), the Et_2O layer washed (NaOH, water) and evaporated to give 0.42 g of **5** (94%), mp 71-73°C.

DIELS - ALDER Cyclohexene synthesis

4+2 Thermal cycloaddition between a diene and an activated alkene or alkyne, sometimes catalyzed by Lewis acids.

2 + 1 → (25°, 17 d) → 3 (80%)

4 + 5 (EtO_2C, CO_2Et) → (170°) → 6 (68%) (CO_2Et, CO_2Et)

1	Diels, O., Alder, K.	*Liebigs. Ann.*	**1928**	*460*	98
2	Jakobs, T.L.	*J. Org. Chem.*	**1962**	*27*	87
3	House, H.O.	*J. Org. Chem.*	**1963**	*28*	27
4	Johnson, C.R.	*J. Org. Chem.*	**1987**	*52*	1493
5	Boger, D.L.	*Chem. Rev.*	**1986**	*86*	781
6	Oppolzer, W.	*Angew. Chem.*	**1984**	*96*	840
7	Wenkert, E.	*Org. Prep. Proc. Int.*	**1990**	*22*	131

cis-3a,4,7,7a-Tetrahydroindane-1,3-dione (3).[3] A solution of 4-cyclopentene-1,3-dione **1** (75 g, 0.78 mol), butadiene **2** (58 g, 1.1 mol) and 2,5-di-t-butylhydroquinone (free radical inhibitor) (0.3 g) in PhH (220 mL) was allowed to stand in an autoclave for 12 days at 20-25°C. A second portion of **2** (26 g, 0.48 mol) was added and the mixture was allowed to stand for 5 days. Filtration gave 94.4 g of crude **3** (80.6%), mp 157-161°C. Recrystallized, mp 160-161°C.

DIMROTH Rearrangement

Migration of an alkyl or aryl group from a heterocyclic to an exocyclic N.

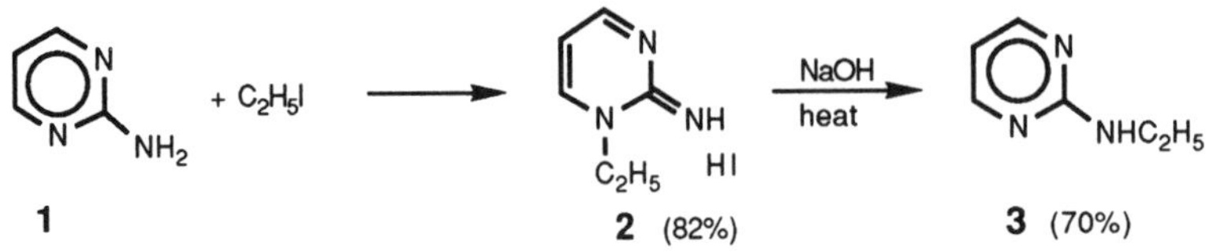

1 + C_2H_5I → 2 (82%) —NaOH, heat→ 3 (70%)

1 Dimroth, O.	*Liebigs Ann.*	**1909**	*364*	183
2 Brown, D.J.	*J. Chem. Soc.*	**1963**		1276
3 Korbonits, D.	*J. Chem. Soc. Perkin I.*	**1986**		2163

2-(Ethylamino)pyrimidine (3).[2] A solution of 2-aminopyrimidine **1** (15.0 g, 0.15 mol) and ethyl iodide (117.0 g, 0.75 mol) in EtOH (60 mL) was refluxed for 28 h. Cooling separated 1-ethyl-1, 2-dihydro-2-iminopyrimidinium iodide **2**, 32.7 g (82.4%).

2 (0.25 g, 1 mmol) in 1N NaOH (10 mL) was heated for 15 min on a water bath. After the pH was brought to 5, the solution was added to a saturated solution of picric acid. The pictrate of **3**, 0.23 g (70%) was obtained, after recrystallization from EtOH, mp 167°C.

DJERASSI - RYLANDER Oxidation

RuO_4 in oxidative cleavage of phenols or alkenes, oxidation of aromatics to quinones, oxidation of alkyl amides to imides or of ethers to esters.

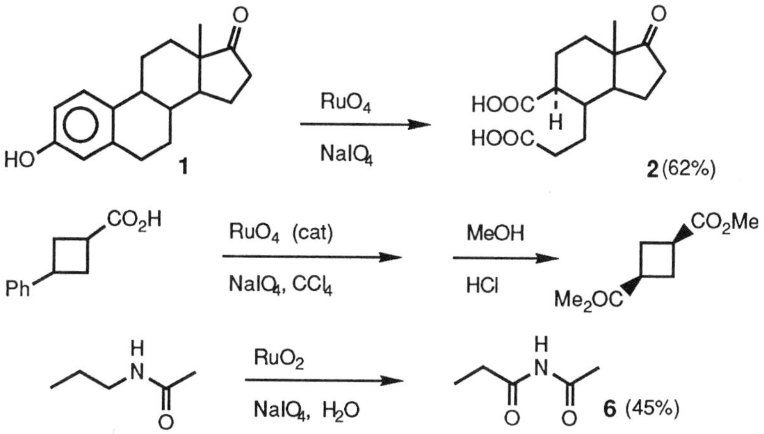

1 → (RuO_4, $NaIO_4$) → 2 (62%)

(RuO_4 (cat), $NaIO_4$, CCl_4) → (MeOH, HCl)

(RuO_2, $NaIO_4$, H_2O) → 6 (45%)

1	Djerassi, C., Engle, R.R.	*J. Am. Chem. Soc.*	**1953**	*75*	3838
2	Pappo, R., Becker, A.	*Bull. Res. Council Isr.*	**1956**	*A5*	300
3	Rylander, P.N.	*J. Am. Chem. Soc.*	**1958**	*80*	6682
4	Caputo, J.A.	*Tetrahedron Lett.*	**1962**		2729
5	Caspi, E.	*J. Org. Chem.*	**1969**	*34*	112,116
6	Tanaka, K.	*Chem. Pharm. Bull.*	**1987**	*35*	364

3-(1-Oxo-8β-methyl-5β-carboxy-trans-perhydroindanyl-4α) propionic acid (2).[3]
Estrone **1** (1.00 g, 3.94 mmol) in acetone (100 mL) was added to stirred RuO_4 (from RuO_2 (400 mg) and $NaIO_4$ (3.00 g) in water (15 mL) and acetone (50 mL)]. The mixture was kept yellow by adding portionwise $NaIO_4$ (11.5 g) in acetone:H_2O (1:1, 115 mL). After 4.5 h a few mL of iPrOH were added. Dilution with an equal volume of acetone, filtration through Celite, evaporation,saturation with NaCl, extraction with Et_2O:AcOEt 1:1, followed by extraction of the acid with aqueous $NaHCO_3$, gave 650 mg of **2** (62%).

DOEBNER - MILLER Quinoline Synthesis

Quinoline synthesis from anilines.

NH2 + CHO / CH3 → (N-ethylidene/crotylidene aniline) —ZnCl2, 250°→ 4-methylquinoline

1 2 3 (32%)

1	Doebner, O., Miller, W.	*Ber.*	**1883**	*16*	2464
2	Leir, C.M.	*J. Org. Chem.*	**1977**	*42*	911
3	Corey, J.E.	*J. Am. Chem. Soc.*	**1981**	*103*	5599
4	Bergstrom, F.W.	*Chem. Rev.*	**1944**	*35*	153

Quinaldine (4-Methylquinoline) (3).[2] To a solution of aniline hydrochloride **1** (42 g, 0.324 mol) in water (100 mL) was added a solution of paraldehyde **2** (25 g, 0.6 mol) in water (100 mL) and the mixture was maintained at rt for a week. The solution was distilled to dryness and the residue was heated with $ZnCl_2$ (21 g, 0.15 mol) at 250°C. The mixture was made alkaline and steam distilled. The organic layer was dried over KOH and distilled to give 15.3 g of **3** (32%), bp 244-245°C (760 mm).

DOERING - LA FLAMME Allene Synthesis

Allene synthesis from olefins via gem-dihalocyclopropanes.

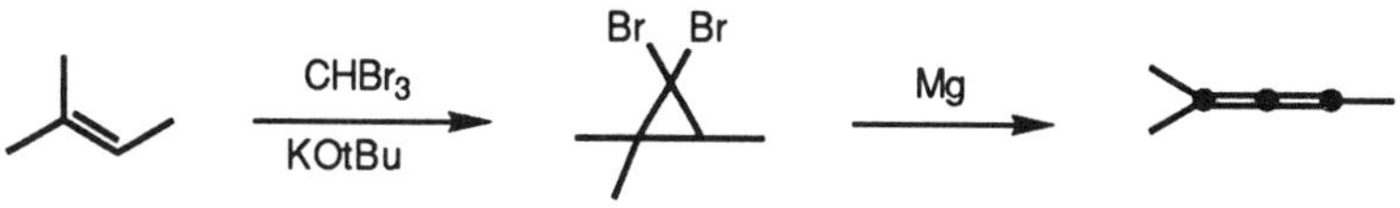

1 **2** (50%) **3** (34%)

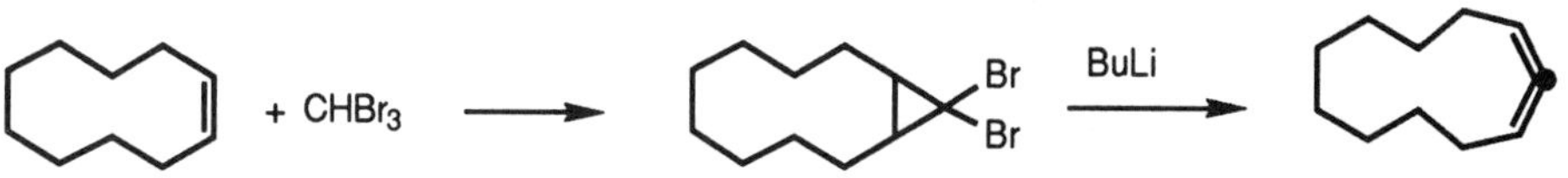

1	Doering, v.W.	*J. Am. Chem. Soc.*	**1954**	*76*	6162
2	La Flamme, P.M.	*Tetrahedron*	**1958**	*2*	75
3	Moore, W.R.	*J. Org. Chem.*	**1962**	*27*	4182
4	Chinoporos, E.	*Chem. Rev.*	**1963**	*63*	235

2-Methyl-2,3-pentadiene (3).[1,2] To a solution of 2-methyl-2-butene **1** (14.0 g, 0.2 mol) in a solution of KOtBu (22.4 g, 0.2 mol) in t-BuOH was added under stirring and cooling $CHBr_3$ (50.6 g, 0.2 mol). The mixture was poured into water, extracted with pentane and distilled to give 24.4 g of **2** (50%), bp 63-66°C (15 mm).

A solution of **2** (24.4 g, 0.1 mol) in THF (50 mL) was added to Mg turningS (4.86 g, 0.2 at g) in THF. Hydrolysis with water and fractionation afforded 2.75 g of **3** (34%), bp 72.5°C, n_D^{25} = 1.435.

DONDONI Homologation

Homologation of aldehydes, ketones, acyl chlorides, two carbon homologation via 2-(trimethylsilyl)thiazole.

Bu Li / TMSC; Ph CHO; 1; 2; (95%); 2 / $Bu_4N^+F^-$; NaH / Bu_4NF, BnBr / MeI -MeCN; 3; 4 (84%); 5 (74%)

1	Dondoni, A.	*J. Chem. Soc. Chem. Commun.*	**1984**		258
2	Dondoni, A.	*J. Org. Chem.*	**1988**	*53*	1748
3	Dondoni, A.	*Tetrahedron*	**1987**	*43*	3533
4	Dondoni, A.	*J. Org. Chem.*	**1989**	*54*	693
5	Dondoni, A.	*Pure & Appl. Chem.*	**1990**	*62*	643

Dibenzyl α-D-glucodialdofuranose (5).[3] To **3** (505 mg, 2.5 mmol) in THF (50 mL) was added dropwise at 20°C **2** (590 mg, 3.7 mmol) in THF (10 mL). After 24 h stirring, the solvent was evaporated and the residue in THF (30 mL) was treated with n-$Bu_4N^+F^-$ (783 mg, 3 mmol) in THF. After 2 h stirring and work up, the residue was chromatographed (petroleum ether:EtOAc 1:1) to give 602 mg of **4** (84%) (diastereomeric ratio 95:5). To **4** (1.32 g, 4.6 mmol) in THF (50 mL) was added NaH (50%) (250 mg, 5.1 mmol) at 20°C. After 20 min reflux were added n-Bu_4NF^- (170 mg, 0.46 mmol) and benzyl bromide (880 mg, 5.1 mmol). After 24 h at 20°C, work up and chromatography (petroleum ether:EtOAc) (7:3), the product was treated with MeI in MeCN and refluxed until disappearance of the adduct. Work up afforded 1.101 g of **5** (74%).

D Ö T Z Hydroquinone synthesis

Hydroquinone synthesis (regiospecific) from alkynes and carbonyl carbene chromium complexes.

$(CO)_5Cr{=}C(OMe)Ph$ **1** + Ph-C≡C-Ph **2** → **3** (62%) + CO

4 + **5** → **6** (81%)

1 Dötz, K.H.	*Angew. Chem. Int. Ed.*	**1975**	*14*	644
2 Dötz, K.H.	*Angew. Chem. Int. Ed.*	**1984**	*23*	587
3 Dötz, K.H.	*Chem.Ber.*	**1988**	*121*	665
4 Hofmann, P.	*Angew. Chem. Int. Ed.*	**1989**	*28*	908
3 Dötz, K.H.	*New J. Chem.*	**1990**	*14*	433

Tricarbonyl(2,3-diphenyl-4-methoxy-1-naphthol)chromium(0) (3)[1] Pentacarbonyl-[methoxy(phenyl)carbene]chromium(0) (0.58 g, 1.86 mmol) and tolan 2 (0.33 g, 1.86 mmol) in Bu_2O (6 mL) is stirred at 45°C for 3 h. At -20°C, pentane (5 mL) was added. The red product was filtered and chromatographed (silica gel, CH_2Cl_2:pentane (1:1), to give after recrystallization from CH_2Cl_2:pentane (1:5) 0.55 g of **3** (62%).

Tricarbonyl[dimethyl-2-(1-hydroxy-4,5,9,10-tetramethoxy-2-anthracenylmethyl)-succinate]chromium (6).[3] A solution of **4** (2.12 g, 5 mmol) and **5** (1.01 g, 5.5 mmol) in t-BuOMe (25 mL) is stirred under Ar at 55°C for 1 h. After cooling (-40°C) the supernatant is decanted. The residue dissolved in CH_2Cl_2 was precipitated with pentane to give 2.26 g of **6** (81%).

DUFF Aldehyde Synthesis

Formylation of phenols and anilines with hexanethylene tetramine.

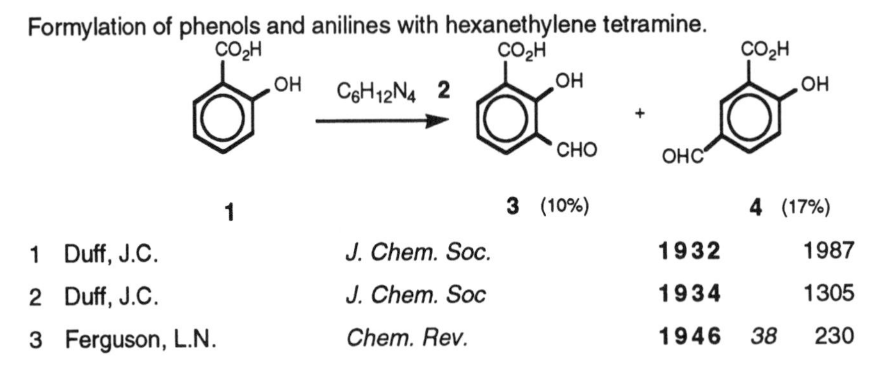

1 Duff, J.C. *J. Chem. Soc.* **1932** 1987
2 Duff, J.C. *J. Chem. Soc* **1934** 1305
3 Ferguson, L.N. *Chem. Rev.* **1946** *38* 230

3- and **5-Formylsalicylic acid (3)** and **(4).**[2] **1** (10 g, 72 mmol) and hexamethylene-tetramine **2** (10 g, 71 mmol) in anhyd. AcOH was heated for 6 h and treated with water (80 mL) and conc. HCl (50 mL)., Filtration gave a mixture of 1.2 g of **3** (10%) and 2 g of **4** (17%).

DUTT - WORMALL Azide formation

Synthesis of aromatic azides from anilines via diazonium salts.

$$\underset{\mathbf{1}}{\text{Ph-NH}_2} \xrightarrow{\text{HNO}_2} \text{Ph-N}_2^+ \xrightarrow[\mathbf{2}]{\text{TsNH}_2} \underset{\mathbf{3}}{\text{Ph-N}_3}$$

1 Dutt, P.K., Wormall, A. *J. Chem. Soc.* **1921** *119* 2088
2 Dutt, P.K. *J. Chem. Soc.* **1924** *125* 1463
3 Bretschneider, H. *Monatsh.* **1950** *81* 970

Phenylazide (3).[1] **1** (4.6 g, 50 mmol) in water (25 mL) and 35% HCl (15 mL) was diazotized with a solution of $NaNO_2$ (3.5 g, 55 mmol). After diazotization (starch-iodine paper), p-toluenesulfonamide **2** (8.5 g, 50 mmol) in water (175 mL) and NaOH (6.0 g, 150 mmol) were added. After 10-15 min the product was filtered to afford 8.0 g of **3** (85%), mp 85°C.

EHRLICH - SACHS Aldehyde Synthesis

Formation of o-nitrobenzaldehydes from o-nitrotoluenes.

O_2N CH_3 O_2N CH_3 **1** + ON–C6H4–N Me_2 **2** Na_2CO_3

O_2N =N-Ph NMe_2 O_2N =N-Ph NMe_2 **3** (57%)

O_2N CHO NO_2 CHO **4** (22%)

1	Ehrlich, P., Sachs, F.	*Chem. Ber.*	**1899**	*32*	2341
2	Sachs, F.	*Chem. Ber.*	**1900**	*33*	959
3	Ruggli, P.	*Helv. Chim. Acta.*	**1937**	*20*	271
4	Adams, R.	*Org. Synth. Coll.*		*II*	214

4,6-Dinitro-1,3-isophthalaldehyde (4).[3] A mixture of 4,6-dinitro-1,3-xylene **1** (100 g, 0.51 mol), p-nitrosodimethylaniline **2** (150 g, 0.98 mol) and Na_2CO_3 (100 g, 0.94 mol) in EtOH (500 mL) was refluxed for 8 h. The product was refluxed in water (1500 mL) and stirred with acetone (3.250 mL) to afford 130-135 g of **3** (57%). To a mixture of PhH (620 mL) and HNO_3, d = 1.12 (620 mL) was added **3** (100 g, 0.43 mol) and all was shaken for 24 h. The nitrate of p-dimethylaminoaniline was filtered and the crystals washed with PhH, the solvent removed in vacuum and the residue after crystallization from PhH gave 25-29 g of **4** (22-24%), mp 129.5-130°C.

EINHORN - BRUNNER Triazole Synthesis

Condensation of hydrazines with diacyclamines (imides) to triazoles.

$C_6H_5-NH-NH_2 + C_6H_5-CO-NH-CO-CH_3 \longrightarrow$ 3 (78%)

1 2 3

1	Einhorn, M.	*Liebig's Ann.*	**1905**	*343*	299
2	Brunner, K.	*Chem. Ber.*	**1914**	*47*	2671
3	Brunner, K.	*Monatsh*	**1915**	*36*	509
4	Potts, K.T.	*J. Chem. Soc.*	**1961**	*61*	103

3-Methyl-1,5-diphenyl-1,2,4-triazole (3).[4] A mixture of phenyl hydrazine hydrochloride **1** (14.5 g, 0.11 mol), N-acetylbenzamide **2** (16.5 g, 0.1 mol) and NaOAc (10 g, 0.12 mol) in AcOH (30 mL) was refluxed for 10 h. The product was made alkaline with 10% NaOH solution and extracted with Et_2O. Evaporation gave a pale yellow oil which slowly solidified, mp 80-81°C. Recrystallization from 90% EtOH and petroleum ether afforded 18.4 g of **3** (78%).

ELBS Condensation

Intramolecular condensation of methyl biarylketones to polynuclear aromatics

Zn

1

2 (44%)

1	Elbs, K.	*Chem. Ber.*	**1884**	*17*	2847
2	Bachmann, W.E.	*J. Am. Chem. Soc.*	**1937**	*59*	2369
3	Fieser, L.F.	*Org. React.*	**1942**	*1*	129

1,2,3,4,5,6-Tribenzantracene (2).[2] 9-(2-Methyl-1-naphthoyl)phenantrene **1** (15.4 g, 62 mmol) and Zn (5 g) were heated for 3 h at 410°C. Distillation of the product at 3 mm pressure and recrystallization from xylene afforded 6.4 g of **2** (44%), mp 225-228°C.

ELBS Oxidation

Oxidation of monophenols to polyphenols or oxidation of aromatic methyl groups by persulfates (Caro's acid)

$K_2S_2O_8$, $FeSO_4$ 100°

1 → 2 (42%)

$K_2S_2O_8$

1	Elbs, K.	*J. Prakt. Chem.*	**1893**	*48*	179
2	Bergmann, E.J.	*J. Am. Chem. Soc.*	**1958**	*80*	3717
3	Sethna, S.M.	*Chem Rev.*	**1951**	*49*	91
4	Dermer, R.	*Chem. Rev.*	**1957**	*57*	103
5	Wallace, T.W.	*Synthesis*	**1983**		1000
6	Minisci,F.	*Acc. Chem Res.*	**1983**	*16*	27

2,5-Dihydroxypyridine (2).[2] To 2-hydroxypyridine **1** (38.0 g, 0.4 mol) and NaOH (80.0 g, 2 mol) in water (1500 mL) at 0°C was added $FeSO_4$ (2.0 g) in water (20 mL) and potassium peroxydisulfate (135 g, 0.5 mol). After 20 h at 20°C and filtration, conc. H_2SO_4 was added (cooling) to pH 0.75 and the mixture was heated to 100°C under N_2 for 30 min. The cooled solution was neutralized by 10 N NaOH to pH 6.5 and evaporated to dryness. Extraction (Soxhlet) with iPrOH and evaporation afforded 19 g of **2** (42%). Recrystallization from iPrOH gave 8.0 g of **2**, mp 250-260°C.

EMMERT Reductive condensation

Condensation of pyridine with ketones in the presence of aluminium amalgam.

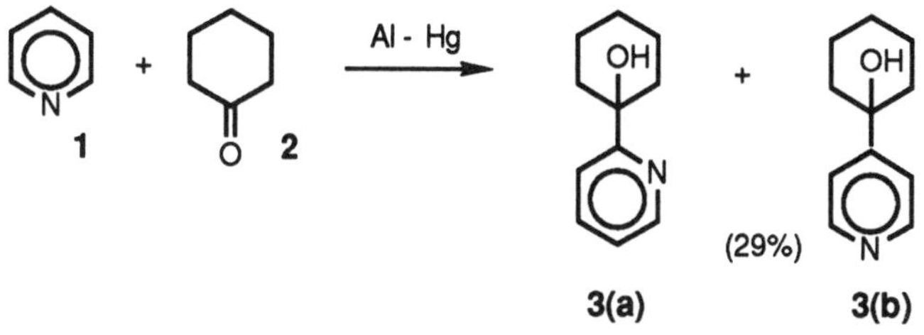

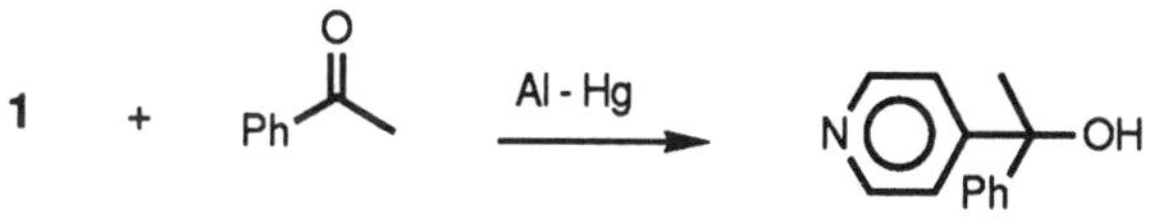

1	Emmert, B.	*Chem.Ber.*	**1939**	*72b*	1188
2	Lochte, H.L.	*J. Am. Chem. Soc.*	**1953**	*75*	4477
3	Russel, C.A.	*J. Chem. Soc.*	**1970**		1406
4	Tilford, C.H.	*J. Am. Chem. Soc.*	**1948**	*70*	4001

1-(2-Pyridyl) and **1-(4-Pyridyl)cyclohexanol (3a)** and **(3b)**[2] A mixture of $HgCl_2$ (1.0 g, 3.7 mmol) and Al turnings (15.0 g, 0.55 at) was heated for 15 min at 120°C. Pyridine **1** (19.0 g, 0.24 mol) and cyclohexanone **2** (17.7 g, 0.18 mol) were added with vigorous stirring. When the reaction began, **1** (75.8 g, 0.96 mol) was added at once and **2** (60.8 g, 0.62 mol) was added at a rate that kept the mixture refluxing. After the addition of all of **2** (1.5 h) the reaction mixture was allowed to cool for 2 h and was poured into a 6N NaOH solution (100 mL) with vigorous stirring. The aqueous layer was extracted with PhH and the PhH solution was extracted with 6N HCl. The HCl solution was basified with NaOH pellets and after distillation afforded a crude mixture of **3a** and **3b** (50.0 g). Recrystallization gave 45.7 g of **3 a+b** (29%), mp 42-43°C.

E N D E R S Chiral reagent

Asymmetric electrophilic substitution of aldehydes and ketones via (S) or (R)-1-amino-2-methoxymethylpyrrolidine (SAMP or RAMP) hydrazones.

1. SAMP
2. LDA, Et_2O
3. n-C_3H_7I

O_3, CH_2Cl_2 or MeI, 4N HCl

(69%, ee = 99.5%)

1 (S,S)-2 (S)-3

1	Enders, D.	*Angew. Chem. Int. Ed.*	**1976**	*15*	549
2	Enders, D.	*Angew Chem. Int. Ed.*	**1979**	*18*	397
3	Enders, D.	*Tetrahedron*	**1984**	*40*	1345
4	Enders, D.	*Org. Synth.*	**1987**	*65*	173,183
5	Enders, D.	*Asymmetric Synthesis*	**1984**	*3*	275

(S)-(+)-4-Isopropyl-3-heptanone (3).[3] **SAMP-Hydrazone** of **1**. SAMP (28 mg, 1 mmol) and 3-pentanone **1** (86 mg, 1 mmol) were stirred at 60°C for 20 h. (TLC or IR). The mixture was poured into CH_2Cl_2:water (6:1) and the dried organic layer on evaporation and short path distillation afforded 170 mg (87%), α_D = +223° (neat).

BuLi (8.4 equiv) was added to diisopropylamine (848 mg, 8.4 mmol) in Et_2O under Ar at 0°C. After 15 min SAMP-hydrazone (1.59 g, 8 mmol) was added and stirring continued at 0°C for 4 h, then the mixture was cooled to -100°C and PrI (0.86 mL, 8.8 mmol) in Et_2O was added. After 3 h at -100° and 12 h at 20°C work up gave 1.91 g of crude **2** (100%). Short path distillation afforded 1.79 g of **2** (91%), bp 63°C/0.1 Torr, α_D = 224° (neat).

2 (1.42 g, 5.16 mmol) in MeI (3.66 g,25.8 mmol) was refluxed for 48 h and the PrI distilled. The residue was stirred in 4 N HCl (25 mL) for 5 min and extracted with pentane. Distillation gave 0.46 g of **3** (69%), 99.5% ee .

ERLENMEYER - PLOCHL - BERGMANN Amino Acid Synthesis

Amino acid synthesis from aldehydes and hydantoin (Bergmann); synthesis of serine derivatives (Erlenmeyer) or of γ-hydroxyaminoacids (Plochl).

Ar-CHO + (hydantoin) —$Ac_2O/AcOH$, NaOAc→ Ar-CH=(hydantoin) —HBr, Δ→ $Ar-CH_2-CH(NH_2)-COOH$

Ar-CHO + $H_2C(NH_2)-COOH$ —KOH, EtOH→ Ar-CH(OH)-CH(N=CH-Ar)-COOH (63%) → Ar-CH(OH)-CH(NH_2)-COOH (79%)

AcO-ketone + oxazolone (Ph) —$Pb(OAc)_2$→ AcO-alkylidene oxazolone —H_2 - Pd/C, HCl→ HO-CH$_2$CH$_2$CH(NH_2)-CO_2H

1	Plochl, J.	*Chem. Ber.*	**1884**	*17*	1616
2	Erlenmeyer, E.	*Chem. Ber.*	**1892**	*25*	3445
3	Bergmann, M.	*Liebigs Ann.*	**1926**	*449*	277
4	Galantay, E.	*J. Org. Chem.*	**1963**	*28*	98
5	Carter, H.E.	*Org. React.*	**1946**	*3*	198
6	Bolhofer, W.A.	*J. Am. Chem. Soc.*	**1953**	*75*	4469
7	Langemann, A.	*Helv. Chim. Acta.*	**1969**	*52*	1095
8	Wieland, Th.	*Angew. Chem.*	**1951**	*63*	13

ESCHENMOSER Fragmentation

Cleavage of epoxy ketones to an acetylene and a carbonyl compound via sulfonylhydrazone.

1. $(NO_2)_2C_6H_3SO_2NHNH_2$
2. $NaHCO_3$

1 → **2**

$TsNHNH_2$

1	Eschenmoser, A.	*Helv. Chim. Acta.*	**1967**	*50*	2101
2	Eschenmoser, A.	*Helv. Chim. Acta.*	**1971**	*54*	2896
3	Corey, E.J.	*J. Org. Chem.*	**1975**	*40*	579
4	Cooper, M.S.	*Tetrahedron Lett.*	**1986**	*22*	5125

Acetylenic aldehyde (2).[3] To a solution of 3,5-dinitrobenzenesulfonylhydrazine (0.276 g, 1.01 mmol) in THF (20 mL) was added at 0°C the ketoepoxide **1** (0.182 g, 1 mmol). The mixture was kept 2 h at 0°C and 2 h at 20°C. The solvent was evaporated, the residue was dissolved in CH_2Cl_2 (15 mL) and the solution was filtered. Upon chilling at -25°C the dinitrobenzenesulfonylhydrazone was filtered to give 0.404 g, mp 102°C. To a solution of hydrazone (0.404 g) in THF (10 mL) was added $NaHCO_3$ (0.25 g) and the mixture was stirred at 25-30°C for 30 min. After filtration through Celite, the clear filtrate was evaporated and the residue chromatographed (silica gel, EtOAc:PhH,1:3) to give **2**, Rf = 0.33

ESCHWEILER - CLARKE Amine methylation

Reductive methylation of amines by a mixture of formaldehyde and formic acid

NH_2 **1** — CH_2O - HCOOH, 95°C → **2** (60%)

Ph, OH, C, pCl Ph, $CH_2CH_2NH_2$ — CH_2O, HCOOH → Ph, OH, C, pCl Ph, $CH_2CH_2N(CH_3)_2$

1	Eschweiler, W.	*Chem. Ber.*	**1905**	*38*	880
2	Clarke, H.T.	*J. Am. Chem. Soc.*	**1933**	*55*	4571
3	Cope, A.C.	*J. Org. Chem.*	**1965**	*30*	2163
4	Borch, R.F.	*J. Org. Chem.*	**1972**	*37*	1673
5	Moore, M.L.	*Org. React.*	**1949**	*5*	301

N,N-Dimethyl-5-amino-1-hexene (2).[3] 1-Methyl-4-pentenylamine **1** (8.5 g, 85.5 mmol) was dissolved in 91% formic acid (24 g) and 37% formaldehyde. The mixture was heated on a steam bath for 6 h, cooled and poured onto ice. The mixture was made strongly basic and extracted with Et_2O. After evaporation of the solvent and distillation of the residue, 6.3 g of **2** (60%), bp 135-136°C was obtained.

ETARD Oxidation

Oxidation (low yield) of aryl or cycloalkyl methyl groups to aldehydes with chromyl chloride.

CH_3 (1) —CrO_2Cl_2→ $CH_3 . CrO_2Cl_2$ (2) —H_2O→ CHO 3 (10%)

1	Etard, A.L.	*Bull. Socl Chim. Fr.*	**1877**	*27*	249(1)
2	Bisagni, E.	*Bull. Soc. Chim. Fr.*	**1968**		637
3	Nenitzescu, C.D.	*Rev. Roum. Chim.*	**1969**	*14*	1553
4	Ferguson,	*Chem. Rev.*	**1946**	*46*	237
5	Hartfort, W.H.	*Chem. Rev.*	**1958**	*58*	1

Cyclohexanecarboxaldehyde (2).[2] A solution of chromyl chloride (77.5 g, 0.5 mol) in CCl_4 (155 mL) was added slowly to a solution of methylcyclohexane **1** (47 g, 0.5 mol) in CCl_4 (100 mL) at 33-38°C. After 15 min of stirring 2-methylbutene (0.3 mL) was added and precipitation of a complex began. The addition of chromyl chloride required 5 h. The complex was filtered after 5 days, 87 g (70%). The chromyl complex was added to 400-500 g of ice and after quenching the mixture was extracted with Et_2O. After evaporation of the solvent the residue was distilled in vacuo to afford 5.5-5.8 g of **2** (10%) bp 75-80°C/20 mm or 161-163°C/760 mm, d=0.925, n_D^{20} = 1.4500.

E V A N S Chiral auxiliary

Enantioselective aldol condensation by means of a chiral auxiliary and boron enolates.

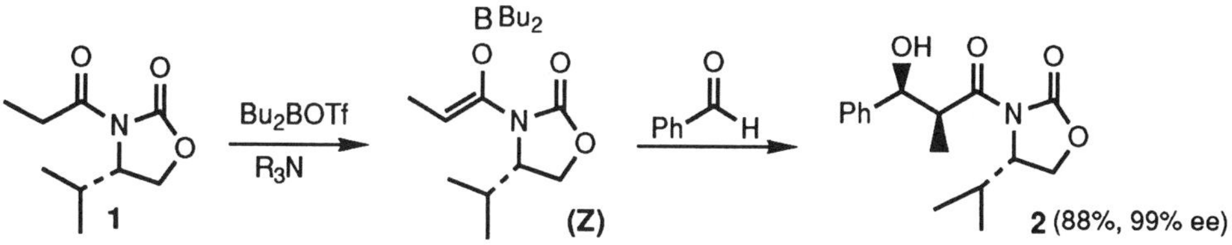

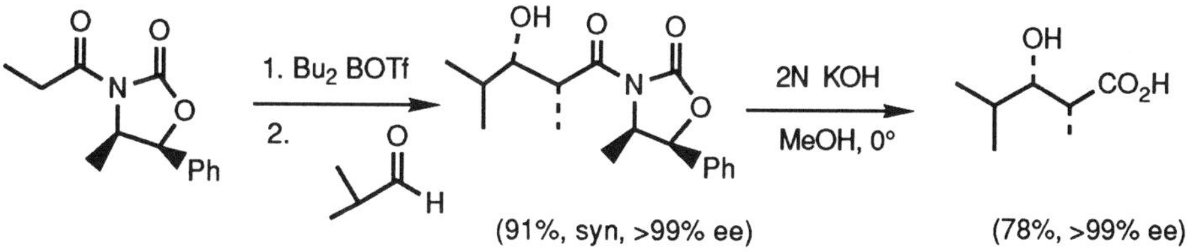

Bu$_2$BOTf, R$_3$N; 1; BBu$_2$; (Z); PhCHO; 2 (88%, 99% ee)

1. Bu$_2$ BOTf; 2. (CH$_3$)$_2$CHCHO; (91%, syn, >99% ee); 2N KOH, MeOH, 0°; (78%, >99% ee)

1	Evans, D.A.	*J. Am. Chem. Soc.*	**1979**	*101*	6120
2	Evans, D.A.	*J. Am. Chem. Soc.*	**1981**	*103*	2127, 2876
3	Newmann, M.S.	*J. Am. Chem. Soc.*	**1951**	*71*	4199
4	Evans, D.A.	*Aldrichimica Acta*	**1982**	*15*	23
5	Evans, D.A.	*Tetrahedron Lett.*	**1987**	*28*	39
6	Evans, D.A.	*Organic Syn.*	**1989**	*68*	89

Syn aldol product(2).[2] To a 0.2-0.5 M solution of oxazolidone **1**, mp 71-72°, a_D = +14.8°, under Ar at 0°C was added 1.1 equivalent of the boron triflate, followed by 1.2 equivalent of diisopropylethyl amine. After 30 min, the boron enolate mixture was cooled (-78°C), 1.1 equivalent of benzaldehyde was added and all was stirred for 30 min at -78°C and 90 min at 25°C. The boron complex was quenched with phosphate buffer (pH 7) and treated with 30% H_2O_2 in MeOH (0°C, 1 h). Extraction with Et_2O gave diastereomerically pure **2** (88%), mp 101°-102°C); **2** (88% erythro, >99% ee).

FAWORSKI - WALLACH Rearrangement

Rearrangement of α-haloketones or α,α'-dihaloketones to acids or acrylic acids (via cyclopropanones).

K_2CO_3 1 → 2 (69%)

NaOMe 3 → 4 (58%)

1	Favorski, A.	*J. Prakt. Chem.*	**1895**	*51*	553
2	Wallach, O.	*Liebigs Ann.*	**1918**	*414*	296
3	Wagner, R.B.	*J. Am. Chem. Soc.*	**1950**	*72*	972
4	Nace, H.R.	*J. Org. Chem.*	**1967**	*32*	3438
5	Kimpe, M.D.	*J. Org. Chem.*	**1986**	*51*	3938
6	Kende, A.S.	*Org. Rect.*	**1960**	*11*	261

Cyclohexanecarboxylic acid (2).[4] 2-Chlorocycloheptanone **1** 5.0 g, 34 mmol) was refluxed in K_2CO_3 (15.0 g, 0.107 mol) water (20 mL) for 6 h with stirring. After washing with Et_2O, the solution was acidified, extracted and the solvent evaporated to give 3.0 g of **2** (69%), mp 22-24°C, bp 232-233°C.

Methyl 3-methyl-2-butenoate (4).[3] To an ice cooled solution of 1,3-dibromo-3-methyl-2-butanone **3** (244 g, 1 mol) in Et_2O (250 mL) was added in small portions (4 h) a suspension of $NaOCH_3$ (111.5 g, 2 mol) in Et_2O (500 mL). After stirring for 30 more min the product was isolated in the usual manner, 66.12 g of **4** (58%), bp 60°C/50 mm, n_D^{20} = 1.4382.

F E I S T - B E N A R Y Furan Synthesis

Synthesis of furans by condensation of an α-halocarbonyl compound with an enol.

$C_2H_5OOC\text{-}C(=O)\text{-}CH_2Br$ (**2**) + $H_2C(COOC_2H_5)\text{-}C(=O)\text{-}CH_2\text{-}COOC_2H_5$ (**1**) → (1. Na; 2. HCl) → 3,4-(HOOC)$_2$-furan-2-CH_2-COOH **3** (65%)

1 Feist, F.	*Ber.*	**1902**	*35*	1539
2 Benary, E.	*Ber.*	**1911**	*44*	489
3 Archer, S.	*J. Am. Chem. Soc.*	**1944**	*66*	1656

3,4-Dicarboxyfuran-2-acetic acid (3).[2] To a cooled suspension of sodium sand (17.2 g, 0.75 at) in dry ether (750 mL), was added under stirring diethyl 1,3-acetondicarboxylate **1** (155 g, 0.7 mol) dropwise (1 h). When all the sodium had reacted, the mixture was cooled and ethyl bromopyruvate **2** (145 g, 0.74 mol) was added over 2 h. After 1 h reflux an equal volume of water was added and the solvent removed from the organic layer. The residue was refluxed with 20% HCl (200 mL) for 2 h, treated with Norite, filtered and cooled overnight to afford **3**, 106 g (65%), mp 200-201°C (monohydrate).

FELKIN Cyclization

Nickel catalyzed steroeselective synthesis of cis and trans methyl vinylcyclopentanes from telemerization of butadiene. Cyclization (ene reaction) of unsaturated allyl Grignard reagents.

$(PPh_3)_2NiCl_2$, n - C_3H_7MgBr; MgBr; 1; 0° - 25°; H; D_2O; CH_2O; H; 3 cis; MgBr; 2; 1. 110°; 2. D_2O; 3 trans; O; H; H; HO

1	Felkin, H.	*Tetrahedron Lett.*	**1972**		1433
2	Felkin, H	*Tetrahedron Lett.*	**1972**		2285
3	Felkin, H.	*J. Chem. Soc. Chem. Comm.*	**1975**		243
4	Oppolzer, W.	*Angew. Chem. Int. Ed.*	**1989**	*28*	32

cis or **trans-2-Methylvinylcyclopentane (3).**[3] A mixture of $(PPh_3)_2$ $NiCl_2$ (32.6 g, 5 mmol), butadiene **1** (12.42 g, 0.23 mol) and a solution of propylmagnesium bromide 1.9 M (0.25 mol) was refluxed (room temperature, condenser cooled with solid CO_2) for 24 h. After deuterolysis one obtains 16.9 g of cis **3** (67%). By heating the reaction mixture, which contained the cis isomer for 24 h in a sealed tube, the thermally more stable trans isomer was obtained. The Ni catalyzed ene cyclization can be performed starting with octadienyl halides and conversion to **2**.[2,4]

FERRARIO - AKERMANN Thiocyclization

A general method for obtaining phenoxthiines, phenothiazines etc., by S insertion

S / $AlCl_3$

2 (74%)

1	Ferrario, E.	*Bull. Soc. Chim. Fr.*	**1911**	*9*	536
2	Akermann, F.	*Germ. Pat.*	**1910**		234.743
3	Sutter, C.N.	*J. Am. Chem. Soc.*	**1936**	*58*	717
4	Deasy, C.L.	*Chem. Rev.*	**1943**	*32*	174
5	Suter, C.M.	*Org. Synth. Coll.*	**1943**	*II*	485

Phenoxthiine (2).[3] A mixture of diphenyl ether **1** (1700 mL, 10.4 mol), sulfur (256 g, 8 mol), and $AlCl_3$ (512 g, 3.84 mol) was heated on a steam bath until the evolution of H_2S ceased. The reaction mixture was poured into ice water and the organic layer was fractionated in vacuo. The fraction bp 185-187°C/23 mm gave 585 g of **2** (74%); after recrystallization from MeOH, mp 56-57°C.

FERRIER Carbohydrate Synthesis

Synthesis of unsaturated carbohydrates by allylic rearrangement.

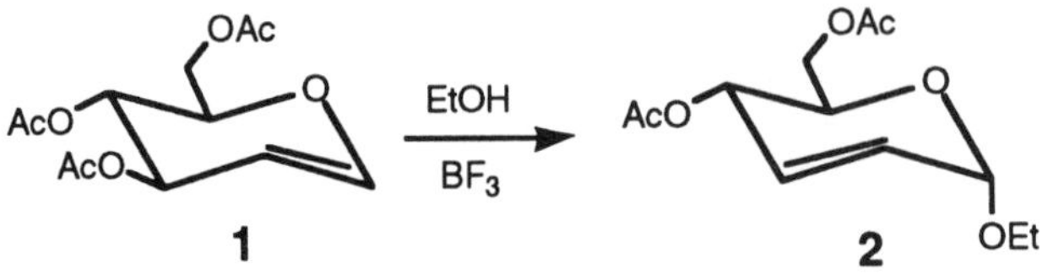

1	Ferrier, R.J.	*Adv. Carbohydrate Chem.*	**1965**	*20*	67
2	Ferrier, R.J.	*J. Chem. Soc. (C)*	**1968**		974
3	Ferrier, R.J.	*J. Chem. Soc. (C)*	**1969**		570

Ethyl 4,6-di-O-acetyl-2,3-dideoxy-α-D-erythro-hex-2-enopyranoside (2).[2] A solution of tri-O-acetyl-D-glucal **1** (5.0 g, 18 mmol) in PhH (20 mL) and EtOH (1.8 mL, 31 mmol) was treated with $BF_3.Et_2O$ (1.0 ml). After 25 min the optical rotation changed from -35° to +20.25°. Neutralization of the catalyst, filtration of the solids and removal of the solvent left a syrup which on trituration with EtOH gave 2.8 g of **2** and a second crop of 0.5 g of **2** (70%). The pure product melted at 78-79°C and had $[\alpha]_D$ = +104° (PhH).

F E R R I E R Chiral cyclohexanone synthesis

Transformation of unsaturated glycosides into cyclohexanone derivatives by heating in aqueous acetone with mercury (II) salts.

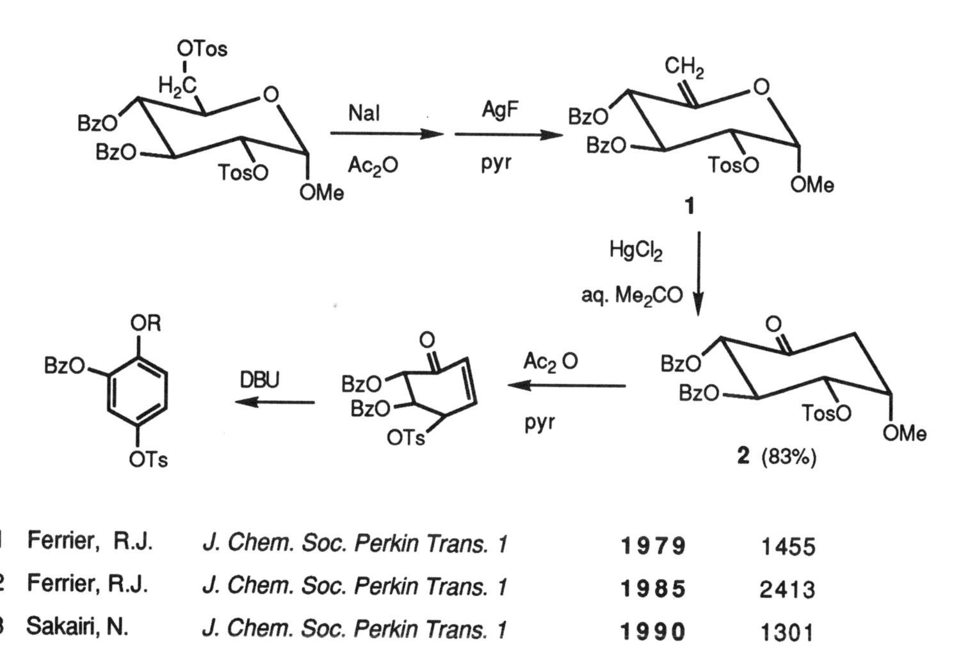

1	Ferrier, R.J.	*J. Chem. Soc. Perkin Trans. 1*	**1979**	1455
2	Ferrier, R.J.	*J. Chem. Soc. Perkin Trans. 1*	**1985**	2413
3	Sakairi, N.	*J. Chem. Soc. Perkin Trans. 1*	**1990**	1301

2,3-Dibenzoyloxy-5-methoxy-4-toluene-p-sulphonyloxycyclohexanone (2).[2]
The olefin **1** (10 g, 18 mmol) and $HgCl_2$ (5 g, 18 mmol) in 1:2 water acetone (250 mL) were heated to reflux for 4.5 h. On cooling **2** separated. A second crop was obtained by dilution with water; total: 8.1 g of **2** (83%), mp 175-176°C (from MeOH), α_D = +10°C (c 2,0 Py).

FINEGAN Tetrazole Synthesis

Tetrazole synthesis from azides by dipolar cycloaddition with activated nitriles or intramolecularly with nitriles in the presence of acids.

$F_3C\text{-}CN$ + $N_3\text{-}C_8H_{17}$ —150°C→ 3 (96%)

2 1 3 (96%)

CN, $N=\overset{+}{N}=\overset{-}{N}$ —$ClSO_3H$, 25°→ (79%)

1 Finegan, W.G. *J. Am. Chem. Soc.* **1956** *80* 3908

2 Carpenter, W.R. *J. Org. Chem.* **1962** *27* 2085

3 Kereszty, von K. *Germ. Pat.* 611.692, C. A. **1935** *29* 5994

1-Octyl-5-trifluoromethyltetrazole (3).[2] Octyl azide **1** (8.66 g, 56 mmol) and trifluoroacetonitrile **2** (3.76 g, 40 mmol) were heated in a sealed tube at 150°C for 17 h. Distillation of the product yielded 2.17 g of **1** in the forerun and 6.20 g of **3** (96%), bp 81.5-85°C/0.25 mm, n_D^{20} = 1.4272.

FINKELSTEIN - GRYSZKIEWICZ - TROCHIMOWSKI - MC COMBIE Halide Displacement

Exchange of halogen in alkyl halides (see also Makosza).

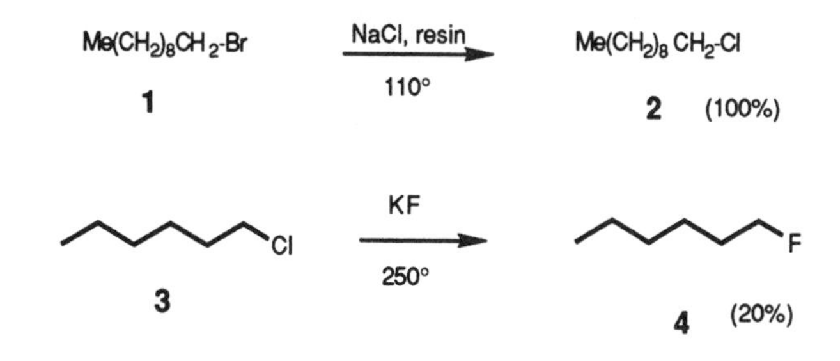

1	Finkelstein, N.	*Chem. Ber.*	**1910**	*43*	1528
2	Regen, S.L.	*J. Org. Chem.*	**1977**	*42*	875
3	Gryszkiewicz-Trochimowski	*Rec. Trav. Chim.*	**1947**	*66*	427
4	McCombie, H.	*Nature*	**1946**	*158*	382
5	Sartori, M.F.	*Chem. Rev.*	**1951**	*48*	237

1-Chlorodecane (2).[2] A mixture of 1-bromodecane **1** (2.3 g, 10.4 mmol) in PhH (20 mL) and NaCl (27 g, 465 mmol) in water (80 mL) was heated with polystyrene ion exchange resin (0.5 g)[2] in a sealed tube at 110°C for 240 h. After cooling, separation from the resin and distillation of the solvent, the residue, 1.8 g was identified as **2** (100%), bp 223°C.

1-Fluorohexane (4).[4] 1-Chlorohexane **3** (250 g, 2 mol) and KF (174 g, 3 mol) were heated under pressure at 250°C. After cooling the products were fractionated through an efficient column to obtain 40 g of **4** (20%), bp 92.5-93°C.

FISCHER Oxazole synthesis

Oxazole synthesis from aldehydes and α-hydroxyamides or cyanohydrins.

CH_3 - CH - $CONH_2$ (OH) + OCH - Ph —TsOH, Δ→ 3 (100%) —$POCl_3$, 80°→ 4 (29%)

1 2

R - CHO + N≡C-CH(OH)—R[1] —HCl→ (R, R')

1	Fischer, E.	*Chem. Ber.*	**1896**	*29*	205
2	Minovici, N.	*Chem. Ber.*	**1896**	*29*	2097
3	Cornforth, J .W.	*J. Chem. Soc.*	**1949**		1028
4	Willey, R.H.	*Chem. Rev.*	**1945**	*37*	410
5	Hill, A.	*J. Am. Chem. Soc.*	**1930**	*52*	769

2-Phenyl-5-methyloxazole (4).[3] A mixture of lactamide **1** (5.0 g, 56 mmol), benzaldehyde **2** (6.0 g, 60 mmol) and TsOH (0.15 g) was refluxed in PhMe in a Soxhlet (with K_2CO_3 in the cartouche) for 12-14 h. After cooling the filtered product gave 4 g of **3** (100%), mp 122-125°C. After recrystallization from water, mp 129-130°C.

3 (2.0 g, 11.3 mmol) in $POCl_3$ (15 mL) was heated at 80-85°C for 15 min, the excess of $POCl_3$ was distilled in vacuum and the residue was steam distilled. From the distillate the oxazole was extracted with Et_2O and the solvent was evaporated to give 0.5 g of **4** (28%), mp 245°C (picrate mp 145-146°C).

FISCHER - BORSCHE - DRECHSEL Indole synthesis

Indole synthesis from phenylhydrazones of ketones (Fischer); in the case of cyclohexanone phenylhydrazones tetrahydrocarbazoles are obtained (Borsche-Drechsel).

HCOOH, 130-135°C

1 → **2** (26%)

3 + **4** → AcOH, reflux 2h → **5** (69%)

1	Fischer, E.	*Chem. Ber.*	**1883**	*16*	2241
2	Iyosuke Simizu	*Chem. Pharm. Bull.*	**1971**	*19*	2561
3	Sarmicole, F.	*Tetrahedron Lett.*	**1984**	*25*	3101
4	Robinson, B.	*Chem. Rev.*	**1969**	*69*	227
5	Welch, W.M.	*Synthesis*	**1977**		845
6	Drechsel, E.	*J. Prakt. Chem.*	**1888**	*38*	69(2)
7	Borsche, W.	*Chem. Ber.*	**1904**	*20*	378
8	Campbell, N.N	*Chem. Rev.*	**1947**	*40*	361

Cis-1,2,3,3a,4,8b-Hexahydro-4-formyl-8b-ethylcyclopent(b)indole. (2).[2]
Phenylhydrazone **1** (850 mg, 4.2 mmol) was treated with 98% HCOOH (902 mg, 19.6 mmol). After a short exothermic reaction the mixture was heated at 130-135°C for 35 min. Evaporation of the HCOOH, extraction with EtOAc and chromatography on alumina (PhH) afforded 246 mg of **2** (25.6%).

FISCHER - HEPP Nitrosamine Rearrangements

Rearrangement of N-nitroso to C-nitroso derivatives.

ON $NC_{19}H_{37}$ (**1**) —HCl→ HCl $HNC_{18}H_{37}$, NO **2** (96%)

1	Fischer, O., Hepp, E.	*Chem. Ber.*	**1886**	*19*	2991
2	Willey, J.	*J. Chem. Soc.*	**1955**		1677
3	Williams, D.L.H.	*J. Chem. Soc. D.*	**1969**		975
4	Hughes, E.D.	*Quart. Rev.*	**1952**	*6*	42

FLOOD Silyl Chloride Synthesis

Synthesis of silyl chlorides from siloxanes.

$[Me_2Si]_2O$ (**1**) —H_2SO_4, 1 h→ $[Me_3Si]_2SO_4$ (**2** (67%)) —HCl, $(NH_4)_2SO_4$→ $2(Me)_3SiCl + H_2SO_4$ (**3** (88%))

1	Flood, E. A.	*J. Am. Chem. Soc.*	**1933**	*55*	1735
2	Whitmore, L.A.	*J. Am. Chem. Soc.*	**1946**	*68*	1881
3	Sommer, L.H.	*J. Am. Chem. Soc.*	**1948**	*70*	445

Trimethylsilyl chloride (3).[1] Hexamethylsiloxane **1** (162 g, 1.00 mol) was treated with conc. sulfuric acid (260 g, 2.6 mol) under stirring and cooling for one hour. The suspension of crystals was extracted with pentane for 24 h. After evaporation of the solvent, **2**, 161 g (67%) was isolated. Dry HCl gas was passed through a mixture of **2** (30 g, 0.124 mol) and ammonium sulfate (20 g, 0.15 mol) in heptane 300 mL. After 2 h the organic layer was distilled to give **3**, 24 g (88%), bp 57°C, d=0.857, n_D^{20} = 1.3870.

FITTIG Pinacolone Rearrangement

Acid catalyzed carbocation rearrangement of 1,2-diols to ketones.

Ph-CH(OH)-C(OH)(Me)-Me (**1**) —HCO_2H, Δ→ Ph-C(Me)(Me)-CHO (**2** (77%)) + Ph-CH(Me)-C(=O)-Me (**3** (7%)) + Ph-C(=O)-CH(Me)-Me (**4** (17%))

5 —H_2SO_4, Δ→ **6** (80%) + **7** (3%) + **8** (17%)

1	Fittig, R.	*Leibig's Ann.*	**1860**	*114*	54
2	Villani, F.J.	*J. Org. Chem.*	**1964**	*27*	3208
3	Sand, E.D.	*J. Org. Chem.*	**1965**	*28*	2690
4	Suzuki, S.	*Tetrahedron Lett.*	**1983**	*24*	4997
4	Olah, G.	*Synthesis*	**1978**		358

1-Phenyl-1,1-dimethylacetaldehyde (2) and **isopropyl phenyl ketone (4).**[2] 1-Phenyl-2-methylpropandiol **1** (0.4 g, 2.4 mmol) in 97% formic acid was heated to reflux for 15 min. The mixture was poured into ice and extracted with Et_2O. **2:3:4** (by GC) 77%:7%:17%.

1-Cyclopenten-1-yl-1-cyclohexene (6); 2-cyclopentylcycloheptanone (7) spirobicyclohexan-2-one (8). 1,1'-Cyclopentylcyclohexyldiol **5** (100 g, 0.657 mol) was refluxed in 25% H_2SO_4 for 2 h. Et_2O extraction and evaporation of the solvent gave 60 g (60%) of a mixture of **6,7 and 8** in a ratio of 80.2%, 2.6% and 17.2%.

FORSTER Diazo Synthesis

Formation of diazo derivatives from oximes.

$NH_4OH/NaOCl$, 0°C

1

2 (50%)

BuONO

=N-OH

NH_4OH / NaOCl

$-N^+_2$

1	Forster, M.C.J.	*J. Chem. Soc.*	**1915**	*107*	260
2	Meinwald, J.	*J. Am. Chem. Soc.*	**1959**	*81*	4751
3	Hassner, A.	*Tetrahedron Lett.*	**1962**		759
4	Kirmse, W.	*Angew. Chem.*	**1957**	*69*	106
5	Rundel, W.	*Angew. Chem.*	**1962**	*74*	469

Diazofluorene (2).[2] A suspension of fluorenone oxime **1** (0.5 g, 2.5 mmol) in 15 N NH_4OH (25 mL) was treated with 5.25% NaOCl (100 mL) at 0°C for a period of 1 h. The reaction mixture was extracted with petroleum ether and the solvent evaporated to leave 0.19 g of **2** (50%). Unreacted oxime **1** was recovered from the aqueous solution (0.25 g).

FORSTER - DECKER Amine Synthesis

Selective monoalkylation of primary amines via imines. An alternative method is the reaction of **1** and **2** in the presence of $NaCNBH_4$ or triacetoxyborohydride (Borch reduction).[4]

$$\underset{\mathbf{2}}{Ph\text{-}CH_2\text{-}CH_2\text{-}NH_2} + \underset{\mathbf{1}}{OCH\text{-}Ph} \xrightarrow{110^\circ} \underset{\mathbf{3}}{Ph\text{-}CH_2\text{-}CH_2\text{-}N{=}CH\text{-}Ph}$$

$$\xrightarrow{NaOH \mid Me_2SO_4} \underset{\mathbf{4}\ (80\%)}{Ph\text{-}CH_2\text{-}CH_2\text{-}NH\text{-}Me}$$

1	Forster, M.O.	*J. Chem. Soc.*	**1899**	*75*	934
2	Decker, H.	*Liebigs Ann.*	**1913**	*395*	362
3	Morrison, A.L.	*J. Chem. Soc.*	**1950**		1478
4	Borch, R.F.	*J. Am. Chem. Soc.*	**1971**	*93*	2897

N-Methyl-N-β-phenethylamine (4).[3] Benzaldehyde **1** (116 g, 1.1 mol), β-phenetylamine (121 g, 1 mol) and PhMe (500 mL) were refluxed with a Dean-Stark water separator untill all water had distilled. Under stirring, dimethyl sulfate (126 g, 1 mol) was added dropwise and reflux was continued for another 3 h. Water and NaOH was added until the pH stabilized at pH=7. After hydrolysis, the solvent and **1** was recovered by steam distillation. The residual water solution was made alkaline to give 114 g of **4** (80-85%), bp 203°C, n_D^{20} = 1.5162, d = 0.930.

F R A N C H I M O N D Cyano Succinic Acid Synthesis

Condensation of α-haloesters to succinic acid derivatives in the presence of CN^-.

CO_2Et Br Br Me CO_2Et — NaCN → CO_2Et CN Me CO_2Et

1 **2** (63%)

1	Franchimond, A.P.N.,	*Chem. Ber.*	**1872**	*5*	1049
2	Fuson, R.C.	*J. Am. Chem. Soc.*	**1930**	*52*	4074

Diethyl 3-methyl-2-cyano-1,2-cyclobutanedicarboxylate (2).[2] Diethyl α,α' dibromo-β-methyladipate **1** (20.0g, 55 mmol) and powdered NaCN (8.0 g, 160 mmol) in abs. EtOH (15 mL) were heated to reflux for 24 h to give after filtration of inorganic salts and evaporation of the solvent 8.0 g of **2** (63%), bp 140-143°C/3 mm, n_D^{21}= 1.44735.

FREUND-GUSTAVSON Cyclopropane Synthesis

Cyclopropane synthesis from 1,3-dihaloalkenes with metals.

Me — CH — CH_2 — Br / Me — C — Br / Me (**1**) + Zn → **2** (86%)

(Br, Cl-substituted cyclobutane) + Na → bicyclobutane (93%)

1	Freund, A.	*Monatsh.*	**1882**	*3*	625
2	Gustavson, G.	*J. Prakt. Chem.*	**1887**	*36*	300
3	Haas, H.B.	*Ind. Eng. Chem.*	**1936**	*28*	1178
4	Bartleson, J.D.	*J. Am. Chem. Soc.*	**1946**	*68*	2513
5	Wiberg, K.	*Tetrahedron Lett.*	**1963**		2173

1,1,2-Trimethylcyclopropane (2).[4] To a suspension of oxygen-free zinc dust (196 g, 3 at) in a mixture of water (100 mL) and n-PrOH (300 mL) cooled in an ice bath, was added 1,2-methyl-2,4-dibromopropane **1** (224 g, 1 mol) over a period of 90 min. The mixture was stirred for about 32 h at 20°C and the hydrocarbon was isolated by distillation. The fraction bp 49-51°C was collected, 78.1 g (86%). After washing with water, freezing out the water and distillation through a column with 100 theoretical plates pure **2** was obtained.

FREUDENBERG - SCHÖNBERG Xanthate Rearrangement

Rearrangement of S-methyl xanthates to S-methyldithiocarbonates (conversion of alcohols to thiols via xanthates, also phenols to thiophenols via thiocarbamates).

HO(CH2)6Cl → $CH_3S-C(=S)-O(CH_2)_6Cl$ **1** ⇌ (Δ) $CH_3S-C(=O)-S(CH_2)_6Cl$ **2** (10%)

1	Freudenberg, K.	*Chem. Ber.*	**1927**	*60*	232
2	Schönberg, A.	*Chem. Ber.*	**1930**	*63*	178
3	Araki, Y.	*Bull. Chem. Soc. Jpn.*	**1970**	*43*	3214
4	Taylor, R.	*J. Chem. Soc. Perkin Trans. 2*	**1988**		183
5	Wiersum, U.E.	*J. Org. Chem.*	**1989**	*54*	5811

S-(6-Chloro-1-hexyl)-S'-methyl dithiocarbonate (2).[5] O-(6-Chloro-1-hexyl)-S-methyl xanthate **1** (60.0 g, 0.265 mol) was distilled at 525°C for 3.5 h into a quartz tube. The crude pyrolizate (45 g) was fractionated, yielding 25.3 g of **1** (81%) and leaving a yellow residue which after distillation gave 5.9 g (10%) of **2**, bp 175-176°C (20 mm).

FRIEDEL - CRAFTS Alkylation-Acylation

Alkylation or acylation of aromatic compounds by means of alkyl halides, alcohols, alkenes, acyl halides in the presence of Lewis acids.

HF, 0°C (74%)

$ZnCl_2$, 110 °C; **1** + **2** → **3** (66%)

1	Friedel, C., Crafts, J.N.	*C.R.*	**1877**	*84*	1450
2	Groggins, P.T.	*Ind.Eng.Chem.*	**1951**	*43*	1970
3	Renfrow, W.B.	*J.Am.Chem.Soc.*	**1951**	*73*	318
4	Kulka, M.	*J.Org.Chem.*	**1986**	*51*	2128
5	Gore, P.	*Chem.Rev.*	**1955**	*55*	229
6	Pearson, D.E.	*Synthesis*	**1972**		533
7	Price, C.C.	*Org.React.*	**1946**	*3*	130

2-Methoxy-5-t-butylbenzophenone (3).[4] A solution of 4-methoxy-t-butylbenzene **1** (100 g, 0.604 mol), benzoyl chloride **2** (85 g, 0.605 mol) and $ZnCl_2$ (0.2 g) in 1,1,2,2-tetrachloroethane (150 mL), was refluxed for 40 h. After removal of the solvent from the reaction mixture, the residue was distilled. The fraction bp 208-210°C (12 mm) gave 132.6 g of **3** which crystallized on standing. Recrystallization from MeOH gave 107 g of pure **3** (66%).

FRIEDLÄNDER Quinoline Synthesis

Quinoline synthesis from condensation of o-aminoaryl aldehydes (ketones) with α-methylene aldehydes (ketones).

CHO, NH_2 **1** + O **2** —(33% KOH, heat)→ Me, N, Me **3** (58%)

1	Friedländer, P.	*Chem. Ber.*	**1882**	*15*	2572
2	Markgraf, J.H.	*J. Org. Chem.*	**1969**	*34*	4131
3	Coffen, D.L.	*J. Org. Chem.*	**1974**	*39*	1765
4	Bergstrom, F.W.	*Chem. Rev.*	**1944**	*35*	151
5	Eckert, K.	*Angew. Chem. Int. Ed.*	**1981**	*20*	208

2,3-Dimethylquinoline (3).[2] A solution of o-aminobenzaldehyde **1** (1.0 g, 8.2 mmol), 2-butanone **2** (0.59 g, 8.2 mmol) and 33% aqueous KOH (5 mL) in 95% EtOH (20 mL) was refluxed for 1 h. The reaction mixture was neutralized with AcOH, concentrated and extracted with Et_2O. The residue after chromatography yielded 0.75 g of **3** (58%), mp 68-69°C.

F R I E S Phenol Ester Rearrangement

Rearrangement of phenol esters to o- or p-ketophenols, Lewis acid catalyzed.

$AlCl_3$, Δ

1 → **2** (18.6%) + **3** (0.5%)

1 Fries, K. *Chem. Ber.* **1908** *41* 4271
2 Cremer, S.E. *J. Org. Chem.* **1961** *26* 3653
3 Martin, A.R. *Tetrahedron Lett.* **1986** *27* 1959
4 Blatt, A.H. *Chem. Rev.* **1940** *27* 429
5 Effenberg *Angew. Chem. Int. Ed.* **1973** *12* 776
6 Blatt, A.H. *Org. React.* **1942** *1* 342

3,5-Dichloro-2-hydroxy-6-methyl acetophenone (2) and **6,8-Dichloro-2,5-dimethylchromone (3).** A mixture of 2,4-dichloro-5-methylphenyl acetate **1** (39.6 g, 0.181 mol) and $AlCl_3$ (32.0 g, 0.24 mol) was heated for 90 min at 135°C. The cooled solution was poured into ice (150 g) and extracted with Et_2O. The Et_2O solution was washed with 10% NaOH and evaporated to give 1.0 g of **3**, mp 125-135°C, after recrystallization from EtOH, 200 mg (0.5%) of **3** and unreacted **1** 3.4 g (8.5%). From the NaOH solution after acidification with 6N HCl and extraction with Et_2O was obtained 8.2 g of **2** (18.6 %), mp 104-105.5°C (recrystallized from petroleum ether).

FRITSCH - BUTTENBERG - WIECHELL Acetylene Synthesis

Alpha elimination from haloethylenes leading via carbene rearrangement to acetylenes.

$$\underset{\mathbf{1}}{(p\text{-}Cl\text{-}C_6H_4)(C_6H_5)C{=}C(Br)H} \xrightarrow{BuLi} \underset{\mathbf{2}\ (22\%)}{p\text{-}Cl\text{-}C_6H_4\text{-}C{\equiv}C\text{-}C_6H_5}$$

1	Fritsch, P.	*Liebigs Ann.*	**1894**	*279*	319
2	Buttenberg, W.P.	*Liebigs Ann.*	**1894**	*279*	327
3	Wiechell, H.	*Liebigs Ann.*	**1894**	*279*	337
4	Curtin, D.Y.	*J. Am. Chem. Soc.*	**1958**	*80*	4599
5	Kobrich, G.	*Chem. Ber.*	**1972**	*105*	1674
6	Kobrich, G.	*Angew. Chem. Int. Ed.*	**1965**	*4*	49

1-Phenyl-2-(p-chlorophenyl)acetylene (2).[4] To a solution of Z-1-phenyl-1-p-chlorophenyl-2-bromoethylene **1** (5.00 g, 17 mol) in ether (100 mL) at -35°C under N_2 was added a 10% solution of butyllithium (26.5 mL) maintaining the temperature at -20°C to -35°C for 30 min. The mixture was allowed to warm to room temperature and was neutralized with 20% HCl (50 mL). Separation of the neutral fraction with ether followed by drying over anhydrous sodium sulfate and removal of the ether gave a residue of 2.24 g. Purification by chromatography on alumina yielded 0.88 g of **2** (22%), mp 82-83°C.

FUJIMOTO - BELLEAU Cyclohexenone Synthesis

Synthesis of fused cyclohexenones from lactones (an alternative to the Robinson annulation).

MeMgI → NaOH →

1 2 3

1 Fujimoto, G.I.	*J. Am. Chem. Soc.*	**1951**	*73*	1856
2 Belleau, B.	*J. Am. Chem. Soc.*	**1951**	*73*	5441
3 Weyl Raynal, J.	*Synthesis*	**1969**		49

Cholestenone (3).[1] Methylmagnesium iodide obtained from CH_3I (142 mg, 1 mmol) was added dropwise to a solution of enol lactone (Turner's lactone) **1** (386 mg, 1 mmol) in Et_2O (3 mL) under N_2 at 0°C. The mixture was let stand for 24 h and was decomposed with NH_4Cl solution. Removal of the solvent gave crude **2**, which on recrystallization from Me_2CO afforded 180-260 mg of **2** (45-65%), mp 174-178°C. Crude **2** was stirred with a solution of 20% NaOH (5 mL) and MeOH (30 mL) for 24 h. The residue, obtained after removing the solvent, was chromatographed on alumina (petroleum ether: PhH 9:1 or 4:1) to give **3**, mp 80-88°C.

FUJIWARA Lanthanide (Yb) reaction

Use of ytterbium or other lanthanoids in substitution, reduction and 1,2 addition.

$$Ph_2C{=}O\ (\mathbf{1}) + Yb\ (\mathbf{2}) \longrightarrow \mathbf{3} \longrightarrow Ph_2C(OH){-}C(OH)Ph_2\ (\mathbf{4})$$

$$\mathbf{1} + \mathbf{2} + Ph{-}CH_2{-}CN\ (\mathbf{5}) \xrightarrow[HMPA]{THF} Ph_2C(OH){-}C(=O){-}CH_2{-}Ph\ \ \mathbf{6}\ (65\%)$$

$$\mathbf{1} + \mathbf{2} + \text{cyclohexene oxide} \longrightarrow \text{2-(hydroxydiphenylmethyl)cyclohexanol}$$

$$Ph{-}I \xrightarrow{\mathbf{2}} Ph{-}Yb{-}I \xrightarrow{Ph{-}CH{=}CH{-}Br} Ph{-}CH{=}CH{-}Ph$$

1	Fujiwara, Y.	*Chem. Lett.*	**1981**		1771
2	Fujiwara, Y.	*J. Org Chem.*	**1984**	*49*	3237
3	Fujiwara, Y.	*J. Org. Chem.*	**1988**	*53*	6077
4	Fujiwara, Y.	*J. Org. Chem.*	**1987**	*52*	3524

2-Oxo-1,1,3-triphenylpropan-1-ol (6).[3] Yb powder (173 mg, 1 mmol) under N_2 was treated with a drop of MeI and was heated to activate the Yb. THF (2 mL) was added, followed by HMPA (1 mL). Under stirring benzophenone **1** (182 mg, 1 mmol) in THF (2 mL) was added, followed by phenylaceto-nitrile **5** (117 mg, 1 mmol). After 4 h stirring at 20°C the mixture was quenched with 2N HCl, extracted with Et_2O and the product separated by medium pressure L.C. to afford 187 mg of **6** (65%) and 50 mg of **4** (35%).

F U J I W A R A Arylation, Carboxylation

A mild Pd catalyzed arylation or carboxylation of a Pd activated double bond.

C_6H_6 + $Pd(OAc)_2$ ⟶ PhPdOAc + HOAc

Ph-CH=CH-C(=O)-Ph (1) + PhPdOAc ⟶ Ph CH(Ph)-CH_2-C(=O)-Ph

1 — 2 (50%)

Thiophene + CO_2 —($Pd(OAc)_2$, tBuOOH)⟶ thiophene-2-CO_2H

Cyclohexane + CO (30 atm.) —($Pd(OAc)_2$ $K_2S_2O_8$, CF_3CO_2H 80°)⟶ cyclohexane-CO_2H

1	Moritari, I., Fujiwara, Y.	*Tetrahedron Lett.*	**1967**		1119
2	Yamamure, K.	*J. Org. Chem.*	**1978**	*43*	724
3	Fujiwara, Y.	*J. Organomet. Chem.*	**1984**	*266*	C44
4	Fujuwara, Y.	*Chem. Lett.*	**1989**		1687

1,1-Diphenyl-2-benzoylethane (2).[2] A mixture of benzylideneacetophenone **1** (2.08 g, 10 mmol) and palladium acetate (2.0 g, 10 mmol) in benzene (150 mL) and AcOH (40 mL) was refluxed under stirring until the precipitation of Pd metal ceased. The insoluble Pd was filtered off and the filtrate after washing with water was concentrated in vacuum. The residue, chromatographed on silica gel (hexane) gave a small amount of biphenyl, followed by 1.40 g of **2** (50%), eluted with hexane:benzene (1:1).

FUJIWARA - HECK Coupling

Cross-coupling reactions of aromatic or vinylic halides and olefins catalyzed by palladium.

$ClCOCH=CHCOCl$ + $PhCH=CH_2$ → ($Pd(OAc)_2$; R_3N, 140°) → $Ph(CH=CH)_4Ph$ (44%)

4-vinylpyridine **2** + Br–C_6H_4–NBu_2 **1** → ($Pd(OAc)_2$; TTP; TEA) → **3** (86%)

1,2-dibromocyclohexene + $CH_2=CHCO_2Et$ → ($Pd(OAc)_2PPh_3$; Et_3N; DMF) → bis($CH=CHCO_2Et$)cyclohexene (98%)

1 Fujiwara, Y. *Tetrahedron Lett.* **1967** 1119

2 Heck, R.F. *J. Am. Chem. Soc.* **1968** *90* 5518

3 Heck, R.F. *J. Am. Chem. Soc.* **1974** *96* 1133

4 Hassner, A. *J. Org. Chem.* **1984** *49* 2546

5 Fujiwara, Y. *Bull. Chem. Soc. Japan* **1990** *63* 438

6 de Meijere, A. *Synlett* **1990** 405

7 Heck, R.F. *Org. React.* **1982** *27* 345

Trans-4-[p(Di-n-butylamino)styryl]pyridine (3).[4] A mixture of p-bromo-N,N-dibutyl aniline **1** (5.68 g, 20 mmol), 4-vinylpyridine **2** (2.63 g, 25 mmol), $Pd(OAc)_2$ (45 mg, 0.2 mmol), tri o-tolylphosphine (120 mg, 4 mmol) and triethylamine (10 mL) was heated at 110°C for 72 h. To the cooled mixture was added water and $CHCl_3$ (all solids dissolved). The water was extracted with $CHCl_3$ (2x100 mL) and the combined organic solution was washed, dried and evaporated. The residue recrystallized from cold hexane gave 5.29 g of **3** (86%), mp 80-81°C.

GABRIEL Amine Synthesis

Synthesis of primary amines from alkyl halides.

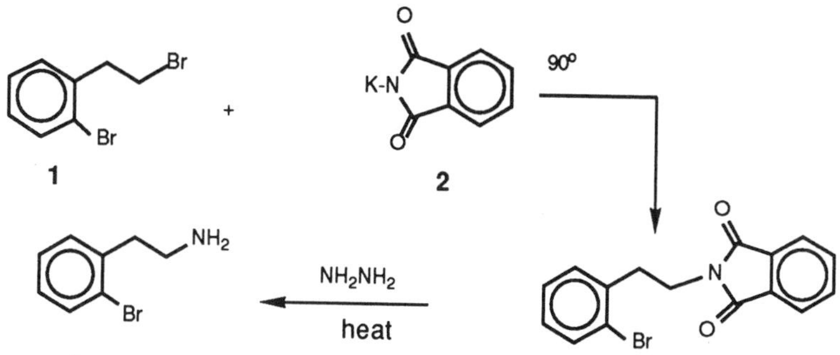

1	Gabriel, S.	*Chem. Ber.*	**1887**	*20*	2224
2	Bradsher, C.H.	*J. Org. Chem.*	**1981**	*46*	327
3	Gibson, J.S.	*Angew. Chem. Int. Ed.*	**1968**	*7*	919
4	Ragnarsson, A.	*Accts. Chem. Res.*	**1991**	*24*	285

2-(2-Bromophenyl) ethylamine (4). 2 A mixture of 2-(2-bromophenyl))ethyl bromide **1** (90.65 g, 0.34 mol) and potassium phthalimide **2** (66.64 g, 0.36 mol) in DMF (280 mL) was stirred for 15 h at 90°C. Dilution with $CHCl_3$ (400 mL), washing with 2N NaOH and evaporation gave 94.27 g of **3** (83%), recrystallized from MeOH/petroleum ether (3:7), mp 96.5-98°C.

A mixture of purified **3** (83.64 g, 0.253 mol), MeOH (250 mL) and hydrazine hydrate (30.43 g, 0.506 mol) was refluxed for 1 h. The residue obtained after evaporation of the solvent was refluxed with conc. HCl (120 mL) for 1.5 h. The solid was filtered and the filtrate was made alkaline with 2N NaOH and cooled to 5°C for 3 h. Extraction with Et_2O and evaporation of the solvent gave 41.87 g of **4** (83%), bp 82-86°C (0.3 torr).

GABRIEL - COLMAN Rearrangement

Synthesis of isoquinolines by rearrangement of phthalimides.

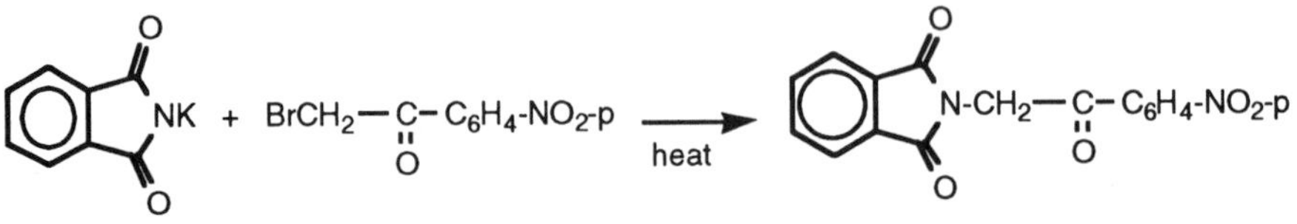

1	Gabriel, S., Colman, J.	*Chem. Ber.*	**1900**	*33*	980
2	Hill, J.H.M.	*J. Org,. Chem.*	**1965**	*30*	620
3	Allen, C.F.H.	*Chem. Rev.*	**1950**	*47*	284

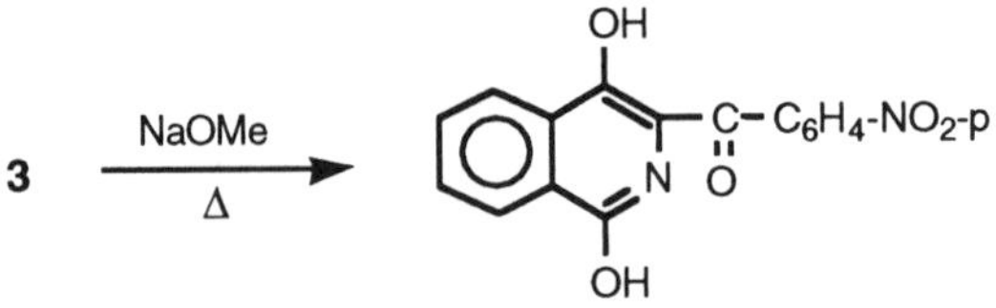

3-(p-Nitrobenzoyl)-1,4-dihydroxyisoquinoline (4).[2] p-Nitro-α-bromoacetophenone **2** (18.2 g, 75 mmol) and potassium phtalimidate **1** (9.3 g, 50 mmol) in Me_2CO (150 mL) were refluxed for 72 h. The brown mixture was poured into water, the solid was filtered and triturated with Et_2O to remove unreacted **2**. The tan residue after recrystallization from DMF gave 8.7 g of **3** (56%), mp 243-244°C.

A solution of **3** (1.95 g, 4 mmol) in 2N MeONa (50 mL) was refluxed under N_2 for 2 h. The cooled mixture was neutralized with dilute HCl, the solid filtered and washed with water. Recrystallization from AcOH afforded **4**, mp 278-280°C.

GAREGG - SAMUELSSON Olefin Synthesis

Conversion of vic trans-diol groups into a double bond by iodine-triphenylphosphine-imidazole reagent.

Ph_3P-I_2, imidazole

1 → **2** (56%) + **3** (17%)

1 Garegg, P.J., Samuelsson, B. *Synthesis* **1979** 469; 813

2 Zamojski, A. *Carbohydrate Research* **1990** *205* 410

Benzyl 2-benzyloxycarbonylamino-2,3,4,6-tetradeoxy-6-iodo-α-D-erythrohex-3-enopyranoside (2).[2] To a mixture of triphenylphosphine (15.72 g, 59.9 mmol) and iodine (15.2 g, 59.9 mmol) in PhMe (200 mL) stirred at 20°C for 30 min, was added a solution of imidazole (4.5 g, 66.1 mmol) in MeCN (50 mL) and stirring was continued for 10 min at 50°C. Benzyl 2-benzyloxycarbonyl-amino-2-deoxy-α-D-glucopyranoside **1** (4.04 g, 10 mmol) was added to the reagent prepared above and the mixture was stirred for 3 h at 80°C. The mixture was cooled to 20°C, filtered through Celite and concentrated in vacuum. The residue was extracted with PhMe and the combined extract evaporated to dryness. The residue chromatographed on silica gel (PhMe:EtOAc 95:5) gave: 2.68 g of **2** (56%), mp 129-130°C, α_D^{23} = +22° (c 0.8 $CHCl_3$), 1.03 g of **3** (17%), mp 119-121°C, α_D^{23} = +133° (c 1 $CHCl_3$) and 330 mg of β–**2** (7%), mp 138-140°C, α_D^{20} = - 151° (c 2 $CHCl_3$).

G A S S M A N Oxindole Synthesis

Synthesis of oxindoles from anilines

1 → (Me$_3$C-OCl **2**, TEA **4**; CH_3-S-CH_2-$COOC_2H_5$ **3**) → **5** (SCH_3, $CHCO_2C_2H_5$, NH_2) →

6 (84%) (SCH_3, H, O, NH) → (Raney nickel) → **7** (76%) (O, NH)

1	Gassman, P.G.	*J. Am. Chem. Soc.*	**1973**	*95*	2718
2	Gassman, P.G.	*J. Am. Chem. Soc.*	**1974**	*96*	5508
3	Johnson, P.D.	*J. Org. Chem.*	**1990**	*55*	1374

Oxindole (7). [2] To a stirred, cooled (-65°C) solution of aniline **1** (4.09 g, 44 mmol) in CH_2Cl_2 (150 mL) was added dropwise t-butyl hypochlorite **2** (4.77 g, 44 mmol) in CH_2Cl_2 (20 mL). After 10 min, ethyl methylthioacetate **3** (5.89 g, 44 mmol) in CH_2Cl_2 (20 mL) was added (exothermic) and stirring was continued for 1 h. TEA **4** (4.44 g, 44 mmol) in CH_2Cl_2 (20 mL) was added. The mixture was allowed to warm to room temperature, water (50 mL) was added and the organic layer was evaporated. The residue was redissolved in Et_2O (150 mL) and was stirred with 2N HCl (20 mL) for 24 h. Filtration afforded 6.61 g of **6** (84%). A solution of **6** (2.00 g, 11 mmol) in anh. EtOH (50 mL) was stirred and refluxed with W-2 Raney nickel (12 g) for 2 h. The supernatant and the washing solution were evaporated to dryness, the residue was dissolved in CH_2Cl_2 (20 mL), the solution dried ($MgSO_4$), filtered and evaporated to give 1.13 g of **8** (76%), mp 116-117°C.

GASTALDI Pyrazine Synthesis

Pyrazine synthesis from α-oximinoketones via α-aminoketones.

1 → (1. NaOH; 2. HNO_2) → 2 (44%) → ($Na_2S_2O_5$, KCN) → → (KCN, 50°) → 3 (8%)

1 Gastaldi, G.	*Gazz. Chim. Ital.*	**1921**	*51*	233
2 Sharp, W.	*J. Chem. Soc.*	**1948**		1862
3 Krems, I., Spoerri, P.	*Chem. Rev.*	**1947**	*40*	301

2,5-Dicyano-3,6-diethylpyrazine (3).[2] To ethyl 3-oxopentanoate **1** (40 g, 0.312 mol) in 2.5% NaOH (600 mL) kept for 24 h, were added $NaNO_2$ (21 g, 0.304 mol) in water and after cooling at 5°C a 20% solution of H_2SO_4 (172 g) dropwise. The mixture was stirred for 60 min at 5°C, extracted with Et_2O and after removal of solvent, the residue was distilled to give 15.7 g of **2** (44.6%). To $Na_2S_2O_5$ (183 g) and water (317 mL) was added **2** (40 g, 0.396 mol). The mixture was heated, then kept 24 h at 20°C and treated with EtOH (1500 mL) and AcOH (7.5 mL). After 24 h the precipitate was filtered. From the mother liquor a second and third crop were obtained. The crude product (50 g) was treated with KCN (22 g) in water (100 mL) with stirring at 40-50°C and the red brown solution was heated for 1 h at 45-60°C with 20% HCl (27 mL) to give a solid. The latter was extracted with EtOH to give 2.82 g of **3** (8%), mp 115°C.

GATTERMAN - KOCH Carbonylation

Synthesis of aromatic aldehydes or ketones using cyanide salts or CO - HCl and Lewis acids.

HO OH → ($Zn(CN)_2$, HCl) → HO OH CHO

1 → 2 (50%)

3 + 4 ($Cl\text{-}CH_2$, Cl_2C, CH_3) + CO → ($AlCl_3$) → 5 (58%) (C_6H_5, CH_3, O)

1	Gattermann, L.	*Chem. Ber.*	**1898**	*31*	1149
2	Adams, R.	*J. Am. Chem. Soc.*	**1923**	*45*	2373
3	Kreuzenberg, A.	*Angew. Chem.*	**1967**	*79*	978
4	Truce, W.E.	*Org. React.*	**1957**	*9*	37
5	Gattermann, L., Koch, J.	*Chem. Ber.*	**1897**	*30*	1622
6	Brunson, H.R.	*J. Org. Chem.*	**1967**	*32*	3359
7	Gore, P.M.	*Chem. Rev.*	**1955**	*55*	235
8	Grounse, N.	*Org. React.*	**1949**	*5*	291

Resorcinaldehyde (2).[2] To resorcinol **1** (20 g, 0.18 mol) in Et_2O (150 mL) was added $Zn(CN)_2$ (37 g, 0.27 mol) and dry HCl gas was bubbled through for 2 h. After decantation the residue was crystallized from water (100 mL) to afford 12.5 of **2** (50%), mp 135-137°C.

2-Methyl-2-phenylindanone (5).[6] To vigorously stirred benzene **3** (140 g, 1.8 mol) and $AlCl_3$ (42 g, 0.3 mol), was added 1,2,2-trichloropropane **4** (44.5 g, 0.3 mol) over 3 h at 24-27°C while CO was rapidly bubbled in. The organic layer after quenching with water, was dried and distilled in vacuum. Crystallization from EtOH gave 39 g of **5** (58%), mp 111°C.

GEWALD Heterocycle Synthesis

Synthesis of 2-aminofurans (thiophenes, pyrroles) by condensation of malononitriles with ketones.

1	Gewald, K.	*Chem. Ber.*	**1965**	*98*	3571
2	Gewald, K.	*Chem. Ber.*	**1966**	*99*	94;1002
3	Gewald, K.	*Z. Chem.*	**1961**	*1*	349
4	Peet, P.N.	*J. Heterocycl. Chem.*	**1968**	*23*	129

2-Amino-4,5,6,7-tetrahydrobenzo-(b)-thiophen-3-carboxylic acid ethyl ester (4).[4] A mixture of ethyl cyanoacetate **1** (56.6 g, 0.5 mol) cyclohexanone **2** (49.1 g, 0.5 mol), morpholine **3** (43.5 g, 0.5 mol) and S_6 (16 g, 0.5 at) in EtOH (150 mL) was stirred and the temperature rose to 50°C. After 1 h a thick precipitate was present. After standing 24 h, the mixture was poured into water and the solid filtered. The crude product, 115 g, recrystallized from EtOH gave 95.2 g of **4** (82%), mp 114-115°C.

GIRARD - SANDULESCU Reagent

Reagents "T" and "P" for separation of aldehydes or ketones by forming a water soluble hydrazide derivative.

Cl^- (pyridinium) N^+ - $CH_2CONHNH_2$ "P"

Cl^- $Me_3N^+CH_2CONHNH_2$ "T"

"T" + **1** (o-nitrobenzaldehyde, NO_2, CHO) $\underset{H_3O^+}{\overset{heat}{\rightleftharpoons}}$ **2** (NO_2; $CH{=}N{-}NH{-}C(=O){-}CH_2{-}N^+Me_3$ Cl^-)

1	Girard, A., Sandulescu, G.,	*Helv. Chim. Acta.*	**1936**	*19*	1095
2	Lederer, E.	*Bull. Soc. Chim. Fr.*	**1949**		1400
3	Wheller, O.H.	*Chem. Rev.*	**1962**	*62*	205
4	do Amaral, L.	*J. Org. Chem.*	**1991**	*56*	1419

(Carboxymethyl)trimethylammonium chloride hydrazide (Reagent "T").[1] A cooled solution of ethyl chloroacetate (984 g, 6.65 mol) was treated with trimethylamine (200 g, 3.39 mol). After 12-24 h a second portion of trimethylamine (200 g, 3.39 mol) was added. After complete consumption of ethyl chloroacetate, hydrazine (400 g) was added. The product was filtered and dryed under vacuum (H_2SO_4) to afford 1100 g of "T" (90%).

Separation of aldehydes.[2] o-Nitrobenzaldehyde (302 mg) and reagent "T" (502 mg) was stirred in EtOH (5 mL) under reflux. After cooling the water soluble hydrazone **2** (406 mg), mp 229-230°C was obtained. Acid hydrolysis gave back the aldehyde.

GLASER - CHODKIEWCZ Acetylene Coupling

Polyacetylenes from monoacetylenes in the presence of copper salts.

$$\underset{\mathbf{1}}{Me_2\underset{OH}{C}-C\equiv CH} \xrightarrow[Pyr - O_2]{Cu_2Cl_2} \underset{\mathbf{2}\ (90\%)}{Me_2\underset{OH}{C}-C\equiv C-C\equiv C-\underset{OH}{C}Me_2}$$

$$\underset{\mathbf{3}}{Et_3Si-C\equiv C-C\equiv C-Br} \xrightarrow[Cu_2Cl_2,\ EtNH_2]{Ph-C\equiv CH} \underset{\mathbf{4}\ (30\%)}{Ph-C\equiv C-C\equiv C-C\equiv C-SiEt_3}$$

1	Glaser, C.	*Chem. Ber.*	**1869**	*2*	422
2	Stansbury, H.A.	*J. Org. Chem.*	**1962**	*27*	320
3	Chodkiewcz, W.	*Ann. Chim. (Paris)*	**1957**	*2*	819(13)
4	Walton, D.R.M.	*Synthesis*	**1974**		890

2,7-Dimethyl-3,5-octadiyne-2,7-diol (2).[2] 2-Methyl-3-butyn-2-ol **1** (84 g, 1 mol), pyridine (20 g, 0.25 ml), MeOH (78 g) and Cu_2Cl_2 (1.2 g, 12 mmol) was stirred at 30-35°C for 2.5 h while 7.0 L (0.28 mol) of O_2 was absorbed. After treatment with saturated NH_4Cl (400 mL), extraction with Et_2O, evaporation and recrystallization from PhH (1200 mL), 75 g of **2** (90%), mp 127-129°C was obtained.

1-Phenyl-6-triethylsilylhexatriyne (5).[4] Phenylacetylene (2.6 g, 25 mmol) was added dropwise to $EtNH_2$ (1.1 g), Cu_2Cl_2 (0.2 g) and $NH_2OH.HCl$ (0.5 g) in DMF (25 mL) followed by bromobutadiynyl triethylsilane **3** (6.1 g, 25 mmol). After 2 h the mixture was acidified at 0°C with 2N H_2SO_4 (100 mL). Extraction with Et_2O and chromatography on acidic alumina (petroleum ether) gave 1.98 g of **4** (30%).

G O L D Reagent

Reagent for dialkylaminomethynilation of activated methylenes or NH_2 groups.

Cl, N, Cl, N, N, Cl **1** + 6 Me_2NCHO **2** $\xrightarrow{65^o}$ Cl^- $Me_2\overset{+}{N}$=CH—N=CH—NMe_2 **3** (86%)

p-Br-C_6H_4-CO-CH_3 **4** + **3** $\xrightarrow[heat]{NaOMe}$ p-Br-C_6H_4-CO-CH=CH-N Me_2 **5** (74%)

Ph-NH_2 $\xrightarrow{3}$ Ph-N=CH-NMe_2 Ph—NH-NH_2 $\xrightarrow{3}$ Ph-N, N=, N

1	Gold, H.	*Angew. Chem.*	**1960**	*72*	959
2	Eschenmoser, A.	*Angew. Chem. Int. Ed.*	**1971**	*10*	330
3	Kunst, G.	*Angew. Chem. Int. Ed.*	**1977**	*15*	239
4	Bryson, T.A.	*J. Org. Chem.*	**1980**	*45*	524
5	Gupton, J.T.	*J. Org. Chem.*	**1980**	*45*	4522

3-(Dimethylamino)-2-azaprop-2-en-1-ylidene dimethylammonium chloride (3).[5] **1** (148 g, 0.808 mol) in dioxane (800 mL) at 25°C, with DMF **2** (385 g, 5.28 mol) was heated for 45 min at 65°C (CO_2 evolution). After 2-3 h at 68-85°C and cooling, **3** crystallized 339 g (86%), mp 94-96°C.

Ketone (5).[5] To MeONa (from 1.5 g Na and 100 mL MeOH) was added p-bromoacetophenone **4** (10 g, 50 mmol) and then **3** (10.6 g, 65 mmol). The solution was refluxed for 24 h, and after evaporation the residue taken up in $CHCl_3$ washed and the solvent evaporated. The residue was stirred in CCl_4:hexane (1:9) (50 mL) and filtered to give 9.4 g of **5** (74%), mp 75-76°C.

GOMBERG - BACHMANN - GRAEBE - ULLMANN Arylation

Aryl-aryl bond formation via diazonium salts. Carbazole synthesis by intramolecular aryl-aryl bond formation.

NO_2 NH 2 + S $NaNO_2$ - HCl NO_2 S

1 2 3 (51%)

N2+ N H N H

1	Gomberg, M., Bachmann, W.E.	*J. Am. Chem. Soc.*	**1924**	*42*	2339
2	Smith, P.A.S.	*J. Am. Chem. Soc.*	**1951**	*73*	2452,2626
3	Dermer, O.C.	*Chem. Rev.*	**1957**	*57*	77
4	Graebe, C., Ullman, F.	*Liebigs Ann.*	**1896**	*291*	16
5	Ashton, B.W.	*J. Chem. Soc.*	**1957**		4559
6	Campbell, N.	*Chem. Rev.*	**1948**	*40*	360

2-(2'-Nitrophenyl)thiophene (3).[2] A cooled solution of 2-nitroaniline **1** (28.0 g, 0.2 mol) in water (80 mL) and HCl (45 mL), was diazotized with $NaNO_2$ (14.5 g, 0.21 mol) in water (50 mL). The diazonium salt was filtered and treated under cooling with thiophene **2** (500 g, 5.9 mol) and a solution of $NaOAc.3H_2O$ (80 g, 0.588 mol) in water (200 mL) was added dropwise. After stirring for 3 h at 5-10°C and for 24 h at 20°C, the organic layer and the Et_2O extract of the water layer were vacuum distilled to give 25.6 g of **3** (80%), bp 126-132°C (0.3-0.5 mm). Redistillation and recrystallization from MeOH gave 21.0 g of **3** (51%), mp 51-52°C.

GRIECO Reagent

Pyridinium p-toluenesulfonate (PPTS) reagent **1** for protection of alcohols as the tetrahydropyranyl ethers as well as for cleavage of ethers (in warm ethanol).

$$CH_3\text{-}C_6H_4\text{-}SO_3H.H_2O + C_5H_5N \longrightarrow CH_3C_6H_4\text{-}SO_3^- \; C_5H_5N^+H$$

1

2 + dihydropyran → (**1** CH_2Cl_2, 20°) **3**; **3** → (**1**, EtOH 55°) **2**

1 Grieco, P.A. *J. Org. Chem.* **1977** *42* 3772

2 Pinnick, H.W. *Tetrahedron Lett.* **1978** *44* *4261*

Pyridinium p-toluenesulfonate (PPTS) (1).[1] TsOH monohydrate (5.70 g, 30 mmol) was added to pyridine (12.1 mL, 150 mmol) with stirring at 20°C. After 20 min the excess of pyridine was removed in vacuum (60°C). Recrystallization from Me_2CO gave 6.8 g of **1** (90%), mp 120°C.

Geraniol tetrahydropyranyl ether (4).[1] Geraniol **2** (154 mg, 1 mmol) and dihydropyran (126 mg, 1.6 mmol) in dry CH_2Cl_2 (7 mL) containing PPTS (25 mg, 0.1 mmol) is stirred for 4 h at 0°C. Dilution with Et_2O, washing with brine to remove the catalyst, evaporation of the solvent and vacuum distillation afforded 326 mg of **3** (99%), bp 140°C/10 mm.

Geraniol (2) A solution of **4** (119 mg, 0.5 mmol) and PPTS (12.6 mg, 0.05 mmol) in EtOH (4 mL) was stirred at 55°C for 3 h. Chromatography on silica gel afforded 77 mg of **2** (100%).

GRIESS Deamination

Deamination of aromatic amines via diazonium salts, by means of alcohols (Griess), or PO_2H_3, $Sn(OH)_2$, etc.

1. $NaNO_2$-HCl
2. PO_2H_3

1 → **2**

1	Griess, P.	*Phil. Trans.*	**1864**	*154*	683
2	Griess, P.	*Chem. Ber.*	**1897**	*21*	547
3	Fletcher, T.L.	*Synthesis*	**1973**		610
4	Kornblum, N.	*Org. React.*	**1944**	*2*	262

1-Bromophenantrene (2).[3] 2-Amino-1-bromophenantrene **1** (0.8 g, 3 mmol) was stirred with THF (5 mL), 32% HCl (30 mL) and water (10 mL). Under stirring and cooling (5-10°C) a solution of $NaNO_2$ (0.3 g, 4.5 mmol) in water (1.5 mL) was added dropwise over 5 min. After 30 min stirring at 0-5°C, 50% hypophosphorous acid (15 mL) was added and the mixture was stirred at 0°C for another 30 min, then at room temperature for 17 h. Water was added and the product collected and chromatographed on alumina, eluent PhH, then crystallized from MeOH, to afford 0.5 g of **2** (66%), mp 109.5-110 5°C.

GRIGNARD Reagents

Organomagnesium reagent capable of reacting with active "H" compounds or in additions to C=X bonds; also nickel catalyzed coupling.

CH_3MgI

1 → **2** (89%)

Ph–CH=CH–Br + t Bu Mg X —Ni (II)→ Ph–CH=CH–tBu

+ t Bu MgCl → tBu–CH₂CH₂–COOH

1	Barbier, P.	*C.R.*	**1899**	*128*	110
2	Grignard, V.	*C.R.*	**1900**	*130*	1322
3	Kirmse, W.	*Synthesis*	**1983**		994
4	Sonntag, N.O.V.	*Chem. Rev.*	**1953**	*53*	372
5	Bogdanovichi, B.	*Angew. Chem.*	**1983**	*95*	749
6	Hayashi, T.	*Chem. Lett.*	**1980**		1209
7	Dubois, J.F.	*Ann. Chim. Fr.*	**1951**	*6*	407
8	Walborsky, H.M.	*Acc. Chem.Res.*	**1990**	*23*	286
8	Walling, C.	*Acc. Chem.Res.*	**1991**	*24*	255

exo-2-Methylbicyclo[3.2.0]heptan-endo-2-ol (2).[3] To MeMgI prepared from MeI (2.3 g, 16 mmol), Mg turnings (0.4 g, 17 mmol) in Et_2O (60 mL) was added bicyclo [3.2.0] heptan-2-one **1** (1.7 g, 15 mmol) in Et_2O (10 mL) . After 1 h reflux the mixture was hydroly-zed (25 mL water) and extracted with Et_2O (2x25 mL). Evaporation gave 1.7 g of **2** (89%), purity 98% by GLC, purified by preparative GLC (Carbowax + KOH,110°C), mp ~ 25°C.

GUARESKY - THORPE Pyridone Synthesis

Synthesis of pyridones from beta diketones and activated amides.

5 + 6 → (Piperidine, EtOH) → 7 (65%)

1	Guaresky, A.	*Mem. Real. Accad. Sci. Torino*	**1896**	*46*	25(II)
2	Thorpe, J.F.	*J. Chem. Soc.*	**1911**	*99*	422
3	Katritzky, A.R.	*Adv. Heterocycl. Chem.*	**1963**	*1*	347
4	Kellog, R.M.	J. *Org. Chem.*	**1986**	*45*	2856

3-Cyano-4,6-dimethyl-2-pyridone (7).[3] Pentane-2,4-dione **5** (4.85 g, 48.5 mmol), cyanoacetamide **6** (4.5 g, 53 mmol) and piperidine (2 mL) in EtOH (50 mL) were heated to reflux for 45 min. The product crystallized on cooling. Recrystallization from EtOH gave 5-5.4 g of **7** (63-68%), mp 298-300°C.

GUY - LEMAIRE - GUETTE Phenol Chlorination

Regioselective ortho or para halogenation of phenols and the corresponding ethers (only para) by hexachloro cyclohexadiene **2** or **3** respectively.

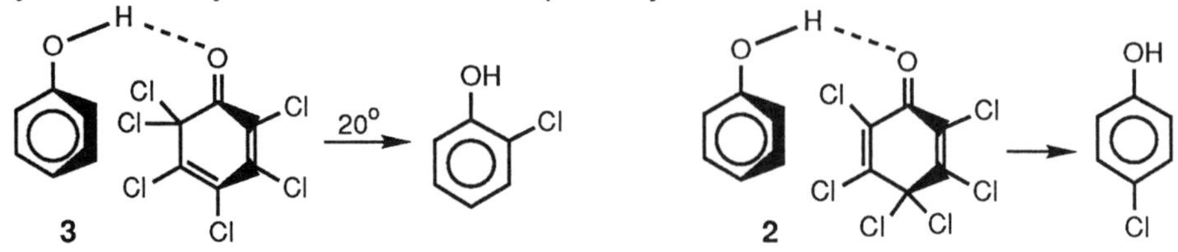

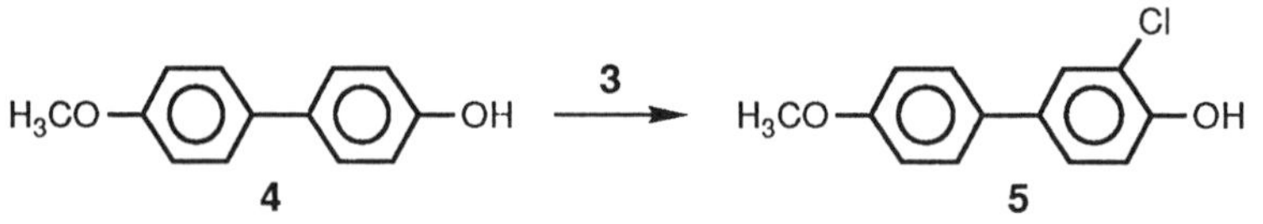

1	Guy, A., Lemaire, M., Guette, J.P.	*J. Chem. Soc. Chem. Commun*	**1980**		8
2	Guy, A., Lemaire, M., Guette, J.P.	*Tetrahedron*	**1982**	*38*	2339
3	Guy, A., Lemaire, M., Guette, J.P.	*Synthesis*	**1982**		1018
4	Guy, A., Lemaire, M., Guette, J.P.	*C.R. Acad. Sci. Ser. 2*	**1984**	*299*	693

2,3,4,5,6,6-Hexachlorocyclohexa-2,4-diene-1-one (3).[2] To stirred Cl_2 (0.3 mol) in CCl_4 (300 mL), was added sodium pentachlorophenolate **1** (57.5 g, 0.2 mol) at 0-5°C. After 3 h, filtration and evaporation (below 40°C) left a residue. This was stirred in petroleum ether (200 mL) at -10°C for 24 h and filtered to give 34 g of **2** (56%).

2,3,4,4,5,6-Hexachlorocyclohexa-2,5-diene-1-one (2).[2] Cl_2 was passed through a suspension of $AlCl_3$ (5 g, 37 mmol) in pentachlorophenol for 4 h. After 48 h, filtration, evaporation and recrystallization from petroleum ether (200 mL) gave 28 g of **3** (46%).

4-(4-Methoxyphenyl)-2-chlorophenol (5).[2] To **4** (0.4 g, 2 mmol) in 25% DMF in CH_2Cl_2 (8 mL) was added **3** (0.6 g, 2 mmol). After 48 h at 20°C, workup gave 0.37 g of **5** (80%).

HADDADIN-ISSIDORIDES Quinoxaline synthesis

Synthesis of quinoxaline dioxides from benzofurazan oxide and ketone enolates or enamines (also known as Beirut reaction).

1 + 2 (Et_3N, 20°) → 3, R: Ph-CO; 4, R: H

1 + (1-morpholinocyclohexene) → 48%

1	Haddadin, M.J.,Issidorides, C.H.	*Tetrahedron Lett.*	**1965**		3253
2	Issidorides, C.H., Haddadin, M.J.	*J. Org. Chem.*	**1966**	*31*	4067
3	Haddadin, M.J.,Issidorides, C.H.	*Tetrahedron*	**1974**	*30*	659
4	Haddadin, M.J.,Issidorides, C.H.	*US Pat. 4,343,942 ; CA*	**1984**	*101*	171227
5	Lin, S. K.	*Yonji Huaxue*	**1991**	*11*	106

2-Phenyl-3-benzoylquinoxaline-N,N'-dioxide(3).[2] A solution of benzofurazan-N-oxide **1** (3.4 g, 25 mmol) and dibenzoylmethane 2 (5.6 g, 26 mmol) in warm Et_3N (25 mL) was allowed to stand at 20°C for 24 h. The mixture was diluted with Et_3N, filtered to give 2.5 g of **3**. The filtrate after standing another 30 h afforded a second crop of crystalls. Total yield: 3.6 g of **3** (42%), mp 234°C (from MeOH). Heating of 1 g of **3** in 45 mL of 2% KOH in MeOH until all dissolved gave 0.65 g (95%) of **4**, mp 205-206°C.

HALLER - BAUER Ketone Cleavage

Cleavage of ketones, lacking α-hydrogen, with sodium amide.

Ph Ph
O
C —Ph
CH_3
1
$NaNH_2$
Ph Ph
CH_3
2 (79%)

1	Haller, A., Bauer, E.	*C.R.*	**1909**	*148*	127
2	Impastato, F.I.	*J. Am. Chem. Soc.*	**1962**	*84*	4838
3	Paquette, L.A.	*J. Org. Chem.*	**1988**	*53*	704
4	Hamlin, K.E.	*Org. React.*	**1957**	*9*	1

1-Methyl-2,2-diphenylcyclopropane (2).[2] A mixture of $NaNH_2$ (3.0 g, 75 mmol) and 1-benzoyl-1-methyl-2,2-diphenylcyclopropane **1** (9.3 g, 30 mmol) in PhMe (80 mL) was refluxed for 5 h. The cooled reaction mixture was treated with cracked ice (50 g) and the separated organic layer, after washing with brine was distilled. The fraction bp 106-107°C (2.5 mm) was collected. There were obtained 4.9 g of **2** (79%).

H A N T S C H Pyridine Synthesis

One step synthesis of substituted pyridines from a beta keto ester, an aldehyde and ammonia.

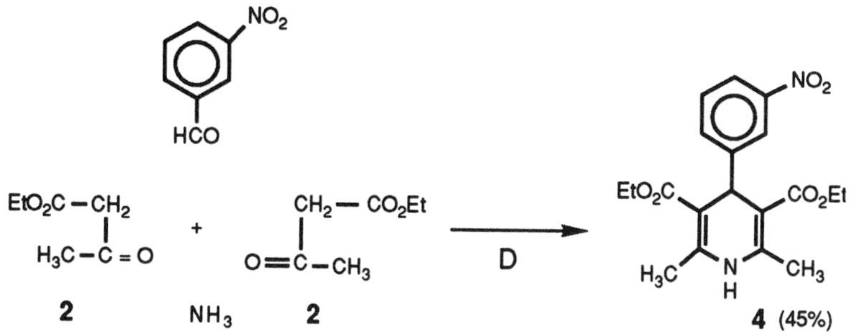

1	Hantsch, A.	*Liebig's Ann.*	**1882**	*215*	172
2	Phillips, A.P.	*J. Am. Chem. Soc.*	**1949**	*71*	4003
3	Svetlik, J.	*J. Chem. Soc. Perkin I*	**1987**		563
4	Eisner, U.	*Chem. Rev.*	**1972**	*72*	1

3,5-Di(ethoxycarbonyl)-1,4-dihydro-2,6-dimethyl-4-(m-nitrophenyl) pyridine (4).[2] m-Nitrobenzaldehyde **1** (15.1 g, 0.1 mol), ethyl acetoacetate **2** (28.6 g, 0.22 mol) and concentrated NH_4OH **3** (8 mL) in EtOH (60 mL) was heated to reflux for 3 h. The hot solution was diluted with water (40 mL), cooled, filtered, washed with 50% EtOH (10 mL) to give 16-18 g of **4** (43-48%), mp 165-167°C.

HASS-BENDER Carbonyl Synthesis

Aldehyde or ketone synthesis by reaction of an alkyl halide with the sodium salt of 2-nitroalkanes.

Me, Me C=NO_2 - Na^+ + (Br) → ($^-$O–N^+=C(Me)(Me), O, H) → (O) + (Me)(Me)C = N - OH

1 2 3 (44%)

1 Hass, H.B., Bender, M.L. *J. Am. Chem. Soc.* **1949** *71* 1767

2 Bersohn, M. *J. Am. Chem. Soc.* **1961** *83* 2136

3 Epstein, W.W. *Chem. Rev.* **1967** *67* 247

2-Cyclohexenone (3).[2] To a solution of sodium (782 mg, 34 mmol), in MeOH (40 mL) was added at room temperature, and under N_2 2-nitropropane **1** (3.94 g, 44 mmol) and then 3-bromocyclohexene **2** (5.5 g, 34 mmol). After 20 h, the solvent and cyclohexadiene were distilled at atm. pressure, the residue was treated with water and ether, the organic layer washed with 10% solution of NaOH, and water. After drying over $MgSO_4$, the solvent was evaporated and the residue distilled in vacuum. There was obtained **3** as a colorless liquid, 1.43 g (44%).

H A S S N E R Azide aziridine synthesis

Stereospecific and regioselective addition of IN_3 (via iodonium ions) or of BrN_3 (ionic or free radical) to olefins and conversion of the adducts to aziridines or azirines.

Ph 1 ICl NaN3 Ph I N3 2 (erythro) LAH Ph Me N H 3trans (95%)

KOtBu Ph N3 4 (86%) hυ or Δ Ph N H 5 (94%) LAH Ph Me N H 3 cis

Ph N3 Br BrN3 pentane (rad.) Ph BrN3 MeCN (ionic) Ph Br N3

1 Hassner, A. *J. Am. Chem. Soc.* **1965** *87* 4203

2 Hassner, A. *J. Am. Chem. Soc.* **1969** *91* 5046

3 Hassner, A. *J. Org. Chem.* **1968** *33* 2686

4 Hassner, A. *J. Am. Chem. Soc.* **1968** *90* 216

5 Hassner, A. *Accts. Chem. Res.* **1971** *4* 9

trans-2-Methyl-3-phenylaziridine (3).[2] To a slurry of 15 g (0.25 mole) of NaN_3 in MeCN (100 mL) below 0°C was added slowly 18.3 g (0.113 mole) of iodine monochloride over 15 min. After 10 min stirring, E-1-phenylpropene (0.1 mole) was added and the mixture stirred at 20°C overnight. The slurry was poured into 300 mL of cold 5% sodium thiosulfite and the orange oil extracted with ether, washed with water (5x200 mL), dried and evaporated. Flash chromatography (Woelm neutral alumina, petroleum ether) gave erythro **2** (100%). Note. *Some S-compounds react explosively with IN_3.* To stirred LAH (2.5 g) in anh. ether (90 mL) was added **2** (10.3 g, 0.035 mol) in ether (15 mL) at 0°C over 20 min. Work up with 20% NaOH (10 mL) stirring, filtration, drying and evaporation gave 4.93 g (85%) of **3** and 5% of **1**.

HASSNER - RUBOTTOM α-Hydroxylation

α-Hydroxylation, iodination, or oximation of carbonyls via silyl enol ethers.

Ph–CH₂–CH=CH–OSiMe₃ (**1**) —mCPBA→ Ph–CH₂–CH(OH)–CH(OCOAr)–OSiMe₃ (**2** (85%)) —Ac_2O→ Ph–CH₂–CH(OAc)–CHO (**3** (42%))

1-(trimethylsilyloxy)cyclohexene —mCPBA / NaI→ [iodonium intermediate, I^+] → 2-iodocyclohexanone

γ-butyrolactone —LDA / TMS-Cl→ 2-(trimethylsilyloxy)-4,5-dihydrofuran —NOCl→ α-oximino-γ-butyrolactone (=NOH)

	Author	Journal	Year	Vol.	Page
1	Hassner, A.	*J. Org. Chem.*	**1974**	*39*	1785, 2558
2	Rubottom, G.M.	*Tetrahedron Lett.*	**1974**		167
3	Hassner, A.	*J. Org. Chem.*	**1975**	*40*	3427
4	Rubottom, G.M.	*J. Org. Chem.*	**1979**	*44*	1731
5	Ching-Kang, Sho	*J. Org. Chem.*	**1987**	*52*	3919

3-Phenyl-2-acetoxypropanol (3).[3] To 3-phenyl-1-trimethylsilyloxypropene **1** (1.03 g, 5 mmol) in CH_2Cl_2 was added slowly m-chloroperbenzoic acid (mCPBA) (1.1 g, 5.5 mmol). After 1 h, aqueous Na_2SO_3 was added. Work up gave 1.6 g of **2** (85%) erythro:threo (7:3).

2 (2.12 g, 5.6 mmol) in Et_2O (10 mL) was treated with Ac_2O (1 mL), triethylamine (2 mL) and 4-pyrrolidinopyridine (0.02 g) and the mixture stirred for 15 min. MeOH was added to destroy Ac_2O, the mixture was washed ($NaHCO_3$, 1.5 M HCl, water). Bulb-to-bulb distillation afforded 0.45 g of **3** (42%), bp 110°C (0.15 torr).

HAUSER - BEAK Ortho Lithiation

Ortho-alkylation of benzamides.

1	Hauser, C.R.	*J. Heterocycl. Chem.*	**1969**	*6*	475
2	Beak, P.	*J. Org. Chem.*	**1977**	*42*	1823
3	Hauser, C.R.	*J. Chem. Eng. Data*	**1978**	*23*	183
4	Beak, P.	*Acc. Chem. Res.*	**1982**	*15*	306
5	Katritzky, A.R.	*Org. Prep. & Proced. Int.*	**1987**	*19*	263

2-n-Butylbenzanilide (3).[5] To benzanilide **1** (1.97 g, 10 mmol) in THF (28.5 mL) and HMPA (1.5 mL) was added 2.5 M n-butyllithium (4 mL), dropwise at -70°C. The mixture was warmed to 20° and CO_2 was passed through for 5 min. After removal of the solvent under vacuum, THF (30 mL) was added under Ar and at -70°C 1.7 M tert butyllithium (6.5 mL), was added slowly. The mixture was maintained for 20 min at -20°C and recooled to -70°C. n-Butyl bromide **2** (1.37 g, 10 mmol) was added. After warming to 20° the mixture was stirred for a few hours. The solvent was removed and 2N HCl was added to the residue at 0°C. The precipitate was collected and recrystallized to give 1.85 g of **3** (73%), mp 72-73°C.

HAYASY Rearrangement

Rearrangement of o-benzoylbenzoic acids.

$$\mathbf{1} \xrightarrow[\text{1h/65°C}]{H_2SO_4} \mathbf{2}$$

1	Hayasy, M.	*J. Chem. Soc.*	**1927**		2516
2	Sandin, R.B.	*J. Am. Chem. Soc.*	**1955**	*78*	3817
3	Caspar, M. L.	*J. Org. Chem.*	**1968**	*33*	2020

2-Methoxy-6(-2-chlorobenzoyl)-benzoic acid (2).[2] 3-Methoxy-2(3-chlorobenzoyl)-benzoic acid **1** (1.0 g, 3.4 mmol) was heated in conc. H_2SO_4 (6 mL) for 1 h at 65°C. The cooled reaction mixture was poured into ice and neutralized. The cyclized product was filtered off (0.2 g) and the filtrate was acidified to give **2**, mp 205°C.

HELL - VOLHARDT - ZELINSKI Bromination

α–Bromination of carboxylic acids.

COOH $\xrightarrow{PBr_3,\ Br_2}$ $\xrightarrow{MeOH}$ Br COOCH$_3$

1 **2** (88%)

1	Hell, C.	*Chem. Ber.*	**1881**	*14*	891
2	Volhardt, J.	*Liebigs Ann.*	**1887**	*242*	141
3	Zelinski, Y.	*Chem. Ber.*	**1887**	*20*	2026
4	Gibson, Th.	*J. Org. Chem.*	**1981**	*46*	1003
5	Haworth, C.	*Chem. Rev.*	**1962**	*62*	99

Methyl 2-(1,5-Dimethylbicyclo [2.1.1] hexanyl-2-bromoacetate) (2).[4] To a mixture of 1,5-dimethylbicyclo [2.1.1] hexane-2-acetic acid **1** (2.92 g, 12.4 mmol) in PBr_3 (7.94 g, 29.3 mmol) maintained for 1 h at room temperature, was added Br_2 (7.94 g, 57 mmol) in two batches under Ar. The reaction mixture was heated on a steam bath for 3 h, cooled, quenched with anh. MeOH, diluted with Et_2O and the organic layer was washed with 5% $NaHCO_3$ solution. Evaporation of the solvent and distillation of the residue gave 4.0 g of **2** (88%), bp 58-59°C (0.33 mm).

HENBEST Iridium Reagent

Reagent for selective reduction of ketones by means of an iridium hydride.

H_2IrCl_6

94 h, 20°

1

2 (94%)

1	Henbest, H.B.	*J. Chem. Soc.*	**1962**		954
2	Blicke, T.A.	*Proc. Chem. Soc.*	**1964**		361
3	Hirschmann, H.	*J. Org. Chem.*	**1966**	*31*	375
4	Hill, J.	*J. Chem. Soc. (C)*	**1967**		783
5	Kirk, D.M.	*J. Chem. Soc. (C)*	**1969**		1653

5-α-Androstan-3α-hydroxy-17-one (2).[5] A stock solution of reagent was prepared from chloroiridic acid (0.1 g), trimethyl phosphite (5 mL) and 2-propanol containing 10% water (250 mL).

5α-Androstane-3,17-dione **1** (20 mg, 0.07 mmol) was heated with 5 mL of stock reagent solution for 94 h. The cooled mixture was poured into water and extracted with ether:benzene. The organic layer was washed with water and $NaHCO_3$ solution and the solvent was removed in vacuum to leave a crude product which contained 94% axial, 2% equatorial alcohol and 3% unchanged **1**. Pure **2** (axial OH) was obtained by chromatography.

HENRY Nitro Condensation

Aldol condensation of nitroalkanes with aldehydes.

CH_3 - CH_2 - NO_2 + (cyclohexyl-CHO) → (Al_2O_3) → (product with OH, NO_2)

1 **2** **3** (75%)

1	Henry, L.	*C.R.*	**1895**	*120*	1265
2	Barker, R.	*J. Org. Chem.*	**1964**	*29*	869
3	Rosini, G.	*Synthesis*	**1983**		1014
4	Hass, H.B.	*Chem. Rev.*	**1943**	*32*	406
5	Lichtentaler, F.W.	*Angew. Chem. Int. Ed.*	**1964**	*3*	211

1-Cyclohexyl-2-nitro-1-propanol (3).[3] To ice cooled nitroethane **1** (3.75 g, 50 mmol) was added cyclohexylcarboxaldehyde **2** and the mixture was stirred for 2-3 min. Chromatographic alumina (Carlo Erba RS, activity I according to Brockmann, 10 g) was added and stirring was continued for 1 h at room temperature. After standing for 23 h, the alumina was washed with CH_2Cl_2 (3x40 mL). The filtered extract was evaporated at reduced pressure to give the crude **3**. At a temperature below 80°C, unreacted **1** and **2** are distilled off. Then the temperature is carefully raised to afford 5.96 g of **3** (75%), bp 90-92°C (0.6 mm).

HERBST - ENGEL - KNOOP - OESTERLING

Aminoacid Synthesis

Alpha amino acids (and aldehydes) synthesis by reaction of an alpha keto acid with another amino acid (Herbst-Engel) or by reaction of a keto acid with ammonia under reducing conditions (Knoop-Oesterling).

$$\underset{\mathbf{1}}{C_6H_5\text{-}CH(NH_2)\text{-}COOH} + \underset{\mathbf{2}}{CH_3\text{-}C(=O)\text{-}COOH} \longrightarrow \underset{\mathbf{3}}{CH_3\text{-}CH(NH_2)\text{-}COOH} + \underset{\mathbf{4}}{C_6H_5\text{-}CHO}$$

$$\underset{\mathbf{5}}{HOOC\text{-}CH_2\text{-}CH_2\text{-}C(=O)\text{-}COOH} \xrightarrow[PtO_2 / H_2]{NH_4Cl} \underset{\mathbf{6}}{HOOC\text{-}CH_2\text{-}CH_2\text{-}CH(NH_2)\text{-}COOH}$$

1	Herbst, R.M., Engel, W.	*J. Biol. Chem.*	**1934**	*107*	505
2	Herbst, R.M.	*J. Am. Chem. Soc.*	**1936**	*58*	2239
3	Mix, H.	*Z. physiol. Chem.*	**1961**	*325*	106
4	Wieland	*Angew. Chem.*	**1942**	*55*	147
5	Knoop, F., Oesterling, H.	*Z. Chem.*	**1925**	*148*	294
6	Wieland	*Chem. Ber.*	**1944**	*77*	34

Glutamic acid (6).[6] 2-Ketoglutaric acid **5** (5.0 g, 36 mmol), NH_4Cl (1.0 g, 20 mmol) and NaOH (0.5 g, 12 mmol) in water (50 mL) was hydrogenated in the presence of PtO_2. After 5 h the catalyst was filtered and the product purified by chromatography on alumina to give **6**, mp 194°C.

H E R Z Benzothiazole Synthesis

Reaction of aromatic amines with sulfur monochloride and an acyl chloride in the presence of Zn salts to give 1,3-benzothiazoles.

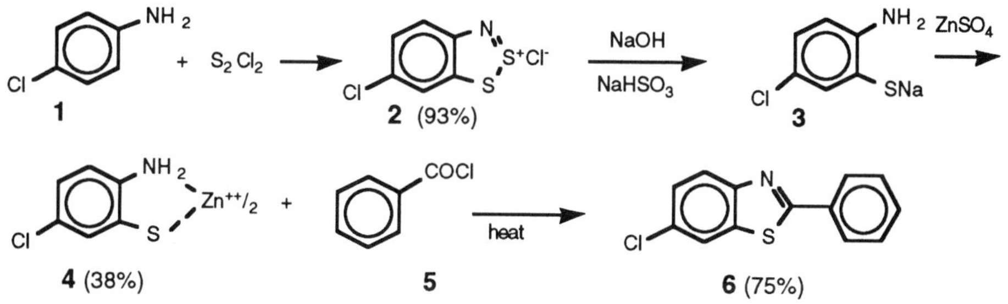

1 Herz, R. *Chem. Zent. Bl.* **1922** *4* 948

2 Herz, R. *U.S. Patent* 1.637.023

3 Huestins, L.D. *J. Org. Chem.* **1965** *30* 2763

4 McChenard *J. Org. Chem.* **1984** *49* 1224

5 Warburton, W.K. *Chem. Rev.* **1957** *57* 1011

6-Chloro-2-phenylbenzothiazole (6).[3] **1** (5.7 g, 0.045 mol) in AcOH (7 mL) was added to S_2Cl_2 (42. g, 0.31 mol; 25 mL) stirred for 3 h at 25°C then for 3 h at 70-80°C. The cooled mixture was stirred with PhH (50 mL) and filtered to give 9.3 g (93%) of **2**, mp 210-225°C (dec.). A vigorously stirred suspension of **2** (8.3 g, 37 mmol) in ice-water (500 mL) was made alkaline with 6N NaOH. Then $NaHSO_3$ (5.0 g, 40 mmol) was added and after 1 h heating, the mixture was treated with Norite and filtered. To the solution of **3** was added excess $ZnSO_4$ to precipitate the zinc mercaptide **4** (2.65 g, 38%). To a suspension of **4** (1.3 g, 3.4 mmol) in AcOH (40 mL) was added benzoyl chloride **5** (2.0 g, 14 mmol). After 30 min reflux, decomposition with water and crystallization from MeOH, gave 1.25 g of **6** (75%), mp 156-157°C.

HILBERT - JOHNSON Nucleoside Synthesis

Nucleoside synthesis from bromosugars and methoxypyrimidines (see Vorbrueggen).

1 + 2 ⟶ 3 (25%)

1	Johnson, T.B.	*Science*	**1929**	*69*	579
2	Hilbert, G.E.	*J. Am. Chem. Soc.*	**1930**	*52*	2001
3	Wolfrom, P.H.	*J. Org. Chem.*	**1965**	*30*	3058
4	Ulbricht, P.L.	*Angew. Chem.*	**1962**	*74*	767

2-Oxo-6-methoxy-3-(tetraacetylglucosido) pyrimidine (3).[2] Acetobromoglucose **1** (2.0 g, 5 mmol) and **2** (1.96 g, 14 mmol) was heated to 50°C for 48 h. The mixture was triturated with water and 3-methyluracyl (1 g) separated. The residue solidified upon addition of Et_2O (5 mL), the solid was filtered and washed with Et_2O to remove unreacted **1**. Two recrystallizations from 50% EtOH gave 0.5-0.7 g of **3** (23-32%), mp 220-221°C.

HINSBERG Thiophene synthesis

Synthesis of thiophenes from α–diketones.

1 + 2 —t BuOK⟶ 3 (93%)

1	Hinsberg, O.	*Chem. Ber.*	**1910**	*42*	901
2	Wynberg, N.	*J. Org. Chem.*	**1964**	*29*	1919
3	Wynberg, N.	*J. Am. Chem. Soc.*	**1965**	*87*	1739

HINSBERG - STOLLÉ Indole-Oxindole Synthesis

Indole synthesis from anilines and glyoxal (Hinsberg), oxindole synthesis from anilines and α-haloacyl halides (Stolle).

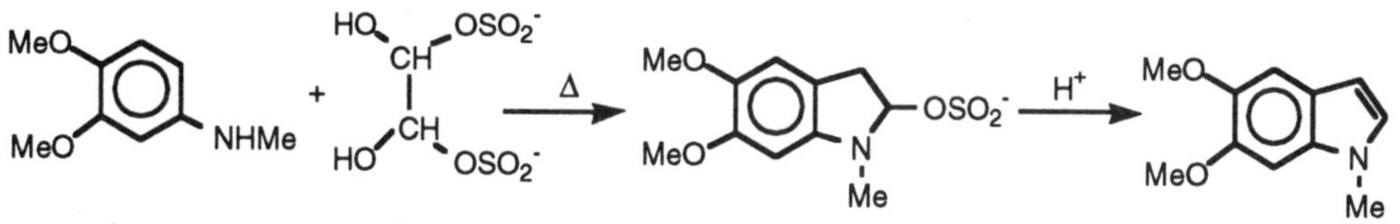

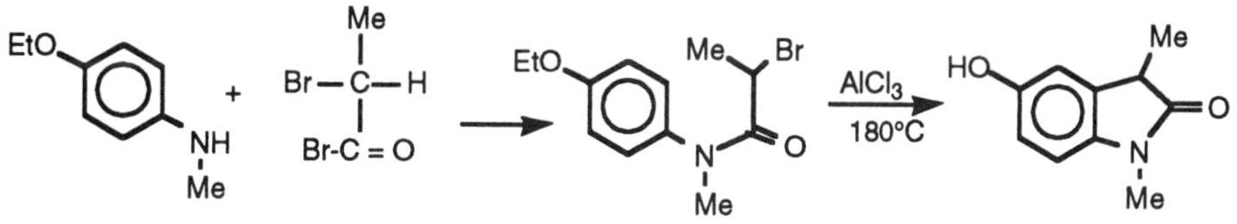

1	Hinsberg, O.	*Chem. Ber.*	**1888**	*21*	110
2	Burton, H.	*J. Chem. Soc.*	**1932**		546
3	Stollé, R.	*Chem. Ber.*	**1913**	*46*	3915
4	Stollé, R.	*J. Prakt. Chem.*	**1930**	*128*	1
5	Julian, P.L.	*J. Am. Chem. Soc.*	**1935**	*57*	563,2026
6	Sumter, W.	*Chem. Rev.*	**1944**	*34*	396

5,6-Dimethoxy-1-methylindole (4).[2] Glyoxal sodium bisulfite adduct **2** (32 g, 0.158 mol) in water (250 mL) and 4-methylaminoveratrole **1** (20 g, 0.12 mol) in EtOH (150 mL) were refluxed for 48 h. The hot solution was filtered, cooled (0°C) and the sodium salt filtered to give 19 g of **3** (83%), mp 187°C. **3** (5 g, 25 mmol) was refluxed in water:HCl (1:2) for 30 min, filtered and extracted with Et_2O; the solution was basified to give **4**, mp 120-121°C.

HIYAMA Aminoacrylate synthesis

Synthesis of 3-aminoacrylic acids or derivatives from nitriles by aldol type condensation.

$$CH_3\text{-}CN + CH_3\text{-}COOBu(t) \xrightarrow[\text{EtMgBr}]{(iPr)_2NH} (CH_3)(H_2N)C{=}CH(COOBu(t))$$

2 **1** **3** (66%)

$$CH_3\text{-}C(OCH_3)_2\text{-}CN + \mathbf{2} \xrightarrow{\text{BuLi}} [CH_3O\text{-}C(CH_3)(OCH_3)](H_2N)C{=}CH(C{\equiv}N)$$

4 **5** (81%)

1	Hiyama, T.	*Tetrahedron Lett.*	**1982**	*23*	1597
2	Hiyama, T.	*Tetrahedron Lett.*	**1983**	*24*	3509
3	Hiyama, T.	*Bull. Chem. Soc. Jpn.*	**1987**	*60*	2127, 2131, 2139

t-Butyl 3-Amino-2-butenoate (2).[1,3] Diisopropylamine (0.61 g, 6 mmol) was added to a solution of EtMgBr (4.3 mL, 3 mmol) in Et_2O at 0°C, and the mixture was stirred for 1 h. t-Butyl acetate **1** (0.35 g, 3 mmol) was added at 0°C and after 30 min MeCN **2** (62 g, 1.5 mmol), and the mixture allowed to react for 3 h at 0°C. Quenching with NH_4Cl, extraction with Et_2O and distillation gave 0.156 g of **3** (66%).

4,4-Dimethoxy-3-amino-2-pentenonitrile (5).[2] To a well stirred solution of **2** (0.14 g, 10 mmol), in THF (20 mL) was added dropwise 1.6 mol/L of BuLi (6.3 mL, 10 mmol) at -78°C and stirring continued for 15 min. 2,2-Dimethoxy propannitrile **4** (1.15 g, 10 mmol) was added. Work up as above and recrystallization gave 1.26 g of **5** (81%).

HIYAMA - HEATHCOCK Stereoselective allylation

Stereoselective synthesis of anti homoallylic alcohols by Cr promoted allylation of aldehydes.

1 + 2 —$CrCl_2$→ 3(75%) —O_3 / H_2O_2→

+ Ph–CHO —$CrCl_3$ - LAH (2:1)→

1	Hiyama, T.	*J. Am. Chem. Soc.*	**1977**	*99*	3179
2	Heathcock, C.H.	*J. Am. Chem. Soc.*	**1977**	*99*	247, 8109
3	Heathcock, C.H.	*Tetrahedron Lett.*	**1978**		1185
4	Hiyama, T.	*Bull. Chem. Soc. Jpn.*	**1982**	*55*	562
5	Mulzer, J.	*Angew. Chem. Int. Ed.*	**1990**	*29*	679

Anti - 4-Hydroxy-3-methyl-1-octene (3).[3] To a suspension of 3.7 g (23 mmol) of anhyd. $CrCl_3$ in 50 mL of THF at 0°C, 440 mg (12 mmol) of $LiAlH_4$ is added in portions followed by 8.6 mmol of aldehyde **2.** Crotyl bromide **1** (2.32 g, 17.2 mmol) in 50 mL of THF is added over 30 min. The mixture is stirred at 20°C for 2 h, poured into water and extracted with ether. Removal of solvent gives **3** in 75% yield.

HOCH - CAMPBELL Aziridine Synthesis

Aziridines from oximes or from α-haloimines (via azirines).

Me Mg Br (70%)

LAH heat

1 **2** (84%)

1	Hoch, J.	*C.R.*	**1934**	*198*	1865
2	Campbell, K.N.	*J. Org. Chem.*	**1943**	*8*	103
3	De Kimpe, N.	*J. Org. Chem.*	**1980**	*45*	5319
4	Laurent, A.	*Bull. Soc. Chim. Fr.*	**1973**		2680
5	Kotera, K.	*Tetrahedron*	**1968**	*24*	3681, 5677

Cis-1,2-dimethyl-3-phenylaziridine (2).[3] To a cooled and well stirred suspension of LAH (3.8 g, 0.1 mol) in Et_2O (100 mL) was added dropwise a solution of 2,2-dichloro-N-methyl propiophenone imine **1** (10.8 g, 0.05 mol) in Et_2O (100 mL). After overnight reflux, the mixture was poured with caution onto ice water and the product extracted with Et_2O. Evaporation of the solvent gave 6.21 g of **2** (84%), bp 79-82°C.

HOFMANN Amide Degradation

Degradation of amides to amines by means of hypohalides.

$$CH_3\text{-}(CH_2)_4\text{-}CONH_2 \;(\mathbf{1}) + NaOCl \xrightarrow{HCl} CH_3\text{-}(CH_2)_3\text{-}CH_2\text{-}NH_2.HCl \;\; \mathbf{2}\ (95\%)$$

1	Hofmann, A.W.	*Chem. Ber.*	**1881**	*14*	2725
2	Magnieri, E.	*J. Org. Chem.*	**1958**	*23*	2029
3	Applequist	*Chem. Rev.*	**1954**	*54*	1083
4	Cohen, L.A.	*Angew. Chem.*	**1961**	*73*	260
5	Wallis, E.S.	*Org. React.*	**1946**	*3*	268

1-Pentylamine hydrochloride (2).[2] A suspension of caproamide **1** (11.5 g, 0.1 mol) in NaOCl solution was heated to 45°C under stirring then at 75°C for 1 h and the product was steam distilled into excess of HCl. Evaporation of the distillate left 11.7 g of **2** (hydrochloride) (95%), mp 137-139°C.

HOFMANN Isonitrile Synthesis

Isonitrile synthesis from primary amines and dichlorocarbene or dibromocarbene.

$$C_2H_5\text{—}NH_2 \;(\mathbf{1}) + CHBr_3 \;(\mathbf{2}) \xrightarrow{R_4N^+OH^-} C_2H_5\text{-}N\equiv C \;\; \mathbf{3}\ (47\%)$$

1	Hofmann, A.W.	*Liebigs Ann.*	**1868**	*146*	107
2	Smith, P.A.S.	*J. Org. Chem.*	**1958**	*23*	1599
3	Ugi, J.K.	*Angew. Chem. Int. Ed.*	**1972**	*11*	530

Ethylisocyanide (3).[2] To a solution of 33% **1** (40 mL, 0.3 mol), benzyltriethylammonium chloride (0.6 g) and NaOH (35 g, 0.875 mol) was added dropwise bromoform **2** (50.3 g, 0.2 mol). The mixture was stirred for 24 h at 20°C and distilled (**1** and **3**). Redistillation from KOH gave 5.2 g of **3** (47%), bp 65°C.

HOFMANN Elimination

Olefins by elimination from quaternary ammonium salts (less substituted olefin preferred).

1 H_3C N+ I⁻ H_3C CH_3 OH —Ag_2O→ —150°→ OH N H_3C CH_3 **2** (75%)

1	Hofmann, A.W.	*Chem. Ber.*	**1881**	*14*	659
2	Hinskey, R.G.	*J. Org. Chem.*	**1964**	*29*	3678
3	Cope, A.C.	*J. Org. Chem.*	**1965**	*30*	2163
4	Francke,	*Angew. Chem.*	**1960**	*72*	397
5	Brewster, J.H.	*Org. React.*	**1953**	*7*	137
6	Cope, A.C.	*Org. React.*	**1960**	*11*	317

3-(Dimethylaminomethyl)-2-methyl-6-heptene-2-ol (2).[3] 2-Methyl-5-(dimethylcarbinol)-N,N-dimethylpiperidinium iodide **1** (1.90 g, 10 mmol) was dissolved in distilled water and treated with freshly prepared Ag_2O. The mixture was filtered and concentrated under vacuum to a syrup. The residue was heated at about 150°C and 0.3 mm and distilled to afford 0.85 g (75%) of **2**, bp 50°C (0.5 mm).

HOFMANN - LOEFFLER - FREYTAG Pyrrolidine Synthesis

Synthesis of pyrrolidines or piperidines from N-haloamines (free radical reaction).

NCS, hν; base

1 → 2 (100%)

1	Hofmann, A.W.	*Chem. Ber.*	**1883**	*16*	558
2	Loffler, K., Freytag, C.	*Chem. Ber.*	**1909**	*42*	3427
3	Kimura, M.	*Synthesis*	**1976**		201
4	Wolff, M.E.	*Chem. Rev.*	**1963**	*63*	55

Perhydrodipyridino[1,2-a][1',2'-c]-pyrimidine (2).[3] (a) To an ice-cooled solution of 2-[2-(piperidyl)ethyl] piperidine **1** (3.2 g, 10.2 mmol) in Et_2O (200 mL) was added N-chlorosuccinimide (NCS) (1.7 g, 12.7 mmol). Under stirring, the reaction mixture was irradiated with a 300 W high pressure mercury lamp under N_2 for 5 h. The precipitate was filtered, dryed and extracted with n-pentane. Evaporation of the solvent and distillation gave 1.0 g of **2** (50%), bp 140°C (20 torr).

(b) To **1** (1.0 g, 5.1 mmol) and NCS (1.34 g, 10 mmol) in cooled Et_2O (200 mL) TEA (2.0 g, 20 mmol) was added with stirring and irradiation as above. Workup gave 980 mg of **2** (100%).

HOFMANN-MARTIUS-REILLY-HICKINBOTTOM

Aniline Rearrangement

Thermal or Lewis acid catalyzed rearrangement of N-alkylanilines to o-(p-) alkylated anilines.

H CH3 N Ph 1 → 200° ZnCl2 → NH2 CH3 Ph 2 (42%) + CH3 Ph H2N 3

1	Hofmann, A.W., Martius, C.A.	*Chem. Ber.*	**1871**	*4*	742
2	Hart, H.	*J. Org. Chem.*	**1962**	*27*	116
3	Ogatta, Y.	*Tetrahedron*	**1964**	*20*	2717
4	Reilly, J., Hickinbottom, W.	*J. Chem. Soc.*	**1920**	*117*	103
5	Cripps, R.W.	*J. Chem. Soc.*	**1943**		14
6	Fischer, A.	*J. Org. Chem.*	**1960**	*25*	463

o- and **p-α-Phenetylaniline (2)** and **(3).** (a) By thermal rearrangement. N-α-Phenethylaniline **1** (hydrochloride) (23.3 g, 0.1 mol) was heated in a sealed tube at 200-230°C for 6 h. The cooled mixture was extracted with 15% HCl and Et_2O. The Et_2O layer was extracted with HCl. The acid extract after neutralization with 5% NaOH, extraction with PhH and distillation gave 2.3 g of aniline and 4.2 g of **2** and **3** (24%), bp 165-170°C in a ratio of 3:7.

(b) By catalytic rearrangement. A mixture of **1** (19.7 g, 0.1 mol) and $ZnCl_2$ (13.6 g, 0.1 mol) was heated in a sealed tube as above. The same work gave up 3.6 g of aniline and 6.8 g of **2** and **3** (41%), o:p ratio 10:90.

HOFFMAN - YAMAMOTO Stereoselective allylations

Synthesis of syn or anti homoallylic alcohols from Z or E crotylboronate and aldehydes (Hoffman) or of syn homoallylic alcohols from crotylstannanes, BF_3 and aldehydes - (Yamamoto).

C_9H_{19} CHO + Me...B(O)(O)...CO_2i Pr, CO_2i Pr → (-78°) C_9H_{19}...OH...Me
1 **2** **3** (90%)

MeCHO + ...B(O)(O)...COO iPr, COO iPr → Me...OH

...SnBu₃ + Ph CHO →(BF_3) Ph...OH

1	Hoffman, R.W.	*Angw. Chem. Int. Ed.*	**1979**	*18*	326
2	Hoffman, R.W.	*J. Org. Chem.*	**1981**	*46*	1309
3	Yamamoto, Y.	*J. Am. Chem. Soc.*	**1980**	*102*	7107
4	Yamamoto, Y.	*Aldrichimica Acta.*	**1987**	*20*	45
5	Brown, H.C.	*J. Am.Chem.Soc.*	**1985**	*107*	2564
6	Roush, W.R.	*Tetrahedron Lett.*	**1988**	*29*	5579

Homoallyl alcohol (3).[6] Metalation of (E)-butene (1.05 equiv) with n BuLi (1 equiv) and KOtBu (1 equiv) in THF at -50°C for 15 min followed by treatment of (E)-crotyl potassium salt with $B(OiPr)_3$ at -78°C gave, after quenching with 1 N HCl and extraction with Et_2O containing 1 equiv of diisopropyl tartarate, the crotyl boronate **2**. A solution of decanal **1** (156 mg, 1 mmol) was added to a toluene solution of **2** (1.1-1.5 equiv) (0.2 M) at -78°C containing 4Å molecular sieves (15-20 mg/L). After 3 h at -78°1 N NaOH was added, followed by extraction and chromatography to afford 208 mg of **3** (90%), anti:syn 99:1.

H O L L E M A N N Pinacol Synthesis

Dimerization of ketones to 1,2-diols by means of Mg and other metals.

2 p-C_6H_4-CO-C_6H_4-Cl p —Mg-Hg→ (p Cl-C_6H_4)$_2$-C-OH / (p Cl-C_6H_4)$_2$-C-OH **2** (94%)

1

3 —Mg-Hg, $TiCl_4$→ OH HO **4** (93%)

1	Holleman, M. A. F.	*Rec. Trav. Chim.*	**1906**	*25*	206
2	Goth, H.	*Helv. Chim. Acta.*	**1965**	*48*	1395
3	Corey, E.J.	*J. Org. Chem.*	**1976**	*41*	260
4	Zimmermann, H.E.	*J. Org. Chem.*	**1986**	*51*	46044

4-Chlorobenzopinacol (2).[2] To magnesium turnings (2.5 g, 100 mat) and iodine (9.0 g, 35 mmol) in PhH (100 mL) and Et_2O (30 mL) were added p-chlorobenzophenone **1** (10.82 g, 50 mmol) with efficient stirring (exothermic). After 1 h, hydrolysis gave crude **2**, which was digested with hot petroleum ether (30 mL) to afford 10.32 g of **2** (94%), mp 172-173°C (EtOH).

Cyclohexylpinacol (4).[3] To $HgCl_2$ (0.044 g, 0.16 mmol) in THF (3 mL) was added 70-80 mesh Mg (0.144 g, 6 at) and the mixture stirred under Ar for 15 min. The supernatant was removed, the amalgam washed (THF) and THF (5 mL) was added. At -10°C, $TiCl_4$ (330 L, 0.570 g, 3 mmol) was added, followed by cyclohexanone **3** (0.196 g, 2 mmol) in THF and the purple solution was stirred for 30 min at 0°C. Quenching with 0.5M Na_2CO_3, extraction with Et_2O and evaporation of the solvent afforded 0.186 g of **4** (93%), mp 124-125°C.

HONZL - RUDINGER Peptide Synthesis

Peptide synthesis by coupling of acyl azides with amino esters.

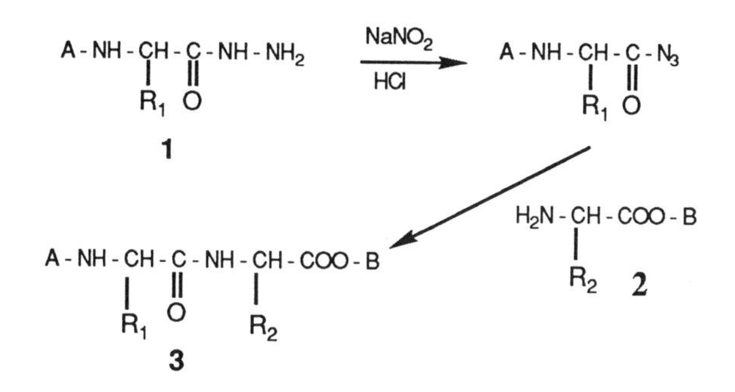

1	Honzl, I., Rudinger, I.	*Coll. Czech. Chem. Comm.*	**1961**	*26*	2333
2	Siebel, F.	*Helv. Chim. Acta.*	**1970**	*53*	2134
3	Medzihradsky, K.	*Acta Chim. Acad. Sci. Hung.*	**1962**	*30*	105
4	Ondetti, M.A.	*J. Am. Chem. Soc.*	**1968**	*90*	4711
5	Klausner, Y.S.	*Synthesis*	**1974**		554

General procedure for azide coupling. [5] Protected peptide hydrazide **1** (26 mmol) in DMF (84 mL) at -20°C was treated with 3.35N NCl in dioxan (70 mmol). t-Butyl nitrite (3.85 mL) (or $NaNO_2$ in water) was added at -15°C and the mixture was maintained at -10°C for 10 min. The amino component **2** (18 mmol) in DMF (320 mL) was added dropwise at -15°C, then ethyldiisopropylamine (70 mmol) and the mixture kept at 0°C for 24 h. During the first 6 h, 0.5 mL of the base was added ever hour. Finally the product **3** was precipitated in ice-cold 1% AcOH (1800 mL).

HORNER - KNOWLES - KAGAN Asymmetric Hydrogenation

Enantioselective hydrogenation of prochiral olefins with chiral Rh catalysts.

Ph(Et)C=CH₂ **1** + RhCl (hexadiene)$_2$ + Ph - P* - Pr (Me) **2** $\xrightarrow{H_2}$ (S) **3**

4 $\xrightarrow{\text{RhCl (cyclooctene)}_2\ H_2,\ \mathbf{5}}$ (R) **6** (90%) 81% ee

1	Horner, L.	*Angew. Chem. Int. Ed.*	**1968**	*7*	942
2	Knowles, W.S.	*J. Chem. Soc. Chem. Commun.*	**1968**		1445
3	Kagan, H.B.	*J. Am. Chem. Soc.*	**1972**	*94*	6429
4	Kagan, H.B.	*J. Organomet. Chem.*	**1975**	*91*	105

(S)-(+)-2-Phenylbutane (3).[1] The catalyst **2** was obtained by adding [Rh-(1,5-hexadiene)Cl]$_2$ (0.5x10^{-4} moles) in PhH (20 mL) to (S)-(+)-methylphenyl-n-propylphosphine (2.2x10^{-4} moles) under H_2. α-Ethylstyrene **1** (10^{-2} moles) was injected. The reaction took 6-8 h (by GC). Chromatography on Al_2O_3 gave **3** (7-8% ee).

N-Acetyl-(R)-phenylalanine (6).[3] The rhodium catalyst was obtained by adding (-) diop **5** (from diethyl tartrate) to a benzene solution of [RhCl(cyclooctene)$_2$]$_2$ under Ar, and stirring for 15 min. A solution of the Rh catalyst (1 mM in EtOH:PhH 4:1) was introduced under H_2 to a solution of α-N-acetylamino-β-phenylacrylic acid **4** (molar ratio Rh:**4** = 1:540). The solvent was evaporated, the residue dissolved in 0.5 N NaOH, the catalyst was filtered and the solution acidified and concentrated to dryness to give **6** (81% ee) in 90-95% yield.

HORNER - WADSWORTH - EMMONS Olefination

Wittig type reaction of phosphonate stabilized carbanions with aldehydes or ketones to form olefins.

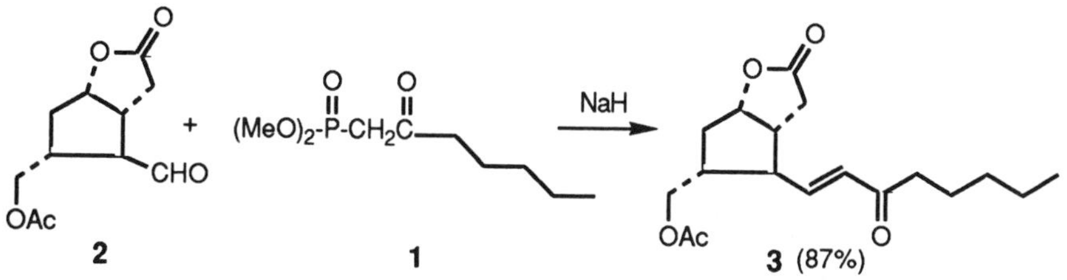

1	Horner, L.	*Chem. Ber.*	**1958**	*83*	733
2	Wadsworth, W.S., Emmons, W.,D.	*J. Am. Chem. Soc.*	**1961**	*83*	1733
3	Berkowitz, W.F.	*J. Org. Chem.*	**1982**	*47*	824
4	Tamizawa, T.K.	*Synthesis*	**1985**		887
5	Sampson, C.R.	*J. Org. Chem.*	**1986**	*52*	2525
6	Tsuge, O.	*Bull. Chem. Soc. Jpn*	**1987**	*60*	4091
7	Boutagy, J.	*Chem. Revs.*	**1974**	*79*	87

Unsaturated ketone (3).[2] To a suspension of NaH (21.4 mg, 0.883 mmol) in DME (4 mL), under N_2 was injected a solution of dimethyl 2-oxoheptylphosphonate **1** (210 mg, 0.95 mmol) in DME (1 mL). After stirring for 1 h (voluminous precipitate) and ice cooling, aldehyde **2** (100 mg, 0.442 mmol) in DME (1 mL) was injected. Stirring was continued for 30 min under ice cooling followed by 2.5 h at 20°C. The mixture was neutralized with AcOH (0.12 mL) and concentrated. Chromatography on silica gel (45 g) and elution with EtOAc:hexane 1:1, gave 125 mg of **3** (87%).

HOUBEN - HOESCH Phenol Acylation

Synthesis of ketones (or aldehydes) by acylation of phenols with nitriles (or ortho formates).

1 + 2 —HCl / $ZnCl_2$→ 3 (28%)

Me-phenol-OH + HC (OEt)$_3$ —$AlCl_3$→ Me, OH, CHO

1	Houben, J.	*Chem. Ber.*	**1913**	*46*	2447
2	Hoesch, K.	*Chem. Ber.*	**1915**	*48*	1122
3	Trucare, J.	*J. Org. Chem.*	**1963**	*28*	3206
4	Roger, R.	*Chem. Rev.*	**1961**	*61*	184
5	Spoerri, P.E.	*Org. React.*	**1949**	*5*	387
6	Gross, H.	*Chem. Ber.*	**1963**	*96*	308

6-(Phenylacetyl)dihydro-β-tubanil (3).[3] A solution of dihydro-β-tubanil **1** (0.8 g, 4.5 mmol), phenylacetonitrile **2** (0.8 g, 7 mmol) and $ZnCl_2$ (2.5 g) in Et_2O (28 mL) and $CHCl_3$ (13 mL) was treated with dry HCl for 10 min under cooling and stirring. The mixture was allowed to stand at 25°C overnight, the Et_2O layer was decanted and the oily residue was treated with alkali and heated on a water bath for 2 h. Extraction with Et_2O, washing the extract with 5% NaOH and evaporation of the solvent gave a residue which was recrystallized from Et_2O:pentane to give 0.38 g of **3** (28%), mp 81-83°C.

HUNSDIECKER - BORODIN - CRISTOL - FIRTH

Decarboxylation - Bromination

Substitution of carboxylic groups by halogen via Ag salts (Hunsdiecker, Borodin) or Hg salts (Cristol-Firth).

COOH —(HgO, Br_2; 80°)→ Br

1 **2** (69%)[3]

$Cl\text{-}C_6H_4\text{-}COOH + Br_2$ —(HgO hυ)→ $Cl\text{-}C_6H_4\text{-}Br$

3 **4** (80%)

1	Borodin, A.	*Liebigs Ann.*	**1861**	*119*	121
2	Hunsdiecker, H.C.	*Chem. Ber.*	**1942**	*75*	291
3	Luh, Tien-Yan Luh	*J. Org. Chem.*	**1981**	*49*	5328
4	Johnson, R.G.	*Chem. Rev.*	**1956**	*56*	219
5	Wilson, C.V.	*Org. React.*	**1957**	*9*	332
6	Cristol, S.J., Firth, W.C.	*J. Org. Chem.*	**1961**	*26*	280
7	Meyers, A.J.	*J. Org. Chem.*	**1979**	*44*	3405

4-Chlorobromobenzene (4).[7] A suspension of p-chlorobenzoic acid **3** (1.56 g, 10 mmol) and HgO (3.25 g, 15 mmol) in CCl_4 (50 mL) was heated to reflux and irradiated with an 100 W bulb. Br_2 (2.40 g, 15 mmol) was added via syringe and the mixture heated and irradiated for 3 h. 5% $NaHCO_3$ (30 mL) was added and the mixture stirred vigorously for 15 min. Work up and distillation gave 1.53 g of **4** (80%), mp 66-68,C.

IMAMOTO Cerium alkylation

Cerium mediated or samarium mediated alkylation of ketones and aldehydes in the presence of esters; see also Kagan-Molander.

$$CH_2{=}CH_2{-}CH_2I \; + \; CH_3{-}\underset{\underset{O}{\|}}{C}{-}C_6H_4{-}Br \xrightarrow[HgCl_2]{Ce} p\,Br{-}C_6H_4{-}\underset{OH}{\overset{CH_3}{C}}{-}CH_2{-}CH{=}CH_2$$

1 + **2** → **3** (95%)

Cyclopropyl methyl ketone **4** + CH_2I_2 **5** $\xrightarrow[0°]{Sm}$ **6** (77%) (cyclopropyl–C(OH)(CH₃)–CH_2I)

1	Imamoto, T.	*J. Chem. Soc. Chem. Commun.*	**1984**		163
2	Imamoto, T.	*J. Org. Chem.*	**1984**	*49*	3904
3	Imamoto, T.	*Tetrahedron Lett.*	**1985**	*26*	4763
4	Imamoto, T.	*Tetrahedron Lett.*	**1986**	*27*	3243

4-Hydroxy-4-(p-bromophenyl)-1-pentene (3)[2] Ce turnings (280 mg, 2 at) were treated with $HgCl_2$ (250 mg, 0.9 mmol) in EtOH (2 mL) for 2 min under Ar. The solvent was removed and the residue washed with EtOH and dried in vacuum. To the Ce amalgam was added allyl iodide **1** (336 mg, 2 mmol) in THF (3 mL) and **2** (700 mg, 2mmol) under Ar at 0°C. After reaction, the mixture was treated with aq. NH_4Cl and extracted with $CHCl_3$. Evaporation and preparative TLC (silica gel) gave 459 mg of **3** (95%).

2-Cyclopropyl-1-iodo-2-propanol (6). To freshly scraped samarium powder (300 mg, 2 at) at 0°C was added a few drops of a solution of **4** (84 mg, 1 mmol) and **5** (803 mg, 3 mmol), in THF (6 mL). After initiation, the rest was added with stirring during 20 min, and after 20 min at 0°C, the mixture was treated with 1 N HCl and extracted with ether. The combined extracts were washed with aq. $Na_2S_2O_3$ solution and brine, and dried over $MgSO_4$. Evaporation and preparative TLC gave 175 mg of **6** (77%).

ISAY Pteridine Synthesis

Synthesis of pteridines from diaminopyrimidines and α-diketones or α–ketoaldehydes.

1	Isay,	*Chem. Ber.*	**1906**	*39*	250
2	Shiman, R.x	*J. Org. Chem.*	**1971**	*36*	3925
3	Albert, A.	*Quart. Rev. Chem. Soc.*	**1949**		2077

2-Amino-4-hydroxy-6-phenylpteridine (3).[2] A suspension of 2,4,5-triamino-6-hydroxypyrimidine sulphate **2** (2.9 g, 8.4 mmol) in water (30 mL) was treated with $BaCl_2$, $2H_2O$ (2.0 g, 8.4 mmol), and after 10 min stirring was filtered. The pH of the filtrate was adjusted to 4.0 with NaOAc, phenyl glyoxal **1** (1.1 g, 8.2 mmol) in MeOH (10 mL) was added and the mixture was heated on a steam bath for 3 h. The precipitate was dissolved in 1M NaOH and reprecipitated by 10M NaOH. By neutralization with AcOH and drying in high vacuum one obtains 0.85 g of **3** (41%), λ_{max} (0.1M NaOH): 270 mμ.

JACOBSEN Rearrangement

Alkyl group migration in aromatic systems.

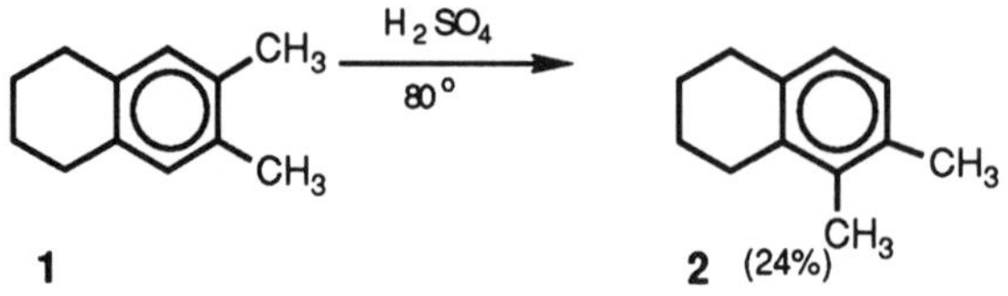

1	Jacobsen, C.	*Chem. Ber.*	**1886**	*19*	1209
2	Smithe, L.I.	*J. Am. Chem. Soc.*	**1948**	*70*	2209
3	Marwell, E.N.	*J. Org. Chem.*	**1965**	*30*	4014
4	Smith, L.I.	*Org. React.*	**1942**	*1*	371

5,6-Dimethyltetralin (2).[2] A mixture of 6,7-dimethyltetralin **1** (18.0 g, 0.11 mol) and H_2SO_4 was heated under stirring to 80°C and maintained at this temperature for 15 min, then at 25°C for 24 h. The mixture was diluted with water and distilled with superheated steam. From the distillate by extraction with Et_2O and distillation there was obtained 4.5 g of **2** (24%), bp 110-115°C, n_D = 1.5530.

J A P P Oxazole Synthesis

Oxazole synthesis from benzoin and nitriles or ammonium formate.

Ph–CH–OH / Ph–C=O **1** + K C≡N **2** —(H_2SO_4)→ **3** (30%)

n-Pr–CH–OH / n-Pr–C=O + $HCOO^- NH_4^+$ → (61%)

1	Japp, F.R.	*J. Chem. Soc.*	**1893**	*63*	469
2	Bredereck, H.I.	*Chem. Ber.*	**1954**	*87*	726
3	Willey, R.H.	*Chem.Rev.*	**1945**	*37*	420

4,5-Diphenyloxazole (3).[1] A mixture of benzoin **1** (30.0 g, 0.14 mol) and KCN **2** (60.0 g, 0.9 mol) was added to conc. H_2SO_4 (800 mL) under stirring and cooling (CARE! HCN is formed). After a few hours at 25°C the mixture was poured into water, the oily product was heated with Na_2CO_3 and extracted with PhH. Vacuum distillation afforded 9.0 g of **3** (30%), bp 192-195°C.

JAPP - KLINGEMANN Hydrazone Synthesis

Synthesis of hydrazones from diazonium salts and an activated methylene group (or enamine).

$+N_2-C_6H_4-NO_2$ -p

1 **2** **3** (82%)

$Ph-N_2^+$

1	Japp, F.R., Klingemann, F.	*Chem. Ber.*	**1887**	*20*	2492
2	Frank, R.L.	*J. Am. Chem. Soc.*	**1949**	*71*	2804
3	Jackman, A..	*Chem. Commun.*	**1967**		456
4	Phillips, R.R.	*Org. React.*	**1959**	*10*	143
5	Robinson, R.	*Chem. Rev.*	**1969**	*69*	233

2-Formylpyridine p-nitrophenylhydrazone (3).[2] Ethyl 2-pyridylacetate **1** (1.06 g, 6.3 mmol) was stirred for 3 h with KOH (0.4 g, 70 mmol) in water (15 mL), then diluted with water (10 mL) and acidified with HCl. Diazonium salt **2** (from p-nitroaniline 0.869 g 6.3 mmol and $NaNO_2$ (0.434 g, 6.3 mmol) was added followed by a cold solution of NaOAc (2.0 g, 24 mmol) in water (4 mL) and kept for 20 h at 0-5°C. Filtration gave 1.25 g of **3** (82%), mp 231-234°C.

JAROUSSE - MAKOSZA Phase transfer

Phase transfer (PT) catalysis by quaternary ammonium salts of substitution, addition, carbonyl formation, oxidation, reduction

$$CH_3\text{-}CO\text{-}CH_2COOCH_3 + CH_3I \xrightarrow[NaOH\text{ - }CHCl_3]{R_4N^+HSO_4^-} CH_3\text{-}CO\text{-}CH(CH_3)\text{-}COOCH_3$$

1 **2** **3** (80%)

$$C_6H_5OH + n\text{-}C_4H_9Br \xrightarrow[NaOH]{PT} C_6H_5\text{-}O\text{-}C_4H_9(n)$$

4 **5** **6** (90%)

1	Jarousse, M.I.	*C.R. ser. C.*	**1951**	*232*	1424
2	Makosza, M.	*Rocz. Chem.*	**1965**	*39*	1977
3	Dockx, J.	*Synthesis*	**1973**		411
4	Weber, W.P.	*Angew. Chem. Int. Ed.*	**1972**	*11*	530
5	Dehmlow, E.V.	*Angew. Chem. Int. Ed.*	**1974**	*13*	170
6	Harris, J.M.	*J. Org. Chem.*	**1985**	*50*	5230
7	Wang, J.X.	*J. Org. Chem.*	**1986**	*51*	275

Methyl acetopropanoate (3).[3] (C-alkylation). Tetrabutylammonium hydrogen sulfate (PT catalyst) (34.6 g, 0.1 mol) and NaOH (8.0 g, 0.2 mol) in water (75 mL) was added to well stirred **1** (11.6 g, 0.1 mol) and MeI **2** (28.4 g, 0.2 mol) in $CHCl_3$ (75 mL). The reaction is exothermic and becomes neutral after a few min. The $CHCl_3$ layer was evaporated. Et_2O was added to filter the PT catalyst. Evaporation of the Et_2O gave a mixture of **3** (80%) (monoalkylated) and 20% dialkylated product.

Phenyl butyl ether 6 (O alkylation). Phenol **4** (9.4 g, 0.1 mol) and butyl bromide **5** (27.4 g, 0.2 mol) were stirred with NaOH (6.0 g, 0.15 mol) in water (500 mL) in the presence of PT catalyst (1-10 mmol). After all phenol was consumed, usual work up gave 13.5 g of **6** (90%), bp 210-211°C, n_D^{20} = 1.4970.

JEGER Tetrahydrofuran synthesis

Free radical ring closure of alcohols with $Pb(AcO)_4$ to tetrahydrofurans.

Pb (AcO)4

Δ

1

cis **2** (62%)
trans **2** (7%)

4

Pb (AcO)4

1	Jeger, O.	*Helv. Chim. Acta.*	**1959**	*42*	1124
2	Micovic, V.N.	*Tetrahedron*	**1964**	*20*	2279
3	Moon Sung	*J. Org. Chem.*	**1969**	*34*	288
4	Michailovici, M.L.	*Synthesis*	**1970**		209

cis-7-Oxabicyclo[4,3,0]nonane (2).[3] A mixture of 2-cyclohexylethanol **1** (10 g, 78 mmol) and lead tetraacetate (60 g, 135 mmol) in PhH (100 mL) was refluxed for 19 h. The mixture was filtered, the filtrate washed (H_2O, $NaHCO_3$), the solvent evaporated and the residue distilled in vacuum to give 6.19 g of **2** (62%), bp 39-62°C/1.5 mm. Byproducts identified by GC: trans 7-oxabicyclo[4,3,0]octane (7%) and 2-cyclohexylethyl acetate (30%).

J O H N S O N Phenol alkynylation

A base induced rearrangement by conversion of aryl tri- or tetra-haloethyl ethers into o-hydroxyphenylalkynes and into heterocyclic systems.

1 Johnson, F. *J. Org. Chem.* **1985** *50* 5430

2 Johnson, F. *J. Org. Chem.* **1986** *51* 5040

1,1-Difluoro-2,2-dichloroethyl aryl ether (2) Phenol **1** (4.7 g, 50 mmol) was dissolved in 40% KOH (14 mL) and 40% $Bu_4N^+OH^-$ (2 mL). To this solution at 0°C was added 1,1-dichloro-2,2-difluoroethylene (9.98 g, 75 mmol) and CH_2Cl_2 (35 mL) and shaken vigorously at 25°C for 16 h. Isolation afforded 10.56 g (90%) of **2**.

2-Hydroxyphenyl trimethylsilyl acetylene (3).[4] The ether **2** (2.39 g, 10 mmol) in THF (30 mL) at -78°C under N_2 was treated with n-BuLi (41 mmol, 20 mL of 2 molar solution) in hexane over 20 min with stirring. After overnight heating to 20°C, quenching in Me_3SiCl (1 mL), treating with petroleum ether and filtration, the solvent was evaporated and the residue chromatographed (petroleum ether: EtOAc 9:1) to afford 1.52 g of **3** (80%).

J O N E S Oxidation Reagent

Oxidation of alcohols to aldehydes or ketones with CrO_3-H_2SO_4 in acetone.

Ph - CH(OH) - C≡CH **1** → (CrO_3-H_2SO_4, Me_2CO, 5°) → Ph - C(=O) - C≡CH **2** (68%)

CH_3O ... OH → (") → CH_3O ... O

1 Jones, E.R.H. *J. Chem. Soc.* **1946** 39

2 Dauben, W.C. *J. Org. Chem.* **1980** *45* 4413

Phenyl ethynyl ketone (2).[1] To a stirred solution of phenyl ethynyl carbinol **1** (342 g, 2.59 mol), in Me_2CO (500 mL) was added slowly a solution of CrO_3 (175 g, 1.75 mol), in water (500 mL) and 98% H_2SO_4 (158 mL) under N_2 at +5°C in 4-5 h. After stirring for a further 30 min, the mixture was diluted with water and the product extracted with Et_2O. Evaporation of the solvent and recrystallization of the residue from MeOH:H_2O afforded 258 g of **2** (68%), mp 50-51°C.

JULIA-BRUYLANTS Cyclopropyl Carbinol Rearrangement

Synthesis of allyl halides (usually E) by rearrangement of cyclopropyl carbinols (acid catalyzed).

HBr

OH

1

Br

2 (80%)

Br

$ZnBr_2$

O

Br

OH

1	Bruylants, P.	*Bull. Acad. Royale Belge*	**1928**	*28*	160
2	Julia, M.	*Bull. Soc. Chim. Fr.*	**1960**		1072
3	Julia, M.	*Bull. Soc. Chim. Fr.*	**1961**		1849
4	Julia, M.	*Tetrahedron Suppl.*	**1966**		443
5	Kocienscky, P.J.	*Chem. & Ind.*	**1981**		549
6	Nakamura, H.	*Tetrahedron Lett.*	**1973**		111

Trans 1-Bromo-2-pentene (2).[3] Methyl cyclopropyl carbinol **1** (43.5 g, 0.5 mol) was stirred at room temperature with 48% HBr (200 mL) for 10-15 min. After extraction with petroleum ether (40-60°C) and washing the extract with $NaHCO_3$ solution, the solvent was removed in vacuum and the residue distilled to give 52 g of **2** (80%), bp 125-126°C (757 mm) or 68-69°C (102 mm), n_D^{20} = 1.4683.

JUNG - OLAH - VORONKOV Ether cleavage

Cleavage of ethers or esters, carbamates, phosphonates with trimethylsilyl iodides. Deoxygenation of sulfoxides.

$$\underset{\mathbf{1}}{C_6H_5\text{-}COOCH_3} + \underset{\mathbf{2}}{Me_3SiI} \xrightarrow[50^\circ]{CCl_4} \underset{\mathbf{3}}{PhCO_2SiMe_3} \longrightarrow \underset{\mathbf{3a}\ (85\%)}{Ph\,CO_2H}$$

$$\underset{\mathbf{4}}{C_6H_{11}\text{-}OMe} + \mathbf{2} \xrightarrow[60^\circ]{pyr} \underset{\mathbf{5}}{C_6H_{11}\text{-}OSiMe_3} \longrightarrow \underset{\mathbf{5a}\ (84\%)}{C_6H_{11}\text{-}OH}$$

1	Jung, M.E.	*J. Am. Chem. Soc.*	**1977**	*99*	968
2	Jung, M.E.	*J. Org. Chem.*	**1977**	*42*	3761
3	Olah, G.A.	*Angew. Chem. Int. Ed.*	**1976**	*15*	774
4	Olah, G.A.	*Synthesis*	**1977**		581
5	Jung, M.E.	*Chem. Commun*	**1978**		315
6	Olah, G.A.	*Tetrahedron*	**1982**	*38*	2225
7	Voronkov, M.G.	*Zh.Obshch.Khim.*	**1976**	*46*	1908

Benzoic acid (3a). Methyl benzoate **1** (136 mg, 1 mmol) trimethylsilyl iodide **2** (0.16 mL, 1.2 mmol) in CCl_4 (0.5 mL) was heated to 50°C for 35 h. (NMR yield of **3** 100%). The reaction mixture was stirred with 10% $NaHCO_3$ (10 mL) for 30 min. Acidification of the aq. layer and extraction with Et_2O, followed by evaporation of the solvent afforded 104 mg of **3a** (95%), mp 118-119°C.

Cyclohexanol (5a). A mixture of cyclohexyl methyl ether **4** (1.722 g, 15.1 mmol), **2** (3.5 mL, 24 mmol), pyridine (0.5 mL, 6 mmol) in $CHCl_3$ (4 mL) was heated at 60°C for 64 h. Quenching with MeOH (2 mL), extraction with Et_2O, filtration of pyridine.HCl, concentration of the filtrate and chromatography (silica gel 70 g, eluent Et_2O) gave 1.274 g of **5a** (84%).

K A B E Chromanone synthesis

Synthesis of 4-chromanones by condensation of salicylaldehydes or o-hydroxyaryl ketones with enamines or ketones.

$COCH_3$ OH + N → O O

1 2 3

CH_3 HO $COCH_3$ H_3C OH CH_3 + O 3 —3 Steps→ CH_3 HO H_3C O CH_3 3

1	Kabe, H.J.	*Liebigs Ann.*	**1976** *511*	511
2	Kabe, H.J.	*Synthesis*	**1978**	887
3	Kabe, H.J.	*Synthesis*	**1978**	888
4	Kelly, S.E.	*J. Org. Chem.*	**1991** *56*	1325

2-Spiro-4-chromanone (3).[2] A mixture of 2-hydroxyacetophenone **1** (600 g, 4.4 mol) and pyrrolidinocyclopentene **2** (630 g, 4.6 mol) in MeOH (1000 mL) was refluxed for 2 h. After work up the product, distilled under vacuum, afforded 770 g of **3** (86%), bp 100-105°C/0.05 torr, mp 38-39°C.

KAGAN-MOLANDER Samarium reagent

Lanthanide reagents, specifically samarium, for generation of free radicals; useful in cyclizations, reductions.

Sm + I-CH$_2$-CH$_2$-I → SmI$_2$
1 2 3

4 —3→ 5 (74%)[4]

—3→ (75%)[6]

—3, THF 20°→ (ref 7)

1	Kagan, H.B.	*J. Am. Chem. Soc.*	**1980**	*102*	2693
2	Kagan, H.B.	*Tetrahedron*	**1986**	*42*	6573
3	Molander, G. A.	*J. Org. Chem.*	**1986**	*51*	2596
4	Molander, G. A.	*J. Org. Chem.*	**1991**	*56*	4112
5	Soderquist, J. A.	*Aldrichimica Acta*	**1991**	*24*	15
6	Molander, G. A.	*Chem. Rev.*	**1992**	*92*	29
7	Hanessian, S.	*Tetrahedron Lett.*	**1992**	*33*	573

Samarium diiodide (3).[1] Sm **1** (3 g, 20mmol) in THF (250 mL) treated with 1,2-diiodoethane **2** or CH_2I_2 gave a $4x10^{-2}$ M blue-green solution of **3**.

2,2,9,9-Tetramethylbicyclo(3.3.1)nonan-1-ol (5).[4] To SmI_2 **3** (18.8 mL, 1.88 mmol) in THF was added iron trisdibenzoylmethane (12.6 mg, 0.018 mmole) in THF (15 mL). Iodoketone **4** (272 mg, 0.845 mmol in THF (10 mL) was added at -78°C in 5 min. After 2 h stirring at 0°C, the mixture was quenched with sat. $NaHCO_3$ (9 mL). Flash chromatography (pentane - Et_2O) and sublimation afforded 122 mg (74%) of **5** , mp 169-172°C.

KAISER - JOHNSON - MIDDLETON Dinitrile cyclization

Synthesis of heterocycles by cyclization of dinitriles by means of HBr.

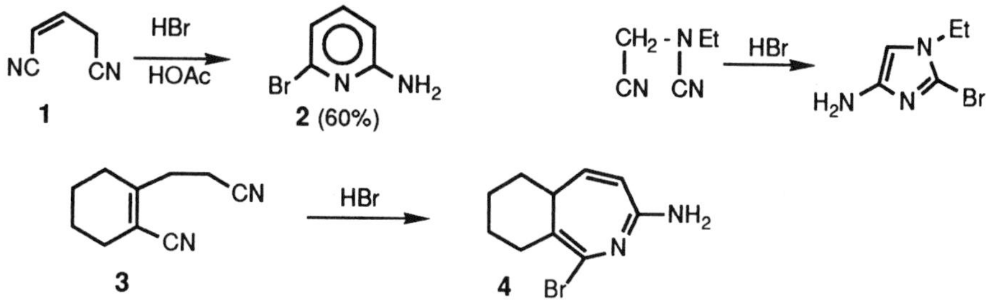

1	Johnson, F.	*J. Org. Chem.*	**1962**	*27*	2241, 2473, 3953
2	Kaiser, A.M., Roemer, J.J.	*U.S. Patent*	**1953**		2,630,433; 2,658,893
3	Middleton, W.J.	*J. Am. Chem. Soc.*	**1958**	*80*	2822, 2832
4	Johnson, F.	*J. Org. Chem.*	**1964**	*29*	153

2-Amino-6-bromopyridine (2).[1] Glutacononitrile **1** (12.5 g, 27 mmol), was added dropwise to a solution of HBr in AcOH (30 g of 30%) in 5 min with cooling and stirring. The yellow precipitate after filtration and washing ($NaHCO_3$ sol), was extracted with Et_2O and recrystallized from Et_2O-petroleum ether to give 2.7 g of **2** (60%), mp 88-89°C.

3-Amino-1-bromo-6,7,8,9-tetrahydro-4H-2-benzazepine(4).[5] 2(β-Cyanoethyl)-1-cyano-cyclohexene **3** (1.2 g 7.5 mmol) in Et_2O was treated with a stream of HBr for 15 min at 0°C. The solvent was removed in vacuum and the residue recrystallized from Me_2CO-Et_2O to give 0.9 g of **4** HBr, mp 212-214°C (MeOH-Et_2O); **4**, mp 150-155°C (CH_2Cl_2).

KAKIS Rearrangement

Formation of ketones by bromination - rearrangement of substituted ethylenes.

$$Ph_2C{=}CHPh + Br_2 \longrightarrow Ph_2C(Br){-}CHPh(Br) \xrightarrow{Ag^+} Ph_2C^+{-}CHPh(Br) \xrightarrow{H_2O}$$

$$Ph_2C(OH){-}CHPh(Br) \xrightarrow{Ag^+} PhC(Ph)(OH)\overset{+}{-}CHPh \quad Br^- \ldots Ag^+ \longrightarrow Ph{-}C({=}O){-}CHPh_2$$

$$\underset{\mathbf{1}}{Ph_2C{=}CPh_2} + Br_2 \xrightarrow[H_2O]{Ag^+} \underset{\mathbf{2}\ (94\%)}{Ph{-}C({=}O){-}CPh_3}$$

1 Kakis, F.J. *J. Org. Chem.* **1971** *36* 4117

2 Fetizon, M. *J. Org. Chem.* **1973** *38* 1733

Phenyl Trityl ketone (2).[1] A solution of tetraphenylethylene **1** (3.3 g, 10 mmol) in $CHCl_3$ (250 mL) cooled in an ice bath was saturated with Cl_2 until a distinct yellow color was present. The stirred mixture was treated with a 9:1 solution $MeOH:H_2O$ saturated with $AgNO_3$ and the stirring was continued at 20°C for 20 h. The inorganic salts were filtered off and the filtrate diluted with water. The organic layer was separated, washed and dried ($MgSO_4$) and the solvent was removed in vacuum to give a residue which crystallized spontaneously. One obtains 3.27 g of **2** (94%), mp 183-184°C.

KALUZA Isothiocyanate synthesis

Formation of isothiocyanates from amines.

$$\underset{\mathbf{1}}{p\text{-}Cl\text{-}C_6H_4\text{-}NH_2} \xrightarrow[Et_3N]{CS_2,\ 0^\circ} \underset{\mathbf{2}}{Ar\text{-}NH\text{-}C(=S)\text{-}S^-\ \overset{+}{N}HEt_3} \xrightarrow[Et_3N,\ 0^\circ]{ClCO_2Et} \underset{\mathbf{3}\ (70\%)}{p\text{-}Cl\text{-}C_6H_4\text{—}N=C=S}$$

1	Kaluza, H.	*Monatsh.*	**1912**	*33*	363
2	Hodgkins, J.E.	*J. Org. Chem.*	**1956**	*21*	404
3	Hodgkins, J.E.	*J. Am. Chem. Soc.*	**1961**	*83*	2532
4	Hodgkins, J.E.	*J. Org Chem.*	**1964**	*29*	3098

KAMETANI Amine oxidation

Oxidation of primary amines to nitriles by Cu(I)Cl - O_2 - pyridine

$$\underset{\mathbf{1}}{p\text{-}MeO\text{-}C_6H_4\text{-}CH_2\text{-}NH_2} \xrightarrow[pyr,\ 60^\circ]{Cu_2Cl_2,\ O_2} \underset{\mathbf{2}\ (99\%)}{p\text{-}MeO\text{-}C_6H_4\text{-}CN}$$

1	Kametani, T.	*Synthesis*	**1977**	245
2	Capdevielle, P	*Synthesis*	**1989**	455

p-Methoxybenzonitrile [2]. p-Methoxybenzylamine **1** (0.137 g, 1 mmol), 4 A molecular sieves (8 g), and Cu_2Cl_2 (0.6 Cu(I) equiv.) in dry pyridine (50 mL) were stirred at 60°C for 4 h under O_2 atmosphere. More Cu catalyst was added and the reaction continued 20 h. The mixture was poured on ice (100 g) and 36% HCl (60 mL) and extracted with CH_2Cl_2 (3x50 mL). The extract was washed ($NaHCO_3$), dried and evaporated to give 0.131 g of **2** (99%), mp 61°C.

KEINAN Silane Reagent

Diiodosilane (DIS) reagent for mild hydrolysis of ketals, acetals or reductive iodination of ketones and aldehydes.

H_2SiI_2

-42°, 1 min

1

2 (99%)

CHO

DIS

20°

3

CH_2I

4 (96%)

DIS

(94%) + (2%)

1	Keinan, E.	*J. Org. Chem.*	**1987**	*52*	4846
2	Keinan, E.	*J. Org. Chem.*	**1990**	*55*	2927
2	Keinan, E.	*Synthesis.*	**1990**		641

p-Bromoacetophenone (2).[2] Ketal **1** (625 mg, 2.58 mmol) in CH_2Cl_2 under N_2 at -42°C was treated with DIS (73 mg, 0.26 mmol) and stirred for 1 min. $NaHCO_3$ (1 mL) was added with vigorous stirring and the mixture allowed to warm up to 20°C. The organic layer was washed with $Na_2S_2O_3$ (5 mL) and dried over $MgSO_4$. Evaporation of the solvent and chromatography (silica gel -CH_2Cl_2) afforded 510 mg of **2** (99%).

Benzyl iodide (4)[2]. DIS (420 mg, 1.5 mmol) was added to **3** (106.2 mg, 10 mmol) in CH_2Cl_2 (5 mL). The mixture was stirred for 5 min at 20°C and quenched with 10% aqueous $NaHCO_3$ (0.5 mL) and 10% $Na_2S_2O_3$ (0.5 mL), diluted with CH_2Cl_2 (20 mL), and washed with water. The organic phase was evaporated and the residue in CH_2Cl_2 filtered through silica gel. Evaporation afforded 210 mg of **4** (96%).

KHAND - PAUSON Cyclopentenone Annulation

Cyclopentenone synthesis from carbon monoxide acetylene and olefins, cobalt carbonyl catalyzed.

HC≡CH , CO

$Co_2(CO)_8$

1

2 (74%)

1	Khand, J.U., Pauson, P.L.	*J. Chem. Soc. Perkin*	**1973**		977
2	Pauson, P.L.	*Tetrahedron*	**1985**	*41*	5855
3	Schore, N.S.	*J. Org. Chem.*	**1987**	*52*	569
3	Schore, N.S.	*J. Org. Chem.*	**1988**	*53*	203
4	Schore, N.S.	*Org. React.*	**1991**	*40*	1

Endo 3a,4,5,6,7,7a-Hexahydro-4,7-methano-2-indene-1-one (2).[1] A solution of octacarbonyldicobalt (1.0 g, 3 mmol) and bicyclo [2.2.1] hept-2-ene **1** (3.0 g, 32 mmol) in isooctane (100 mL) was stirred first with acetylene and then under a mixture of 1:1 of acetylene and carbon monoxyde at 60-70°C until gas absorption ceased (total 1550 mL). The mixture was concentrated and the residue chromatographed on neutral alumina. Petroleum ether: PhH (1:1) eluted acetylene hexacarbonyl dicobalt 70 mg, PhH:$CHCl_3$ (1:1) eluted a yellow oil which after distillation afforded 3.54 g of **2** (74%), bp 101-102°C (15 mm). Colorless crystals from pentane, mp 32°C.

KHARASH - SOSNOVSKY Allylic Oxidation

Allylic or propargylic oxidation with t-butyl peresters.

Et—C≡C—Et (1) + Ph—C(=O)·O·O—t-Bu (2) —(Cu Cl, heat)→ Et—C≡C—CH(Me)—O-CO Ph 3 (57%)

(1-butene) or (2-butene) + Me--C(=O)-O-O-t Bu ——→ (3-acetoxy-1-butene) OAc + (crotyl acetate) OAc

1	Kharash, M., Sosnovky, G.	*J. Am. Chem. Soc.*	**1958**	*80*	756
2	Julia, M.	*Tetrahedron Lett.*	**1976**		2141
3	Kropp, H.	*Synthesis*	**1977**		894

2-Benzoyloxy-3-hexyne (3).[3] Peroxybenzoic acid t-butyl ester **2** (5.83 g, 30 mmol), Cu$^{(I)}$ chloride (10 mg) and 3-hexyne **1** (8.5 g, 130 mmol) was heated for 14 h at 140°C. The cooled reaction mixture was diluted with Et_2O (20 mL), and the organic extract was washed with Tritiplex solution (2x25 mL). Evaporation of the solvent afforded 3.5 g of **3** (57%), bp 75-78°C (0.04 mm).

KILIANI - FISCHER Sugar Homologation

Synthesis of C_{n+1} sugars from C_n sugars.

H-C=O | CH 2 | H-C-OH | H-C-OH | CH 2 OH — KCN / NaOH → (lactone: H, OH, O, O; H-C-OH | CH 2 OH) — Na OO C-COOH / Na BH4 → H-C=O | H-C-OH | CH2 | H-C-OH | H-C-OH | CH 2 OH

1 **2** (100%) **3** (38%)

1	Kiliani, H.	*Chem. Ber.*	**1885**	*18*	3066
2	Fischer, E.	*Chem. Ber.*	**1889**	*22*	2204
3	Wood, H.B.	*J. Org. Chem.*	**1961**	*26*	1969
4	Mowry, D.T.	*Chem. Rev.*	**1948**	*42*	239

3-Deoxy-D-ribohexose (3).[3] To a frozen solution of KCN (2.01 g, 41 mmol) and NaOH (1.49 g, 37 mmol) in water (186 mL), was added a solution of 2-deoxy-D-ribose **1** (5.0 g, 37 mmol) in 0.2M $NaHCO_3$ (186 mL) and the mixture was stirred at 20°C until homogeneous. After 16 h (negative Fehling test) the solution was deionized on Amberlite IR-120-H+, concentrated in vacuum and precipitated with Me_2CO (25 mL) to give 6.87 g of **2** (100%), α_D^{20} = +23.1.

A solution of **2** (1.1 g, 6.8 mmol) in water (50 mL) and sodium acid oxalate (2.0 g) was cooled to 0°C and under stirring was treated with $NaBH_4$ (0.5 g, 13 mmol) in water (10 mL). The pH of the solution was maintained at 4.5-5. After dilution with MeOH to precipitate the salts, the solution was deionized by Amberlite IR-120-H+ and Duolite A-4 and the concentrate was treated with anhydrous EtOH. After several days at +5°C, crystals were filtered, 0.42 g (38%), mp 102-104°C. Pure **3** from EtOH, mp 105-106°C, α_D^{20} = +32° (H_2O).

KISHNER Cyclopropane Synthesis

Cyclopropanes from unsaturated carbonyl compounds via pyrazolines by catalytic pyrolysis.

Ph - CH = CH - CHO (**1**) + $N_2H_4.H_2O$ (**2**) ⟶ **3** (37%)

4 $\xrightarrow[\text{KOH, }\Delta]{\text{Pt}}$ **5** (12%)

1	Kishner, N.	*J. Russ. Phys. Chem. Soc.*	**1911**	*43*	1132
2	Jeger, O.	*Helv. Chim. Acta.*	**1949**	*32*	1817
3	Hammond, G.S.	*J. Am. Chem. Soc.*	**1954**	*76*	4081
4	Van Auken, T.V.	*J. Am. Chem. Soc.*	**1962**	*84*	3736
5	Crawford, R.J.	*Can. J. Chem.*	**1974**	*54*	4033

5-Phenyl-Δ^2-pyrazoline (3).[3] Cinnamaldehyde **1** (30 g, 0.227 mol) and hydrazine hydrate **2** (15 mL, 0.3 mol) in EtOH (60 mL) was refluxed for 4 h. The solvent was evaporated and the residue distilled in vacuum under N_2. The first fraction, bp 120°C/14 mm, was cinnamaldehyde followed by a fraction with bp 142-146°C/14 mm. This fraction on redistillation gave 12 g of **3** (37%), bp 138-139°C/12 mm, $n_D^{19.5}$ = 1.5695, mp 41-42°C; picrate mp 107-107.5°C.

Phenylcyclopropane (5).[3] **3** (118 g, 0.8 mol), powdered KOH (30 g, 0.53 mol) and platinized asbestos (2.5 g) were heated slowly and the heat was shut off at the first sign of reaction. After the exothermic reaction the temperature was again raised and the product was distilled. Both the distillate and the residue were steam distilled, the distillate was extracted with Et_2O, the solvent evaporated and the residue distilled to give 11.5 g of **5** (12%), bp 173.5°C (740 mm), n_D^{20} = 1.5320.

KNOEVENAGEL - DOEBNER - STOBBE Condensation

Base catalyzed aldol condensation of aldehydes or ketones with an activated methylene group of a malonic ester (Knoevenagel - Doebner) or a succinic ester (Stobbe).

2,3-dimethoxybenzaldehyde (**1**) + $H_2C(CO_2Et)(CO_2H)$ (**2**) —piperidine→ ethyl 2,3-dimethoxycinnamate (**3**) (95%)

$Ph_2C{=}O$ (**4**) + $CH_2\text{-}COOEt / CH_2\text{-}COOEt$ (**5**) —NaH, 20°C→ $Ph_2C{=}C(COOEt)CH_2COOEt$ (**6**) (97%)[8]

1	Knoevenagel, E.	*Chem. Ber.*	**1896**	*29*	172
2	Doebner, O.	*Chem. Ber.*	**1900**	*33*	2140
3	Rapoport, H.	*J. Org. Chem.*	**1981**	*46*	5064
4	Grayson, D.H.	*J. Chem. Soc. Perkin I*	**1986**		2137
5	Emden, D.	*Chem. Ber.*	**1987**	*120*	2717
6	Jones, G.	*Org. React.*	**1967**	*15*	204
7	Stobbe, H.	*Chem. Ber.*	**1893**	*26*	2312
8	Daub, G.R.	*J. Am. Chem. Soc.*	**1948**	*70*	418
9	Martelly, J.	*Tetrahedron*	**1974**	*30*	3063
10	Johnson, W.S.	*Org. React.*	**1951**	*6*	1

KNORR Pyrazole Synthesis

Pyrazole synthesis from a β-dicarbonyl compound and a hydrazine.

1 Knorr, L. *Chem. Ber.* **1883** *16* 2587

2 Mosley, M.S. *J. Chem. Soc.* **1957** 3997

3 Katritzky, A.R. *Tetrahedron* **1964** *20* 299

3-(3,5-Dimethylpyrazol-4'-yl)pentane-2,4-dione (3).[2] To a stirred suspension of diacetyl hexanedione **1** (2.0 g, 10 mmol) in boiling EtOH (50 mL) was added hydrazine hydrate **2** (0.6 mL). The clarified (charcoal) hot solution was evaporated in vacuum and the oily residue was crystallized from water to afford 0.9 g of **3** (46%), mp 134-140°C.

KNORR Quinoline Synthesis

Quinoline synthesis by cyclization of acetoacetanilides.

1 Knorr, L. *Liebigs Ann.* **1886** *236* 69

2 Hodgkinson, A. *J. Org. Chem.* **1969** *34* 1709

3 Bergstrom, F.W. *Chem. Rev.* **1944** *35* 157

4 Bergstrom, F.W. *Chem. Rev.* **1948** *48* 47

KNUNYANTS Fluoroalkylation

Fluoroalkylation of aromatics using hexafluoroacetone.

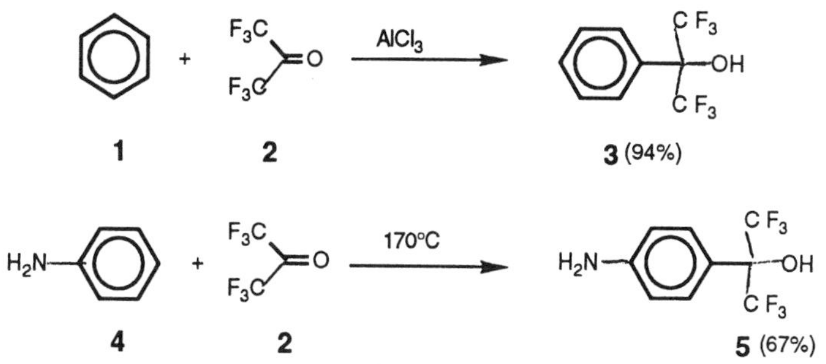

1 + 2 →($AlCl_3$) 3 (94%)

4 + 2 →(170°C) 5 (67%)

1	Knunyants, I.L.	*Zh. Vses. Chim. Obsh.*	**1960**	*4*	114
2	Simmons, H.E.	*J. Am. Chem. Soc.*	**1960**	*82*	2288
3	Gilbert, E.E.	*J. Org. Chem.*	**1965**	*30*	998,1001
4	Knunyants, I.L.	*Zh. Akad.Nauk.SSSR*	**1962**	*4*	682

Bis(Trifluoromethyl)phenylcarbinol (3).[3] To a suspension of $AlCl_3$ (5.0 g, 37 mmol) in PhH **1** (880 g, 11.3 mol) cooled externally, was bubbled hexafluoroacetone **2** (bp -28°C) until it was absorbed - 115 g, 6.72 mol (ca 6 h). The reaction mixture was washed, dried and distilled to give 541 g of **3** (94%).

p-Aminophenyl-bis(trifluoromethyl)carbinol (5). Aniline **4** (93.1 g, 1 mol) was heated at 170°C and **2** (175 g, 1.05 mol) was bubbled in over 2 h. The mixture was heated to 170-180°C for 30 min. The cooled mixture was dissolved in Et_2O and petroleum ether was added. After concentration, there are obtained a first crop of 131 g of **5** (50%), mp 148-150°C. The mother liquor gave 45 g of 5 (17%), mp 80-100°C.

KOCHI Cross Coupling

Cross coupling of organometallics with organic halides catalyzed by iron (III).

Ph - MgBr + BrCH=CH-Ph $\xrightarrow{Fe(DBM)_3}$ Ph-CH=CH-Ph

1 **2** **3**

CH_2=CHBr + BrMg~~~ $\xrightarrow{0.5\%\ FeCl_3}$

4 **5** **6**

1	Kochi, J.	*Synthesis*	**1971**		303
2	Kochi, J.	*J. Org. Chem.*	**1975**	*40*	599
3	Kochi, J.	*J. Org. Chem.*	**1976**	*41*	502
4	Molander, G.A.	*Tetrahedron Lett.*	**1983**	*24*	5449

$Fe(DBM)_3$ Catalyst.[3] To an aqueous solution of $FeCl_3$ (0.6 g) was added a solution of dibenzoyl methane (DBM) (1.85 g, 2.76 mmol) in EtOH. After addition of NH_4OH the solid was filtered, washed, dried and recrystallized from PhH-petroleum ether, 70% yield, mp 270°C, λ_{max} 408 nm.

Trans Stilbene (3).[2] A solution of catalyst $Fe(DBM)_3$ (0.15 mol) in THF was added to PhMgBr **1** (45 mmol) in THF. After 5 min stirring, a solution of β-bromostyrene (12 mmol) in THF was added (ice bath). The mixture was stirred for 1 h, filtered and concentrated to 50%. The solution was extracted with 5% HCl and the solvent removed in vacuum to give **3**.

KÖENIGS - KNORR Glycosidation

Synthesis of glycosides from bromosugars in the presence of Ag^+ or Hg^{2+}.

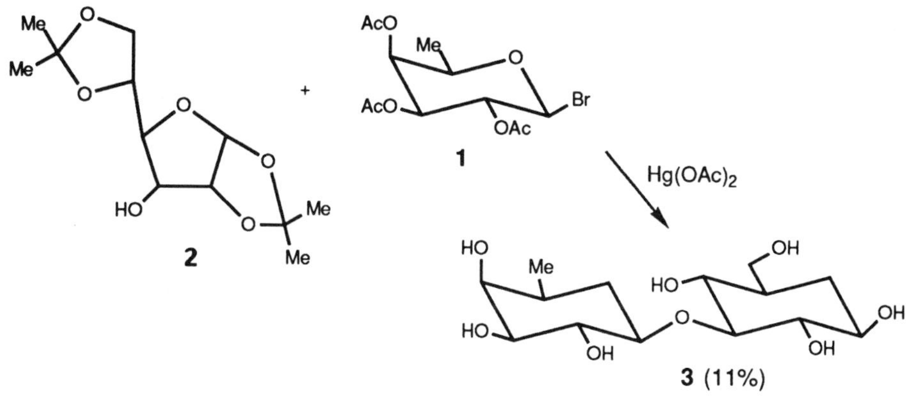

1	Köenigs, W., Knorr, E.	*Chem. Ber.*	**1901**	*34*	957
2	Ice, C.H.	*J. Am. Chem. Soc.*	**1952**	*74*	4606
3	Knochel, A.	*Tetrahedron Lett.*	**1974**		551
4	Israel, M.	*J. Med. Chem.*	**1982**		28

3-0-α-L-Rhamnopyranosyl-D-glucose (3).[2] A mixture of 1,2,5,6-di-0-isopropylidene-α-D-glucofuranose **2** (0.6 , 2.3 mmol), bromorhamnose **1** (0.66 g, 1.9 mmol) and $Hg(AcO)_2$ (0.25 g) in dry PhH (10 mL), was shaken for 4 days at r.t. The mixture was washed with water, distilled to dryness and the residue (0.55 g) was stirred with 0.1 N MeONa (0.5 mL) in MeOH (4 mL) for 6 h. The solution was neutralized with oxalic acid, the solvent removed in vacuum and the residue heated with 0.01 N oxalic acid (10 mL) at 100°C for 3 h. The cooled solution was treated with $BaCO_3$,, filtered and the filtrate deionized by Amberlite IR-100H and IR-4B ion exchange resin. Paper chromatography on Watman 3MM (n-BuOH:AcOH:H_2O 12:3:5) afforded 62 mg of **3** (11%), R_f 0.92.

KOHLER Isoxazole-N-oxide Synthesis

Synthesis of 2-isoxazole-2-oxides from 1,3-dinitroalkanes or 1-halo-3-nitro alkanes.

Ph-CH$_2$—NO$_2$ + Ph—CH=O $\xrightarrow{MeNH_2}$ O$_2$N-CH(Ph)-CH(Ph)-CH(Ph)-NO$_2$ $\xrightarrow{NaOMe}$ 4 (52%)

1 2 3 4 (52%)

1 Kohler, E.P.	*J. Am. Chem. Soc.*	**1924**	*46*	2105
2 Nielsen, A.T.	*J. Org. Chem.*	**1969**	*34*	984
3 Smith, L.I.	*Chem. Rev.*	**1938**	*23*	255

trans-3,4,5-Triphenyl-2-oxazoline-2-oxide (4).[2] Phenylnitromethane **1** (27.4 g, 0.2 mol) and benzaldehyde **2** (10.6 g, 0.1 mol) were mixed with methylamine hydrochloride (0.1 g), K_2CO_3 (0.05 g) and EtOH (1 mL). After a week at 20°C, the solvent was evaporated and PhH added; the mixture was maintained at 25°C for 4 days. The solution was concentrated to a small volume and filtration gave cis-α-nitrostilbene 5.0 g (22%). The filtrate was evaporated to dryness and triturated with EtOH to give 4.5 g of **3** (12%), mp 174-176°C.

A solution of **3** (0.15 g, 0.4 mmol) in MeOH (5 mL) containing NaOMe (0.2 g) was refluxed 3 min, cooled and the solid filtered off. Recrystalllization from MeOH gave 65 mg of **4** (52%), mp 163°C.

KOLBE Electrolysis

Electrochemical decarboxylation-dimerization (via free radicals).

$$2\ Ph_2CH\text{-}COOH \xrightarrow[DMF,\ 17\ h]{electrolysis} Ph_2CH-CHPh_2$$

1 → **2** (24%)[3]

1	Kolbe, H.	*Liebigs Ann.*	**1849**	*69*	257
2	Crum Brown, A., Walker, A.	*Proc. Roy. Soc. Edinburgh*	**1890**	*17*	292
3	Finkelstein, M.	*J. Org. Chem.*	**1969**	*25*	156
4	Rabson, M.	*J. Org. Chem.*	**1981**	*46*	4082
5	Marquet, B.	*Bull. Soc. Chim. Fr.*	**1988**		571
6	Vijh, A.K.	*Chem. Rev.*	**1967**	*67*	625
7	Schaefer, H. J.	*Angew. Chem. Int. Ed.*	**1981**	*20*	911

1,1,2,2-Tetraphenylethane (2).[3] A solution of diphenylacetic acid **1** (21.15 g, 0.1 mol) in DMF was electrolyzed for 17 h and the voltage was increased from 120 to 220 V during electrolysis. The reaction mixture was poured into water and extracted with Et_2O. The extract was filtered, washed and the solvent evaporated. Recrystallization from $CHCl_3$ gave 4 g of **2** (24%), mp 207-208°C.

KOLBE - SCHMIDT Salicylic Acid Synthesis

Carboxylation (usually ortho.) of phenols. Industrial method to obtain salicylic acid derivatives.

1 (m-aminophenol: OH, NH_2) → $KHCO_3$, CO_2, 90 - 95°C → 2 (CO_2H, OH, NH_2) (48%)

1	Kolbe, H.	*Liebigs Ann.*	**1860**	*113*	125
2	Schmidt, R.	*J. prakt. Chem.*	**1885**	*31*	397
3	Doub, L.	*J. Org. Chem.*	**1958**	*23*	1422
4	Lindsey, A.S.	*Chem. Rev.*	**1957**	*57*	583
5	Raecke, B.	*Angew. Chem.*	**1958**	*70*	1
6	Ota, K.	*Bull. Chem. Soc. Jpn.*	**1974**	*47*	2343

p-Aminosalicylic acid (2).[3] A suspension of m-aminophenol **1** (13.1 g, 0.12 mol) and $KHCO_3$ (30.0 g, 0.3 mol) in water was heated under CO_2 pressure at 90-95°C for 4 days. The mixture was cooled to 0°C and after filtration was acidified strongly with conc HCl. The precipitated **2**.HCl (12.8 g, 56%) was treated with water (20 mL) and dissolved by addition of $KHCO_3$. The solution (60 mL) was decolorized by $NaHSO_3$ and charcoal and after filtration was acidified with 50% H_3PO_4, cooled to 0°C, and the product was filtered, washed, and dried to give 8.0 g of **2** (48%), mp 160-162°C.

KONAKA Oxidizing Reagent

Oxidation of alcohols to carboxylic acids (or ketones) with nickel peroxide.

$$NiSO_4 + NaOCl \longrightarrow \underset{\mathbf{1}}{\text{Ni peroxide}}$$

$$\underset{\mathbf{2}}{Ph\text{-}CH_2\text{-}OH} + \mathbf{1} \xrightarrow{NaOH} \underset{\mathbf{3}\ (86\%)}{PhCOOH}$$

$$Ph_2CHOH + \mathbf{1} \longrightarrow Ph_2CO$$

1	Konaka	*J. Org. Chem.*	**1962**	*27*	1660
2	Konaka	*J. Org. Chem.*	**1969**	*34*	1334

Nickel peroxide (1).[1] A mixture of 6% NaOCl (300 mL) and NaOH (42.0 g, 1 mol) was added dropwise to a solution of $NiSO_4 . 6H_2O$ (130 g, 0.5 mol) in water (300 mL) and stirred at 20°C for 30 min. The black nickel peroxide was filtered, washed and dried in vacuum.

Benzoic acid (3). A solution of benzyl alcohol **2** (2.16 g, 20 mmol) and NaOH (1.0 g, 25 mmol) in water (50 mL) was treated with **1** (16.0 g, 1.5 equiv) under stirring at 30°C. After 3 h the solution was filtered and the filtrate was acidified. The precipitate was collected and dried to afford 2.1 g of **3** (86%), mp 122.5°C.

K Ö N I G Benzoxazine Synthesis

Benzoxazine synthesis from quinones.

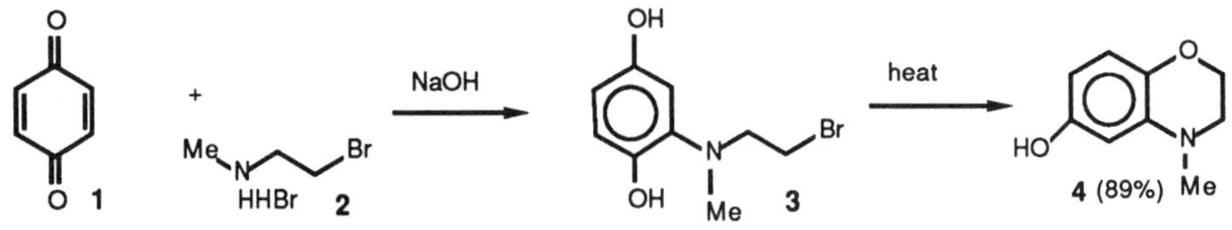

1	Konig, K.H.	*Chem. Ber.*	**1959**	*92*	257
2	Konig, K.H.	*Z. Anal. Chem.*	**1959**	*166*	92
3	Flemming, J.	*Z. Phys. Chem. (Leizig)*	**1964**	*223*	106
4	Day, J.H.	*J. Org. Chem.*	**1965**	*30*	4107
5	McMurtrey, K.D.	*J. Org. Chem.*	**1970**	*35*	4252

3,4-Dihydro-4-methyl-2H-1,4-benzoxazine-6-ol (4).[5] To a solution of 1,4-benzoquinone **1** (30.0 g, 0.277 mol) and 2-bromoethylmethylamine hydrobromide **2** (30.0 g, 0.137 mol) in 50% water- MeOH (2000 mL) at 0°C, was added dropwise 0.2 N NaOH (500 mL). After 2 h the mixture was filtered and the solid washed with water and triturated with Me_2CO. The acetone soluble fraction afforded 6 g of **3** (18%), mp 140-144°C.

A solution of **3** (1.0 g, 4 mmol) in $CHCl_3$ (100 mL) was shaken with aqueous sodium dithionite until the organic layer became colorless. The glassy residue obtained after evaporation was dissolved in dioxane:TEA (1:1) (100 mL) by heating 12 h on a steam bath. Evaporation and chromatography (preparative TLC - PhH:MeOH 9:1) gave 0.6 g of **4** (89%), mp 77-78.3°C.

K O R N B L U M Aldehyde Synthesis

Synthesis of aldehydes from primary alkyl halides or tosylates, using dimethyl sulfoxide (DMSO).

p $Br-C_6H_4-CO-CH_2Br$ —DMSO→ p $Br-C_6H_4-CO-CHO$

1 **2** (91%)

3 —Ag Ts→ R - Ts —DMSO, 150°, Na_2CO_3→ CHO

4 (70%)

1 Kornblum, N. *J. Am. Chem. Soc.* **1957** *79* 6562

2 Kornblum, N. *J. Am. Chem. Soc.* **1959** *81* 4113

p-Bromophenylglyoxal (2).[1] p-Bromophenacyl bromide **1** (15.98 g, 57 mmol) in DMSO (100 mL) was stirred at 20° for 9 h, poured into water and extracted with Et_2O. Evaporation of the solvent and recrystallization of the residue from n Bu-O-Et afforded 11.2 g of **2** (91%), mp 123-124°C.

Heptanal (4).[2] To silver tosylate (11.0 g, 38 mmol) in MeCN (100 mL) were added n-iodoheptane **3**(7.0 g, 30 mmol). The mixture protected from light was maintained 24 h at 20°, poured on ice, the product was extracted with Et_2O and the residue after removal of the solvent, was poured into Na_2CO_3 (20 g) in DMSO (150 mL). After heating 5 min at 150°C under N_2, the aldehyde was separated as its 2,4-dinitrophenylhydrazone (DNPH) to afford 6.9 g of **4** DNPH (70%), mp 106-107°C.

KOSER Tosylation

Vic-bis tosylation of alkenes by means of hydroxytosyloxyiodobenzene.

$$\underset{\mathbf{2}}{H_3C\text{-}CH\text{=}CH\text{-}CH_2\text{-}CH_3} + \underset{\mathbf{1}}{Ph\text{-}I(OTs)(OH)} \xrightarrow[28h]{3°C} \underset{\mathbf{3}\ (44\%)}{H_3C\text{-}CH(OTs)\text{-}CH(OTs)\text{-}CH_2\text{-}CH_3}$$

1	Koser, G.F.	*J. Org. Chem.*	**1977**	*42*	1476
2	Neiland, O.	*J. Org. Chem. USSR (Engl.)*	**1970**	*6*	889
3	Koser, G.F.	*J. Org. Chem.*	**1980**	*45*	1542
4	Koser, G.F.	*J. Org. Chem.*	**1984**	*49*	2462

Erythro(dl)-2,3-bis(tosyloxy)pentane (3).[4] A mixture of hydroxy(tosyloxy)-iodobenzene **1** (3.92 g, 10 mmol), cis 2-pentene **2** (2.5 mL, 1.6 g, 23 mmol) and CH_2Cl_2 (20 mL) was allowed to stand for 28 h at 3°C, to give a yellow solution containing a floating scum. This was washed with water (2x20 mL), dried (Na_2SO_4) and concentrated under vacuum. The oily residue after washing with pentane (15 mL) and recrystallization from a mixture of MeOH (6 mL) and pentane (3 mL) at -20°C gave 827 mg of **3** (40%), mp 82-83°C. A second crop from the mother liquor raised the yield to 44%.

KRIEF - REICH Olefination

Synthesis of olefins by stereospecific reductive elimination of β-hydroxyalkyl selenides (a variant of Peterson olefination) by means of $MeSO_2Cl$, $HClO_4$ or P_2I_4.

$CH_3(CH_3)_5CH(OH)$—$CH(SePh)$-$(CH_2)_5CH_3$ **1** —($HClO_4$, ether, 25°)→ $CH_3(CH_2)_5$ $CH{=}CH$-$(CH_2)_5CH_3$ **2** (85%)

4 —($(PhSe)_2CH_2$ **3**, BuLi)→ OH, CH_2SePh **5** (71%) —($MeSO_2Cl$, TEA)→ **6** (91%)

1	Krief, A.	*Angew. Chem. Int. Ed.*	**1974**	*13*	804
2	Reich, J.R.	*J. Chem. Soc. Chem. Commun.*	**1975**		790
3	Krief, A.	*Tetrahedron Lett.*	**1976**		1385
4	Reich, J.R.	*J. Am. Chem. Soc.*	**1979**	*101*	6638
5	Krief, A.	*Bull. Soc. Chim. Fr.*	**1990**		681

Tetradec-7-ene (2).[3] **1** (125 mg, 0.34 mmol) in 2 mL of ether was stirred with an etheral extract of 7% $HClO_4$ (0.4 mL, 0.14 mmol). After completion (TCL) the mixture was poured into aq. $NaHCO_3$ (10 mL), extracted with Et_2O, dried and purified by preparative TLC (SiO_2 Merck, pentane) to give **2** (85%).

1-Methylene-4-tert-butylcyclohexane (6).[4] To bis(phenylseleno)methane **3** (783 mg, 2.4 mmol) in THF (5 mL) at -78°C under N_2 was added n BuLi (2.4 mmol). After 10 min **4** (309 mg, 2 mmol) in THF (1 mL) was added and after 5 min water (10 mL) and after 5 min the mixture was poured into water and hexane:ether 1:1 (15 mL). After distillation of methyl phenyl selenide, butyl phenyl selenide and **4,** crude **5** (2:8 mixture of diastereomers), was crystallized from 5% Et_2O-hexane to yield 144 mg of **5** (71%), mp 68.5-70°C, and a 2nd and 3rd crop. **5** (309 mg, 0.951 mmol) in CH_2Cl_2 was treated with TEA (480 mg, 4.76 mmol) and $MeSO_2Cl$ (326 mg, 2.85 mmol) to yield of **6** (91% by NMR).

KRÖHNKE - ORTOLEVA Pyridinium Salts

Formation of keto pyridinium salts by reaction of alpha halo ketone derivatives with pyridine and their cleavage to carboxylic acids.

O Br 1 N 2 Br - O + N Ar 3 NaOH heat COOH 4

O CH3 5 + N 2 I2 O N + I - 6

1	Krohnke, F.	*Chem Ber.*	**1933**	*66*	604
2	Ortoleva, G.	*Gazz. Chim. Ital*	**1899**	*25 I*	503
3	King, I.C.	*J. Am. Chem. Soc.*	**1944**	*66*	894,1612
4	Krohnke, F.	*Angew. Chem. Int. Ed.*	**1963**	*2*	380
5	Alvarez, S.I.	*Tetrahedron*	**1986**	*42*	699

1-Naphtoic Acid (4).[1] **1** and pyridine **2** were heated on a water bath for a short time. The product was dissolved in EtOH and precipitated with Et_2O. Recrystallization from 10 volumes of water afforded pure α--naphthacyl pyridinium bromide **3**. A solution of **3** (0.5 g, 1.5 mmol) in water (40 mL), EtOH (12 mL) and 10 N NaOH (2 mL) after heating for 12 min on a water bath, gave on acidification and extraction with Et_2O **4**, mp 160-162°C.

1-Phenacylpyridinium iodide (6).[5] Iodine (25.0 g, 0.1 mol) was added to acetophenone **5** (12.0 g, 0.1 mol) and **2** (25 mL), heated at 95°C for 30 min and at 20°C overnight. Excess **2** was removed and **6** (insoluble in water or EtOH) was separated from pyridinium hydroiodide.

KUCHEROV - DENIGES Hydration

Water addition to a triple bond (Kucherov) or to a double bond (Deniges) under mercury salt catalysis.

$HgSO_4$

1 → 2 (60%)

HgO / H_2SO_4

Ph, OH, OH

3 → 4

HgO / H_2SO_4

CHO

5 → 6 (57%)

1	Kucherov, M.	*Chem. Ber.*	**1881**	*14*	1540
2	Thomas, R.J.	*J. Am. Chem. Soc.*	**1938**	*60*	718
3	Deniges, G.	*Bull. Soc. Chim. Fr.*	**1898**	*19*	494(3)
4	Shearer, D.A.	*Can. J. Chem.*	**1955**	*33*	1002
5	Arzoumanian, N.	*Synthesis*	**1971**		527

2-Hexanone (2).[2] To H_2SO_4 (1.0 g, 10 mmol), $HgSO_4$ (1.0 g, 3.3 mmol), 70% MeOH (150 g), 70% Me_2CO (150 g) and 60% AcOH (50 g) at 60°C ± 1°C, **1** (41.0 g, 0.5 mol) was added under stirring for 1 h. Work up after heating for 3 h gave 30-35 g of **2** (60-70%), bp 127°C.

1,2-Dihydroxy-2-methyl-1-phenylpropane (4).[5] To a suspension of HgO (21.7 g, 0.1 mol) in 1N H_2SO_4 (200 mL, 0.2 mol) was added 2-methyl-1-phenylpropene **3** (6.61 g, 50 mmol). The mixture was shaken for 2 days, filtered and the residue washed with MeOH and Et_2O. The filtrate was extracted with Et_2O and $CHCl_3$ and the solvent was evaportaed to yield 5.8 g of **4** (70%), mp 54-62°C; mp 63-63.5°C after crystallization from Et_2O-petroleum ether.

KUHN - WINTERSTEIN Olefination

Olefin formation from glycols by means of $P_2 I_4$.

$P_2 I_4$ / pyr

1 → 2 (43%)

1	Kuhn, R., Winterstein, A.	*Helv. Chim. Acta.*	**1928**	*11*	87
2	Kuhn, R., Winterstein, A.	*Helv. Chim. Acta.*	**1955**	*27*	309
3	Inhoffen, C.	*Liebigs Ann.*	**1965**	*684*	24

3,4-Bis(4,4-ethylenedioxocyclohexyl)-3-hexene-1,5-diine (2)[3] $P_2 I_4$ (60 g, 0.135 mol) was extracted with CS_2 in a Soxhlett. To this extract was added 3,4-bis(4,4-ethylidendioxycyclohexyl)-3,4-hexanedioldiine **1** (30 g, 0.077 mol) in pyridine (600 mL). The mixture was stirred at 20°C for 2 h and the solvent was distilled. The residue was treated with Et_2O and after work up the product was chromatographed on Woelm Alumina, activ. II. Recrystallization from MeOH afforded 12.9 g of **2** (43%).

KUMADA-FLEMING Stereoselective hydroxylation

Stereoselective conversion of alkyl silanes to alcohols.

Hg(OAc)$_2$, HOOAc / HOAc: 1 → 2

X = O,N

1	Kumada, M., Tamao, K.	*Tetrahedron*	**1983**	*39*	983
2	Kumada, M., Tamao, K.	*J. Org. Chem.*	**1983**	*48*	2120
3	Fleming, I.	*J. Chem. Soc.Chem.Commun.*	**1984**		29
4	Kumada, M., Tamao, K.	*Tetrahedron Lett.*	**1984**	*25*	321
5	Fleming, I.	*Tetrahedron Lett.*	**1987**	*28*	4229
6	Fuchs, P.L.	*Tetrahedron Lett.*	**1991**	*31*	7513

(SR,RS)-4-Hydroxy-3-methyl-4-phenylbutan-2-one (2).[5] To a stirred Metalation of (E)-butene (1.05 equiv) with n BuLi (1 equiv) and KOtBu (1 equiv) in THF at -50°C for 15 min followed by treatment of (E)-crotyl potassium salt with $B(OiPr)_3$ at -78°C gave, after quenching with 1 N HCl and extraction with Et_2O containing 1 equiv of diisopropyl tartarate, the crotyl boronate **2**. A solution of decanal **1** (156 mg, 1 mmol) was added to a toluene solution of **2** (1.1-1.5 equiv) (0.2 M) at -78°C containing 4Å molecular sieves (15-20 mg/L). After 3 h at -78°1 N NaOH was added, followed by extraction and chromatography to afford 208 mg of **3** (90%), anti:syn 99:1.

KUMADA - NEGISHI Cross coupling

Stereospecific alkenyl aluminum-alkenyl halide cross coupling catalyzed by Pd or Ni.

n-Bu-C≡CH —(iBu_2AlH)→ **1** (n-Bu(H)C=C(H)Al(iBu)$_2$) —(5% $Pd(Ph_3P)_2Cl_2$; Br(H)C=C(Me)CO_2Me **2**)→ n-Bu(H)C=C(H)–(H)C=C(Me)CO_2ME

3 (61%, ee 97%)

1	Negishi, E.	*J. Am. Chem. Soc.*	**1976**	*98*	6729
2	Kumada, M.	*Pure. Appl. Chem.*	**1980**	*52*	669
3	Negishi, E.	*Acc. Chem. Res.*	**1982**	*15*	340
4	Negishi, E.	*J. Am. Chem. Soc.*	**1987**	*109*	2393
5	Negishi, E.	*Synthesis*	**1988**		1

Methyl (E,E)-2-methyl-2,4-nonadienoate (3).[1] To dichlorobis(triphenylphosphine)-palladium (0.74 g, 1 mmol) in THF (20 mL) were added sequentially neat diisobutylaluminium hydride (0.37 mL, 2 mmol) (10 min at 25°C), (E)-1-hexenyldiisobutylalane **1** (obtained from 1-hexyne (1.64 g, 20 mmol) in hexane (20 mL) and diisobutylaluminium hydride (3.58 g, 20 mmol)). Methyl (E)-3-bromo-2-methylpropenoate **2** (3.58 g, 20 mmol) was added (25°C) and the mixture was refluxed for 15 min. After addition of 3N HCl, extraction with Et_2O, drying ($MgSO_4$) and evaporation of the solvent, distillation afforded 2.2 g of **3** (61%), bp 78-79°C/1 mm.

KURSANOV - PARNES Ionic Hydrogenation

A non-catalytic hydrogenation of C=C, C=O, C=N bonds and hydrogenolysis of C-OH, C-Hal, etc., under the action of an acid and a hydride ion donor.

$$H_3C\text{-}C(CH_3)\text{=}CH\text{-}(CH_2)_2\text{-}CH_3 \ (\mathbf{1}) \xrightarrow[\text{TFA}]{Ph_2SiH_2\ \mathbf{2}} H_3C\text{-}CH(CH_3)\text{-}(CH_2)_3\text{-}CH_3 \ (\mathbf{3}\ (80\%))$$

$$\text{2-thienyl-}(CH_2)_4\text{-}CO_2H \ (\mathbf{4}) \xrightarrow[BF_3 . Et_2O . TFA]{Et_3SiH\ \mathbf{5}} \text{2-tetrahydrothienyl-}(CH_2)_4\text{-}CO_2H \ (\mathbf{6}\ (90\%))$$

1	Parnes, Z.N.	*Dokl. Akad. Nauk. SSSR*	**1966**	*166*	122
2	Parnes, Z.N.	*Tetrahedron*	**1967**	*23*	2235
3	Kursanov, D.N.	*Synthesis*	**1974**		633
4	Kursanov, D.N.	Ionic Hydrogenation and Related Reactions			
5	Vol'pin, M.E., .Ed.	*Harwood Academic Pub.*	**1985**		

2-Methylhexane (3).[3] To **1** (7.84 g, 80 mmol) and Ph_2SiH_2 **2** (16.21 g, 88 mmol) was added dropwise trifluoroacetic acid (TFA) (91.20 g, 800 mmol). After 1 h stirring at 20°C, it was poured into water, neutralized with $NaHCO_3$ and extracted with Et_2O. The dried organic layer was evaporated. Distillation afforded 6.4 g of **3** (80%), bp 90-92°C, n_D^{20} = 1.3851.

6-(2-Tetrahydrothienyl)-valeric acid (6).[4] To a mixture of 6-(2-thienyl)-valeric acid **4** (5.52 g, 30 mmol) and Et_3SiH **5** (7.19 g, 62 mmol) cooled at 0°C was added dropwise a solution of $BF_3.Et_2O$ (1.15 g, 8 mmol) in TFA (30.78 g, 270 mmol). After 20 min stirring at 20°C the volatiles were removed by distillation and the residue recrystallized from hexane to give 3.95 g of **6** (70%), mp 51-51°C.

LADENBURG Pyridine Benzylation

Cu catalyzed α– and γ-benzylation of pyridine.

.HCl + Cl–CH2–Ph —($CuCl_2$, pyr)→ 2-benzylpyridine + 4-benzylpyridine

1 2 3

1	Ladenburg, A.	*Liebigs Ann.*	**1883**	*16*	1410
2	Croock, K.E.	*J. Am. Chem. Soc.*	**1948**	*70*	416
3	Brewster, J.H.	*Org. React.*	**1953**	*7*	135

2 or 4-Benzylpyridine (3).[2] Pyridine hydrochloride **1** (462 g, 4 mol), $CuCl_2$ (5 g, 37 mmol) and pyridine (60 mL) were treated with benzyl chloride **2** (126.5 g, 115 mL, 1 mol) and refluxed 12 h. The mixture was distilled until the temperature reached 195°C. The cooled residue was treated with water (200 mL) and 32% HCl (15 mL), stirred until homogeneous and made alkaline with 25% NH_4OH. Extraction with petroleum ether: PhH, washing the extract with NH_4OH and evaporation of the solvent afforded **3** as a mixture of two isomers, bp 123-125°C and 175-190°C.

LAPWORTH (BENZOIN) Condensation

Condensation of two molecules of aryl aldehydes to an alpha-hydroxy ketone catalysed by CN^- (via cyanohydrins).

1 + **2** $\xrightarrow[\text{heat}]{\text{KCN}}$ **3** (76%)

Ph-CHO + H-C(=O)-C(=O)-H $\xrightarrow{\text{KCN}}$ Ph-CH(OH)-C(=O)-CH(OH)-CN $\longrightarrow$ **5** (93%)

1	Lapworth, A.	*J. Chem. Soc.*	**1903**	*83*	995
2	Buck, J.S.	*J. Am. Chem. Soc.*	**1931**	*53*	2351
3	Hensel, A.	*Angew. Chem.*	**1953**	*65*	491
4	Ide, V.S.	*Org. React.*	**1948**	*4*	269
5	Dahn, H.	*Helv. Chim. Acta.*	**1954**	*371*	309,1612

p-Dimethylaminobenzpiperoin (3).[2] A solution of piperonal **1** (6.0 g, 40 mmol) and p-dimethylaminobenzaldehyde **2** (5.96 g, 40 mmol) in EtOH (30 mL) was treated with a saturated solution of KCN (4.0 g, 61 mmol) in water. After 2 h reflux and 3 days at 20° the crystalline product was filtered and recrystallized from EtOH to give 9.18 g of **3** (76.7%), mp 132°C.

LAWESSON Thiacarbonylation

Reagents for synthesis of thiaamides, thiaesters from the corresponding carbonyl compounds.

Ph-OMe $\xrightarrow{P_4S_7}$ **2**

1 + **2** $\xrightarrow[\text{reflux}]{\text{PhMe}}$ **3** 98%

1	Lawesson, S.O., Scheibye, S.	*Bull. Soc. Chim. Belg.*	**1978**	*87*	293
2	Lawesson, S.O.,	*Tetrahedron*	**1979**	*35*	1339
3	Heimgartner, H.	*Helv. Chim. Acts.*	**1987**	*70*	1001
4	Hoffmann, R. W.	*Angew. Chem.*	**1980**	*42*	559
5	Moriya, T	*J. Med. Chem.*	**1988**	*31*	1197

Dihydro-2(3H)-furanthione (3).[2] Butyrolactone **1** (0.86 g, 10 mmol) and [2,4-bis(4-methoxyphenyl)-1,3-dithia-2,4-diphosphetane-2,4-disulfide] **2** (2.02 g, 5 mmol) in PhMe (10 mL) were heated to reflux until no more starting material was detected (TLC or GC). After cooling, the solvent was stripped off and the mixture was purified on silica gel using CH_2Cl_2 as eluent to afford **3**, mp 96-97°C.

LEBEDEV Methoxymethylation

Methoxymethyl methyl sulfate **3** as an electrophilic reagent for methoxymethylation of alkenes.

$$(CH_3O)_2CH_2 + SO_3 \longrightarrow CH_3O-CH_2-OSO_2OCH_3 \quad \mathbf{3}$$

1 **2**

3 + methylenecyclobutane (=CH₂) **4** —MeOH→ 1-methoxy-1-(CH₂-CH₂OCH₃)cyclobutane **5**

isoprene —3→ CH_3O-CH_2-allyl cation + —Et_3N / MeOH→ CH_3O…OMe + CH_3O…OMe

1	Lebedev, M.Yu.	*Zh. Org. Khim.*	**1987**	*23*	960
2	Lebedev, M.Yu.	*Zh. Org. Khim. USSR (Eng. trans.)*	**1989**	*25*	391
3	Kal yan, Yu.B.	*Izv. Akad. Nauk. SSSR Ser. Khim.*	**1985**	*9*	2082

Methoxymethyl methyl sulfate (3).[1] To a stirred solution of dimethoxy methane **1** (6.08 g, 80 mmol) in CH_2Cl_2 at 60°C was added freshly distilled SO_3 **2** (6.40 g, 80 mmol) to give a solution of **3**.

1-Methoxy-1-(2-methoxyethyl)cyclobutane 5. To the cooled (-20°C) solution of **3** was added methylenecyclobutane **4** (2.04 g, 30 mmol) in CH_2Cl_2 and the mixture was stirred for 30 min. MeOH (3.2 g, 100 mmol) was added, then at 20°C TEA (1.01 g, 100 mmol) was added. The mixture was diluted with Et_2O and washed with NH_4OH. Evaporation of the solvent and distillation gave 2.19 g of **5** (50%), bp 57°C/ 15 mm, n_D^{20} = 1.4348.

LEHMSTED - TANASESCU Acridone Synthesis

Acridone synthesis from o-nitrobenzaldehyde and aryls.

CHO, NO_2, Cl + Cl → (H_2SO_4, $NaNO_2$ 6d) O, Cl, Cl, N H

1 **2** **3** (53%)

1	Tanasescu, I.	*Bull. Soc. Chim. Fr.*	**1927**	*41*	528
2	Lehmsted, K.	*Chem. Ber.*	**1932**	*65*	834
3	Spalding, D.P.	*J. Am. Chem. Soc.*	**1946**	*68*	1596
4	Silberg, I.	*Rev. Roum. Chim.*	**1965**	*10*	1035

3,6-Dichloroacridone (3)[3] A mixture of 2-nitro-4-chlorobenzaldehyde **1** (18.5 g, 0.1 mol) chlorobenzene **2** (78.7 g, 0.7 mol), conc. H_2SO_4 (37.5 mL) and $NaNO_2$ (0.35 g) was alternatively shaken for 9 h and allowed to stand 15 h, for a total of 6 days. At the end of each two-day period a mixture of H_2SO_4 (10 mL) and $NaNO_2$ (0.1 g) was added. The mixture was poured into water (500 mL) and steam distilled until no further aldehyde solidified in the condenser. The residue from the steam distillation was filtered and digested with PhH, leaving 14 g of **3** (53%).

LEHN Cryptand Synthesis

Synthesis of diaza-polyoxa-macrobicyclic compounds (cryptands).

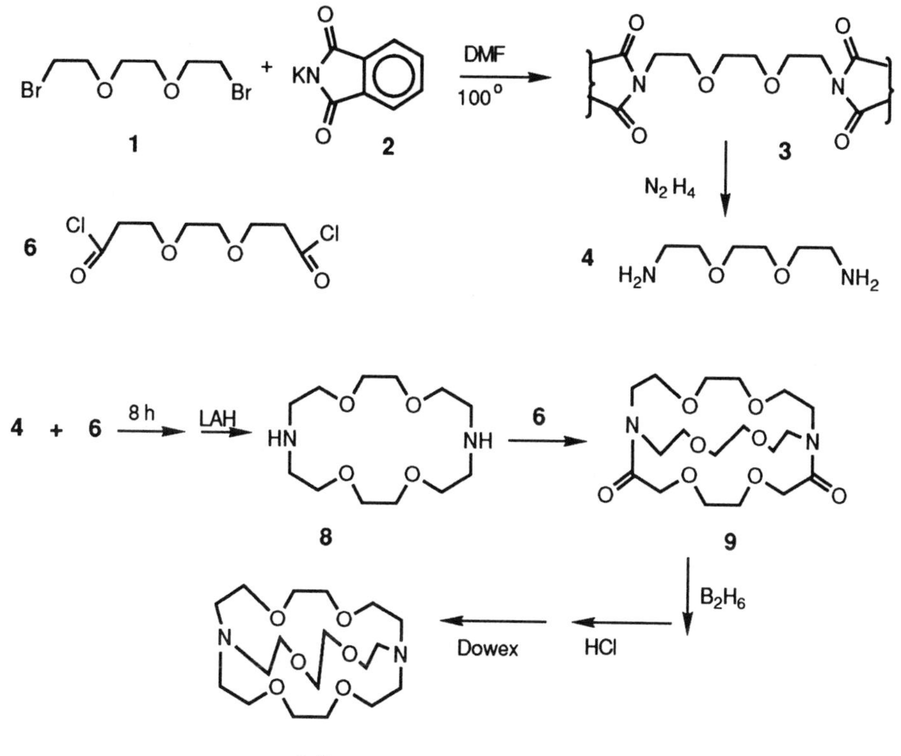

1	Lehn, J.M.	*J. Chem. Soc. Chem. Commun.*	**1972**		1100
2	Lehn, J.M.	*Tetrahedron*	**1973**	*29*	1624
3	Lehn, J.M.	*Angew. Chem. Int. Ed.*	**1974**	*13*	406, 611
4	Echegoyen, L.	*J. Org. Chem.*	**1991**	*56*	1524
5	Lehn, J.M.	*Angew. Chem. Int. Ed.*	**1990**	*29*	1304

cont. on next page

Diamine (4).[2] To a suspension of potassium phtalimide **2** (260 g, 1.5 mol) in DMF (1000 mL) was added bromo ether **1** (217 g, 0.78 mol). After stirring 4 h at 100°C, the mixture was poured into 3 vol of ice and 5% NaOH. The cake was washed (water, Me_2CO). Recrystallization (EtOH) gave 204 g of **3** (85%), mp 104.5-105°C.
3 (200 g, 0.65 mol) in EtOH (500 mL) was refluxed with hydrazine hydrate (50 mL) for 10 h, treated with 6N HCl to pH=1, heated to reflux for 30 min, cooled, filtered and the precipitate was washed with EtOH. The filtrate and washings were concentrated. The solid was dissolved in water, treated with solid KOH and extracted continuously for a few days with PhH to give 67.2 g of **4** (70%), bp 115°C (0.2 mm).

Macrocyclic diamine (8). **4** (14.8 g, 68 mmol) in PhH (500 mL) and **6** (10.8 g, 50 mmol, from oxalyl chloride reaction of the diacid), in PhH (500 mL) were added dropwise simultaneously with stirring to PhH (1200 mL) over 8 h. Evaporation of the solvent and chromatography on alumina (PhH and PhH:$CHCl_3$ 95:5) gave 15.2 g of a diamide **7** (75%). To a suspension of LAH (7.6 g, 0.2 mol) in THF (120 mL) was added dropwise **7** (13.8 g, 47 mmol) via a Soxhlet. Work up afforded 8.02 g of **8** 75%, mp 115-116°C.

Tricyclic diamide (9). From diamide **8** (5.24 g, 20 mmol), acyl chloride **6** (4.4 g, 20 mmol) and TEA (4.5 g) using high dilution as for **7** over 11 h, was isolated 3.67 g of **9** (45%), mp 114-115°C (PhH - petroleum Ether).

[2,2,2]-Cryptand (10). Cooled **9** (10 g, 26 mmol) in THF (100 mL) was treated with 1.5 M diborane (100 mL); the mixture was heated to 25°C over 30 min and refluxed for 2 h. The solvent was removed in vacuum, the crude amine borane complex was refluxed with 6 N HCl (100 mL) for 3 h and evaporated to dryness. The residue was dissolved in a minimum of water and passed through Dowex-1 (50-100 mesh, as HO^-). Concentration to dryness of the alkaline and neutral eluent and azeotropic elimination of water gave 9.28 g of **10** (95%), mp 68-69°C (hexane).

LEUCKART Thiophenol Synthesis

Formation of thiophenols from diazonium salts and xanthates.

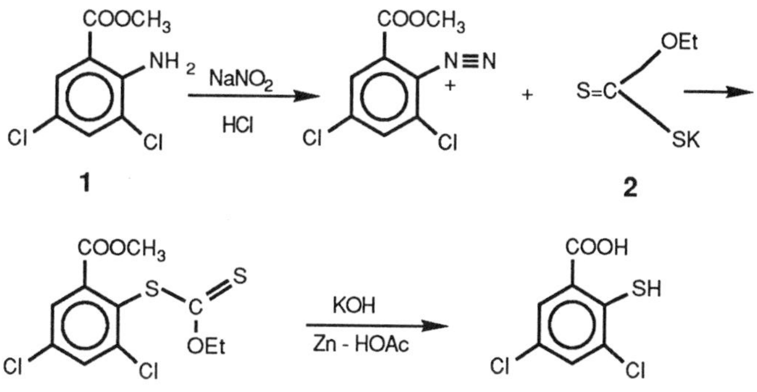

3 (70%)

1 Leuckart, R. *J. Prakt. Chem.* **1890** *41* 187(2)

2 Bourgeoise, E. *Rec. trav. Chim.* **1899** *18* 447

3 Tarbel, D.S. *J. Am. Chem. Soc.* **1952** *74* 48

3,5-Dichlorothiosalicylic acid (3).[3] Methyl 3,5-dichloro-2-aminobenzoate **1** (43.9 g, 0.2 mol) in 32% HCl (50 mL) was diazotized with $NaNO_2$ (17.1 g, 0.25 mol) and stirred for 1 h at 5°C. Water (30 mL) was added, the solution was neutralized with $NaHCO_3$ (350 g) at 0°C and back-titrated with HCl to congo red. The cold solution was added to stirred potassium ethyl xanthate **2** (64.1 g, 0.5 mol) in water (85 mL) at 40-45°C. The mixture was stirred for 30 min, extracted with Et_2O and the solvent removed. The residue was refluxed for 8 h with KOH (46.6 g, 0.83 mol) in EtOH (150 mL). The EtOH was evaporated, the solution washed with Et_2O, the aqueous solution was acidified, the product collected and refluxed for 24 h with Zn/AcOH and finally recrystallized from EtOH to afford 31 g of **3** (70%), mp 207-208°C.

LEUCKARDT - PICTET - HUBERT Phenanthridine synthesis

Amidation of aryls by isocyanates (Leuckardt) or by amides (Pictet-Hubert), catalyzed by Lewis acids and leading to phenanthridines.

NCO $\xrightarrow[70^{o}]{AlCl_3}$ N, OH

1 **2** (73%)

CHO, NH $\xrightarrow[\Delta\ \ 30\ min]{POCl_3,\ SnCl_4}$ N

3 **4** (86%)

1	Leuckardt, R.	*Chem. Ber.*	**1885**	*18*	873
2	Buttler, J.M.	*J. Am. Chem. Soc.*	**1949**	*71*	2578
3	Schmutz, I.	*Helv. Chim. Acta.*	**1965**	*48*	336
4	Pictet, A., Hubert, A.	*Chem. Ber.*	**1896**	*29*	1182
5	Boyer, J.H.	*Synthesis*	**1978**		205
6	Eisch, J.	*Chem. Rev.*	**1957**	*57*	525

Hydroxyphenanthridine (2) To a suspension of $AlCl_3$ (37 g, 0.25 mol) in o-dichlorobenzene (190 mL) was added o-biphenyl isocyanate **1** (48.8 g, 0.20 mol) over 15 min at 70-80°C. Stirring was continued for 1 h at 25°C. The crude product was filtered, washed with Et_2O and dried to give 38 g of **2** (77%), which after recrystallization from HOAc yielded 35.6 g of pure **2** (72.6%), mp 292.5-293-5°C.

LEUCKART - WALLACH Reductive Amination

Reductive amination of carbonyl groups with amines and formic acid or H_2-Ni (Miquonac) or $NaBH_4$ (see Borch).

CHO
1
HN O
HCOOH
N O
2 (95%)

C_6H_5-CHO + NH_3 + H_2 —90 atm / Ni→ C_6H_5-CH_2-NH_2

3 4 (89%)

1	Leuckart, R.	*Chem. Ber.*	**1885**	*18*	2341
2	Wallach, O.	*Liebigs Ann.*	**1892**	*272*	100
3	Marcus, E.	*J. Org. Chem.*	**1960**	*25*	199
4	Moore, M.I.	*Org. React.*	**1949**	*5*	301
5	Miquonac, G.	*C.R.*	**1921**	*172*	223
6	Raudvere, F.	*Ann. farm. bio (Buenos Air)*	**1943**	*18*	81

N-(1-Pyrenylmethyl)morpholine (2).[3] Morpholine (65.5 g, 0.75 mol) and 90% HCOOH (38.5 g, 0.75 mol) were heated slowly to 200°C (to distill water, unchanged amine and acid). To 82 g of the residue were added 1-pyrenecarboxaldehyde **1** (23.0 g, 0.1 mol) and 90% HCOOH (5 mL). After 4 h reflux (182-185°C), N-formylmorpholine was vacuum distilled. The residue was dissolved in Et_2O, filtered and dry HCl was bubbled through to give 32.3 g of **2** .HCl (95%), mp 256-263°C, **2**, mp 90=93°C.

Benzylamine (4).[6] NH_3 (51.0 g, 3 mol), EtOH (300 mL), **3** (318 g, 3 mol) and Raney nickel 10 g) were hydrogenated at 90 atm and 40°C. Filtration and distillation gave 287 g of **4** (89%), bp 70-80°C/8 mm.

LEY - GRIFFITH Oxidation reagent

Oxidation of alcohols to carbonyl compounds with a perruthenate catalyst and N-methylmorpholine - N-oxide (NMO), in the presence of other functional groups.

OH O → NMO, $Pr_4N\ RuO_4$ (TPAP) → O O OTBDPS

1 OTBDPS **2** (70%)

HO Br → NMO, TPAP → O Br

1 Ley, S.V., Griffith, W.P. *J. Chem. Soc. Chem. Commun.* **1987** 1625

2 Ley, S.V., Griffith, W.P. *Tetrahedron Lett.* **1989** *30* 3204

3 Ley, S.V. *Aldrichimica Acta* **1990** *23* 13

Oxirane aldehyde (2).[1] Alcohol **1** (TBDPS = tert butyldiphenylsilyl) (0.5 mmol) in CH_2Cl_2 (5 mL) containing 4Å molecular sieves and 4-methylmorpholine-N-oxide (NMO) (0.1 g, 0.75 mmol) was stirred for 10 min. Tetra-n-propylammonium perruthenate (TPAP) (8.3 mg, 0.025 mmol) was added and the reaction followed by TLC until complete. The mixture was diluted with CH_2Cl_2 , washed with Na_2SO_3 solution (10 mL), brine (10 mL) and with saturated $CuSO_4$ solution (10 mL). The organic layer was filtered and worked up to give **2** (70%).

L I E B E N Hypohalide Oxidation

Oxidation of methyl ketones with hypochloride (or hypobromide) to carboxylic acids and chloroform; with NaOH and iodine iodoform is formed.

1 ($COCH_3$) —KOCl, 60-70°C→ 2 (CO_2H) (87%)

1 Lieben, A.	*Liebigs Ann. Suppl.*	**1870**	*7*	218
2 Fieser, L.F.	*J. Am. Chem. Soc.*	**1936**	*58*	1055
3 Fusson, R.C.	*Chem. Rev.*	**1934**	*15*	275

2-Naphthoic acid (2).[2] To $Ca(OCl)_2$ 65% (250 g) in water (1000 mL) was added K_2CO_3 (175 g, 1.27 mol) and KOH (50 g, 0.9 mol) in water (500 mL) with stirring at 55°C. To this was added **1** (85 g, 0.5 mol) at 60-70°C by cooling. After 1 h the oxidant was destroyed by Na_2S (50 g in water 200 mL) to give **2**, 85 g, mp 181-183°C, from EtOH, 75 g of **2** (87%), mp 184-185°C.

L I E B I G Benzylic Acid Rearrangement

Benzylic acids by rearrangement of diketones (also α-ketol rearrangement).

Ph—C(=O)—C(=O)—Me (1) —10% NaOH, Et_2O 0° 1 h→ Ph—C(=O)—C(Me)(OH)—O⁻ → Ph—C(OH)(Me)—COOH 2 (25%)

1 Liebig, v.J.	*Liebigs Ann.*	**1838**	*25*	27
2 Warren, K.S.	*J. Org. Chem.*	**1963**	*28*	2152
3 Houber, G.	*Angew. Chem.*	**1951**	*63*	501
4 Eastham, J.F.	*Quart. Rev. Chem. Soc.*	**1960**	*14*	221

L O S S E N Rearrangement

Rearrangement of O-acyl hydroxamic acid derivatives with base or heat to amines or urea derivatives (via isocyanates), or rearrangement of carboxylic acids via their hydroxamic acids to amines.

$$\underset{\mathbf{1}}{Ph-\overset{O}{\overset{\|}{C}}-\underset{K}{N}-O-\overset{O}{\overset{\|}{C}}-Ph} \xrightarrow[\mathbf{2}]{H_2N-C_4H_9(n)} Ph-N=C=O \xrightarrow{\mathbf{2}} \underset{\mathbf{3}\ (83\%)}{Ph-NH-\overset{O}{\overset{\|}{C}}-NH-C_4H_9(n)}$$

$$\text{2-naphthalene-COOH} + H_2N\text{-}OH \xrightarrow{PPA} \text{2-naphthalene-}NH_2 \quad (82\%)$$

1	Lossen, W.	*Liebig Ann. Chem.*	**1869**	*150*	314
2	Brend, D.C.	*J. Org. Chem.*	**1966**	*31*	976
3	Popp, F.V.	*Chem. Rev.*	**1958**	*58*	374
4	Cohen, L.A.	*Angew. Chem.*	**1961**	*73*	260
5	Snyder, H.R.	*J.Amer.Chem.Soc.*	**1953**	*75*	2014
6	Ulrich, H.	*J. Org. Chem.*	**1978**	*43*	1544

N-n-Butyl-N'-phenyl urea (3).[2] A solution of potassium O-benzoylbenzhydroxamate **1** (0.73 g, 2.6 mmol) and butylamine **2** (3.0 g, 41.1 mmol) in DMSO (30 mL) was stirred at 30°C. The residue obtained after evaporation of the solvent was extracted with hot PhH to afford 0.435 g of **3** (83%), mp 125-128°C (from PhH).

MACDONALD Porphyrine Synthesis

Porphyrine synthesis from dipyrrolemethanes.

2 + 1 —HI - HOAc, air→ 3 (11%)

1	MacDonald, S.P.	*J. Am. Chem. Soc.*	**1960**	*82*	4384
2	Clesy, P.S.	*Austr. J. Chem.*	**1965**	*18*	1835
3	Chang, C.K.	*J. Org. Chem.*	**1981**	*46*	4610

5,8-Dimethyl-6,7-bis[2-(methoxycarbonyl)ethyl]porfine (3).[3] [5,5-Diformyl-4,4-dimethyl-3,3'-bis[2-(methoxycarbonyl)ethyl]-2,2'-dipyrrylmethane **1** (420 mg), 1 mmol) and 2,2'-dipyrrylmethane **2** (150 mg, 1 mmol) dissolved separately in AcOH (150 mL) were mixed and diluted with AcOH (200 mL) containing 56% HI (3.5 mL). The reaction proceeded in the dark for 1 h. Anhydrous NaOAc (10 g) was added and the mixture was aerated for 30 h in the dark and the solvent evaporated. The residue was refluxed with MeOH (300 mL), ethyl orthoformate (20 mL) and H_2SO_4 (3 mL). After evaporation of the solvent and work up, the product was chromatographed (silica gel, CH_2Cl_2). The red fraction was collected and recrystallized from CH_2Cl_2:MeOH to give 15 mg of **3** (10.8%), mp 205°C.

MADELUNG Indole Synthesis

Indole synthesis by cyclization of N-acyl-o-toluidines.

Cl Me O N H BuLi Cl N H

1 **2** (94%)

CH_2 - CN O N-C-Ph CH_2-Ph BuLi CN Ph N CH_2Ph

1	Madelung, W.	*Chem. Ber.*	**1912**	*45*	1128
2	Pichat, L.	*Bull. Soc. Chim. Fr.*	**1954**		85
3	Hertz, W.	*J. Org. Chem.*	**1960**	*25*	2242
4	Houlihan, W.J.	*J. Org. Chem.*	**1981**	*46*	4511

5-Chloro-2-phenylindole (2).[3] A stirred solution of N-(4-chloro-2-methylphenyl)-benzamide **1** (12.25 g, 50 mmol) in THF (100 mL) was maintained under N_2 at 0°C and treated with n-butyllithium in hexane (0.1-0.15 mol). The stirred mixture was kept at room temperature for 15 h, treated with 2N HCl (60 mL) under ice cooling and extracted with PhH. The combined organic layers were concentrated to dryness to afford 10.6 g of **2** (94%), mp 195-196°C (from Et_2O-PhH).

MAKOSZA Vicarious nucleophilic substitution

Synthesis of alkyl substituted heterocycles by "vicarious" nucleophilic substitution of hydrogen in heteroarenes, nitroderivatives of heteroarenes.

O_2N-pyridine **1** + $CHCl_3$ **2** → (NaOMe; DMF / NH_3 liq.) → O_2N, $CHCl_2$-pyridine **3** (72%)

Cl, Cl-pyridazine → ($ClCH_2$ Ts; t Bu OK/DMF) → Ts-CH_2, Cl, Cl-pyridazine (98%)

O_2N-C6H4-NC → (PhS CH_2CN; NaOH / NH_3) → O_2N, CN-indole (N-H) (50%)

1	Makosza, M	*J. Org. Chem.*	**1983**	*48*	3860
2	Makosza, M.	*J. Org. Chem.*	**1989**	*54*	5094
3	Makosza, M.	*Acc. Chem. Res.*	**1987**	*20*	282
4	Makosza, M.	*Russian Chem. Rev.*	**1989**	*58*	747
5	Makosza, M.	*Synthesis*	**1991**		103

2,2-Dichloromethyl-4-nitropyridine (3).[2] A solution of 3-nitropyridine **1** (372 mg, 3 mmol) and $CHCl_3$ (395 mg, 3.3 mmol) in DMF (2 mL) was added dropwise to a vigorously stirred mixture of NaOMe (650 mg, 12 mmol) in liquid ammonia (10 mL) at -70 to-68°C. After 1 min of stirring, NH_4Cl (1.5 g) was added, ammonia was evaporated, water (50 mL) was added to the residue and usual work up afforded 447 mg of **3** (72%).

MALAPRADE - LEMIEUX - JOHNSON Olefin (diol) cleavage

Oxidative cleavage of 1,2-glycols to two carbonyls (Malaprade) or direct oxidation of olefins by IO_4^- and OsO_4 catalyst (Lemieux-Johnson).

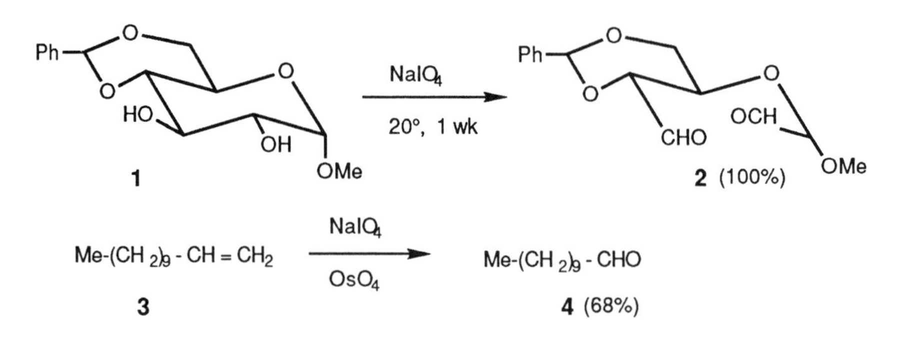

1	Malaprade, L.	*Bull Soc. Chim. Fr.*	**1828**	*43*	683
2	Baddiley, J.	*J. Chem. Soc.*	**1954**		3826
3	Fatiatide, A.J.	*Synthesis*	**1974**		229, 255
3	Jackson, E.I.	*Org. React.*	**1944**	*2*	341
5	Lemieux, R.U.	*Anal Chem.*	**1954**	*26*	920
6	Lemieux, R.U., Johnson, W.S.	*J. Org. Chem.*	**1956**	*21*	478
7	Rapoport, H.	*J. Am. Chem. Soc.*	**1958**	*80*	5767
8	Djerassi, C.	*J. Am. Chem. Soc.*	**1962**	*84*	2990
9	Henbest, N.R.	*Chem. Commun.*	**1968**		1036

Dialdehyde (2).[2] Glucoside **1** (2.8 g, 10 mmol) in water (750 mL) was treated with $NaIO_4$ (2.14 g, 10 mmol) and kept for a week at 20°C. Filtration gave 2.9 g of **2** (100%), mp 142°C.

Undecanal (4). Water (5 mL), dioxane (15 mL), dodecene-1 **3** (0.71 g, 4.2 mmol) and OsO_4 (11.3 mg, 0.044 mmol) were stirred for 5 min. Powdered $NaIO_4$ (2.06 g, 9.6 mmol) was added over 30 min and the slurry stirred for 90 min. The mixture was extracted with Et_2O and **4** was isolated as the 2,4-DNPH, 0.96 g, mp 102-106°C, second crop 0.14 g, total yield 68%.

MANNICH - ESCHENMOSER Methylenation reagent

Aminomethylation of activated methyl or methylene groups by in situ formed $Me_2N^+=CHR$ (Mannich). $Me_2N^+=CH_2\ X^-$ reagent for α-methylenation (Eschenmoser).

$$\mathbf{1} + CH_2O\ (\mathbf{2}) + H_2N^+Me_2Cl^-\ (\mathbf{3}) \xrightarrow[4\,h]{EtOH} \mathbf{4}\ (90\%)$$

Me, O, CH_2-NMe_2

$$Me_3N\ (\mathbf{5}) \xrightarrow{CH_2I_2\ (\mathbf{6})} Me_3\overset{+}{N}\text{-}CH_2I\ I^-\ (\mathbf{7}) \xrightarrow{160°} Me_2\overset{+}{N}=CH_2\ I^-\ (\mathbf{8})\ (81\%) \xrightarrow[LDA,\ MeI]{\mathbf{9}} \mathbf{10}\ (61\%)$$

1	Mannich, C.	*Arch. Phram.*	**1912**	*250*	647
2	House, H.O.	*J.Org. Chem.*	**1964**	*29*	1339
3	Tramontini, M.	*Synthesis*	**1973**		703
4	Dimnoch, D.	*Die Parmazie*	**1986**	*91*	284
5	Blicke, F.F.	*Org. React.*	**1942**	*1*	303
6	Eschenmoser, A.	*Helv. Chim. Acta.*	**1969**	*52*	1823
7	Eschenmoser, A.	*Angew. Chem. Int. Ed.*	**1971**	*10*	330
8	Roberts, J.L.	*Tetrahedron Lett.*	**1977**		1621

Dimethyl(methylene)ammonium iodide (Eschenmoser's salt) (8).[7] **5** (20 g, 0.36 mol), CH_2I_2 **6** (120 g, 0.73 mol) and EtOH were kept closed in the dark for 100 h at 20°C. Filtration, washing and drying for 1 h at 70°C in high vacuum afforded 98 g of **7** (89%), mp 190°C dec. **7** (40 g, 0.122 mol) in sulfolane (120 mL) was heated under N_2 to 160°C and CH_3I was distilled. Filtration, washing (CCl_4) and drying to 50°C in vacuum, gave 18.4 g of **8** (81%), mp 240° (dec).

α-Methylenebutyrolactone (10).[7] iPr_2NH(2.02 g, 20 mmol) in THF (20 mL) at 4°C, and 2.55 M of BuLi in hexane (20 mmol) was stirred for 15 min, at -78°C. Lactone **9** (1.6 g, 19 mmol) and **8** (7.4 g, 40 mmol) was added. At 20°C solvent was removed in vacuum. The residue in MeOH was treated with MeI and stirred for 24 h. Chromatography gave 1.21 g of **10** (61%).

MARKOVNIKOV Regioselectivity

Description of selectivity during addition of unsymmetrical reagents to unsymmetrical olefins. H-X adds selectively with H forming a bond to the less substituted olefin carbon (Markovnikov). Now supplanted by the general term regioselectivity introduced by A. Hassner, denoting selectivity in bond making between an unsymmetrical reagent X-Y and an unsymmetrical substrate (now includes regioselective (o,m,p)-substitution and also applied to bond breaking reactions (regioselective elimination).

$(CH_3)_3$ Si (**1**) + H Br —20°→ $(CH_3)_3$ Si … Br (**2**)

3 + Br N_3 —pentane (free rad.)→ … Br … N_3

3 + Br N_3 —Me CN (ionic)→ C_4H_9 … Br^+ → … N_3 … Br

regioselective additions

… OAc —400°→ …

1	Markovnikov, W.	*Liebigs Ann.*	**1870**	*153*	256
2	Hassner, A.	*J. Org. Chem.*	**1969**	*34*	2628
3	Sommer, L.S.	*J. Am. Chem. Soc.*	**1948**	*70*	2872
4	Nunget, W.A.	*J. Org. Chem.*	**1985**	*50*	5370
5	Kharash, M.	*J. Am. Chem. Soc.*	**1940**	*62*	811
6	Mayo, F.R.	*Chem. Rev.*	**1940**	*27*	351
7	Stasey, F.M.	*Org. React.*	**1963**	*13*	155

MARSCHALCK Aromatic alkylation

Alkylation of quinones or aminoquinones with aldehydes.

1 → 2 ($Na_2S_2O_4$, CH_2O, 40°) (90%)

OCH - CO_2H, MeOH → CH_2CO_2Me (33%)

1	Marschalck, C.	*Bull. Soc. Chim. Fr.*	**1936**	*3*	1545
2	Marschalck, C.	*Bull. Soc. Chim. Fr.*	**1939**	*6*	655
3	Brockmann, H.	*Chem. Ber.*	**1958**	*91*	1920
4	Havlincova, L.	*J. Chem. Soc.*	**1970**		657
5	Krohn, E.	*Angew. Chem. Int. Ed.*	**1979**	*18*	621

1-Amino-2-methylantraquinone (2).[4] 1-Aminoantraquinone **1** (1.56 g, 6.8 mmol), was added to a solution of sodium dithionite (4.42 g, 25 mmol) and 30% NaOH (9.7 mL). The mixture was stirred at 40°C for 30 min, then aqueous 35% formaldehyde (0.55 mL) was added. The mixture was stirred for another 5 hr at 40°C, air was passed through and the product was filtered, washed with water and dried to give 1.45 g of **2** (90%), mp 200°C. (It is advantageous to carry out the reaction under N_2).

MARTIN Dehydrating reagent

Sulfurane reagent for conversion of trans diols to epoxides, generally for dehydration of diols to olefins or cyclic ethers, and as an oxidizing agent.

Cl OH OH $Ph_2S(OC[CF_3]_2Ph)_2$ Cl O

1 **2** **3** (93%)

1	Martin, J.C.	*J. Am. Chem. Soc.*	**1971**	*93*	4327
2	Martin, J.C.	*J. Am. Chem. Soc.*	**1974**	*96*	4604
3	Martin, J.C.	*J. Am. Chem. Soc.*	**1977**	*99*	3511
4	Bartlett, P.D.	*J. Am. Chem. Soc.*	**1980**	*102*	3515
5	Eschenmoser, W.	*Helv. Chim. Acta*	**1982**	*65*	353
6	Burnett, D.A.	*J. Am. Chem. Soc.*	**1984**	*106*	8201
7	Martin, J.C.	*Organic Synthesis*	**1977**	*57*	22

3-β-Chloro-5α,6α-epoxycholestane (3).[2] 3β-Chloro-5α,6β-cholestandiol **1** (465 mg, 1.06 mmol) was added to a solution of bis α,α-bis(trifluoromethyl)benzenemethanolato)-diphenylsulfur **2** (1.2 g, 1.8 mmol) in Et_2O. After stirring for 10-20 min at 20° the phenyl di(trifluoromethyl)carbinol was removed by extraction with 20% KOH, after the Et_2O was replaced by CH_2Cl_2 (because of the solubility of the carbinol in Et_2O). After recovery of the polar sulfoxide (silica gel, Et_2O:pentane 1:8) one obtains 417 mg of **3** (93%), mp 97-97.5°C (EtOH anh).

M A S C A R E L L I Fluorene Synthesis

Synthesis of fluorenes from 2-amino-2'-alkylbiphenyls via diazonium ions.

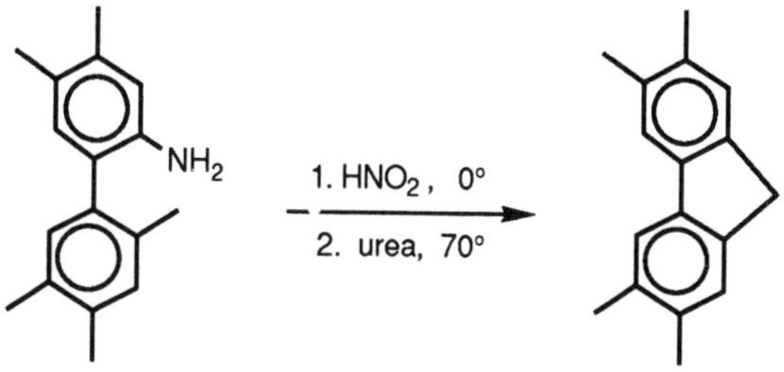

1. HNO_2, 0°
2. urea, 70°

1 → **2** (40%)

1	Mascarelli, L.	*Gazz. Chim. Ital.*	**1936**	*66*	843
2	Cohen, T.	*J. Am. Chem. Soc.*	**1964**	*86*	2514
3	Puskas, I.	*J. Org. Chem.*	**1968**	*3*	4237

2,3,6,7-Tetramethylfluorene (2).[3] A mixture of 2,4,4',5,5'-pentamethyl-2'-aminodiphenyl **1** (5.35 g, 20 mmol) and 3% H_2SO_4 (250 mL) was stirred and warmed to form a fine suspension of amine salt. The suspension was rapidly cooled and was diazotized at -2° to 0°C with a solution of $NaNO_2$ (1.7-2.0 g, 30 mmol) in water (12 mL) and the reaction mixture was stirred for an additional 1 h at 0°C. The diazonium salt solution was filtered and heated with 5 g of urea at 70-75°C. Extraction with Et_2O and evaporation of the solvent gave 1.80 g of **2** (40%), mp 171°C.

MATTOX - KENDALL Dehydrohalogenation

Dehydrohalogenation of α-haloketones with 2,4-dinitrophenylhydrazine or LiCl-DMF.

2,4-DNPH

2,4-DNPH

1

2 (88%)

1	Mattox, V.R., Kendall, E.C.	*J. Am. Chem. Soc.*	**1948**	*70*	882
2	Djerassi, C.	*J. Am. Chem. Soc.*	**1953**	*75*	3500
3	Warnhof, E.W.	*J. Org. Chem.*	**1963**	*28*	887

1-Cholesten-3-one2,4-dinitrophenylhydrazone (2).[2] To a hot solution of 2-chlorocholestane-3-one **1** (0.5 g, 1.5 mmol) in AcOH (20 mL) was added a solution of 2,4-dinitrophenylhydrazine (0.5 g, 1.5 mmol) in AcOH (5 mL). After refluxing for 3 min under N_2 the mixture was cooled and the precipitate was collected to afford 0.57 g of **2** (88%), mp 220-222°C.

McCORMACK - KUCHTIN - RAMIREZ Phosphole Synthesis

Formation of phospholes from butadienes (McCormack) or of dioxaphospholes from 1,2-diketones (Kuchtin-Ramirez).

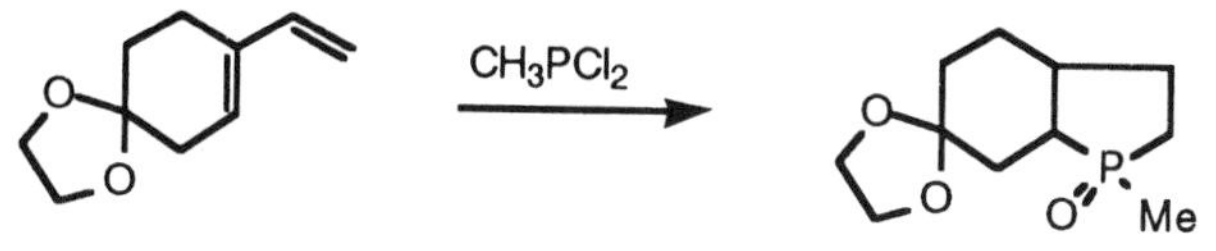

CH_3PCl_2

1 → **2** (72%)

$(EtO)_2$ $POSiMe_3$, **4**, 20°

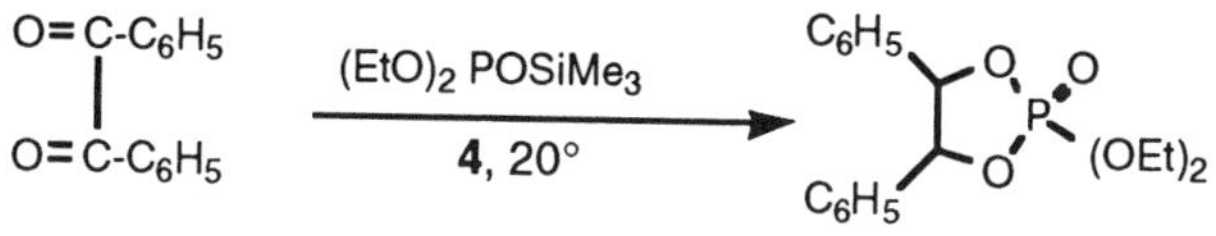

3 → **5** (65%)

1	McCormack, W.B.	*U.S. Pat.*	2.663.736		2.663.737
2	Hajos, A.G.	*J. Org. Chem.*	**1956**	*30*	1213
3	Quin, L.D.	*J. Org. Chem.*	**1981**	*46*	461
4	Kuchtin, V.A.	*Doklad. Akad. Nauk. USSR*	**1958**	*121*	466
5	Ramirez, F.	*J. Am. Chem. Soc.*	**1960**	*82*	2651
6	Mitsuo, S.	*J. Org. Chem.*	**1981**	*46*	4030

6,6-(Ethylenedioxyl)-1-methyl-Δ^3-2,4,5,6,7,7a-hexahydro-1H-phosphindole-1-oxide (2).[3]

A hexane solution of 1-vinyl-4,4-ethylendioxycyclohex-1-ene **1** (20.0 g, 0.12 mol) and freshly distilled CH_3PCl_2 (18.3 g, 0.156 mol) and copper stearate (0.4 g) as a polymerization inhibitor, was allowed to stand at rt for 5 days. The adduct was filtered and the filtrate was kept for 22 days to get more cycloaduct. The first crop was hydrolyzed by $NaHCO_3$, extracted with $CHCl_3$, and after evaporation gave 19.36 g of **2** (72.7%), mp 156-157°C.

McFADYEN - STEVENS Ester Reduction

Reduction of esters to aldehyde via hydrazides.

1 (pyridine-2-COOEt) —NH_2NH_2→ 2 (pyridine-2-CO-NH_2NH_2) —$PhSO_2Cl$→ 3 (pyridine-2-$CONHNHSO_2Ph$) (100%) —Na_2CO_3, 160°, 2 min→ 4 (pyridine-2-CHO) (20%)

1	McFadyen, J.S., Stevens, T.S.	*J. Chem. Soc.*	**1936**		584
2	Nieman, C.	*J. Am. Chem. Soc.*	**1942**	*62*	1681
3	Martin, C.B.	*J. Org. Chem.*	**1974**	*39*	2285
4	Ferguson, L.H.	*Chem. Rev.*	**1946**	*38*	244
5	Mosettig,E.	*Org. React.*	**1954**	*8*	232

Picolinic aldehyde (4).[2] Benzenesulfonyl chloride (6.92 g, 40 mmol) was added with stirring to a chilled solution of picolinic hydrazide **2** (4.62 g, 33 mmol) in pyridine (35 mL). After 1 h the solvent was removed in vacuum, the residue stirred with water, filtered and the solid washed with EtOH and Et_2O to afford 10.5 g of **3** (100%), mp 202-203°C.

A mixture of **3** (25.0 g, 0.18 mol), anhydrous Na_2CO_3 (24.0 g, 0.226 mol) in glycerol (100 mL) was heated to 160°C and maintained for 2 min at this temperature. To the cooled mixture, water (100 mL) was added, the solution was saturated with $NaNO_3$, filtered and the filtrate extracted continuously for 10 h with Et_2O. Evaporation of the solvent afforded 1.99 g of **4** (20%).

MCMURRY Coupling

Formation of olefins by coupling or cross coupling of ketones, mediated by low valent titanium. Also coupling of enol ethers of 1,3-dicarbonyl compounds.

HO
+
Zn, $TiCl_4$
DME
1
2
3 (98%)

EtO
R
Ti
R
O
O
R

1	McMurry, J.E.	*J. Am. Chem. Soc.*	**1974**	*96*	1708
2	McMurry, J.E.	*Acc. Chem. Res.*	**1983**	*16*	405
3	Coe, P.L.	*J. Chem. Soc. Perkin I*	**1986**		475
4	Dormagen, W., Breitmaier, E.	*Chem. Ber.*	**1986**	*119*	1734
5	Breitmann, E.	*Synthesis*	**1987**		96

1-(4-Hydroxyphenyl)-1,2-diphenyl-1-butene (3).[2] To a stirred suspension of powdered Zn (9.0 g, 0.138 at) in dry dimethoxyethane (DME) at -10°C under N_2, $TiCl_4$ (12.9 g, 7.5 mL, 68 mmol) was added via a syringe. The mixture was refluxed for 2 h. To the cold suspension was added **1** (4.5 g, 22 mmol) and **2** (3.0 g, 22 mmol) in dry DME (30 mL). The mixture was refluxed for 4 h, cooled to 20°C and poured into 10% K_2CO_3 solution (300 mL). After Et_2O extraction (3x100 mL) and evaporation, the light brown oil crystallized on standing to afford 6.5 g of **3** (98%), mp 109-111°C. Ratio of E:Z = 7:1.

MEERWEIN Alkylation reagent

$R_3O^+BF_4^-$ reagent for O-alkylation of amides.

$Et_3O^+BF_4^-$

1

OC_2H_5

2 (76%)

$Et_3O^+BF_4^-$

OC_2H_5 (80%)

1	Meerwein, H.	*J. Prakt. Chem.*	**1937**	*147*	17
2	Eschenmoser, A.	*Pure Appl. Chem.*	**1963**	*7*	297
3	Fujita, A.	*Chem. Pharm. Bull. (Tokyo)*	**1965**	*13*	1183
4	Ayers, W.A.	*Can. J. Chem.*	**1967**	*45*	451
5	Curphey, T.J.	*J. Org. Chem.*	**1966**	*31*	1199
6	McMinn, D.G.	*Synthesis*	**1976**		824

2,6-Dimethyl-4-ethoxypyrimidine (2).[6] 2,6-Dimethyl-4-hydroxypyrimidine **1** (23.0 g, 185 mmol) was added in small portions to dry CH_2Cl_2 (100 mL) containing triethyloxonium tetrafluoroborate (Meerwein reagent) (38.5 g, 203 mmol). The mixture was stirred under N_2 overnight and then added to 10% Na_2CO_3 (500 mL). The organic layer was separated, washed with Na_2CO_3 solution (3x50 mL), dried and solvent evaporated. Distillation of the residue afforded 21.4 g of **2** (76%), bp 68°C/2.7 torr.

MEERWEIN - PONNDORF - VERLEY Reduction

Reduction of carbonyl groups to alcohols by means of $Al(iPrO)_3$ and isopropanol.

$Al(OiPr)_3$

iPrOH

1

2 (76%)

$Al(iPrO)_3$

OH

(90%)

1	Meerwein, H.	*Liebigs Ann.*	**1925**	*444*	221
2	Ponndorf, W.	*Angew. Chem.*	**1926**	*39*	138
3	Verley, A.	*Bull. Soc. Chim. Fr.*	**1925**	*37*	537(4)
4	Snyder, C.H.	*J. Org. Chem.*	**1970**	*35*	264
5	Ashby, E.C.	*J. Org. Chem.*	**1986**	*51*	3593
6	Okano, K.	*Chem. Lett.*	**1987**		181
7	Denno, N.C.	*Chem. Rev.*	**1960**	*60*	7
8	Wilds, A.L.	*Org. React.*	**1944**	*2*	178

4-Hydroxy-1,2,3,4-tetrahydrocrysene (2).[4] A mixture of 4-keto-1,2,3,4-tetrahydrocrysene **1** (2.75 g, 11 mmol) and aluminium isopropoxide (8.2 g, 40 mmol) in PhMe (25 mL) was refluxed for 1 h. After cooling, iPrOH (25 mL) was added and the formed acetone was distilled slowly maintaining the volume constant by addition of iPrOH. After all acetone was distilled, the solution was acidified with 10% H_2SO_4. The separated toluene solution was evaporated to give 2.1 g of **2** (76%), mp 156-158°C.

MEISENHEIMER N-Oxide Rearrangement

Chlorination of pyridines via rearrangement of N-oxides.

AcOH / AcOOH → $POCl_3$ / P_2O_5 → +

1 **2** **3** (56%) **4** (43%)

1	Meisenheimer, J.	*Chem.Ber.*	**1926**	*59*	1848
2	Albert, A.	*J. Chem. Soc.*	**1960**		1790
3	Brown, E.V.	*J. Org. Chem.*	**1967**	*32*	241
4	Pandler, W.L.	*J. Org. Chem.*	**1971**	*36*	1720

2-(and-4-)Chloro-1,5-naphthyridine (3) and **(4).**[3] 1,5-Naphthyridine **1** (4.5 g, 34 mmol) was treated with a mixture of AcOH (10 mL) and 40% peracetic acid (5 mL) for 3 h at 70°C. From the mixture of mono and di N-oxides, the mono N-oxide **2** was obtained by recrystallization from methylcyclohexane. **2** (0.77 g, 5 mmol) was heated in $POCl_3$ (30 mL) and P_2O_5 for 30 min. The product was collected and analyzed by GC (15% SE-30 on Chromosorb W 240°C He 40 psi) to be a mixture of **3** (56.8%) and **4** (43.2%).

MEISENHEIMER - JANOVSKY Complexes

The adduct formed from a polynitroaromatic compound in alkaline solution with RO^-, HO^- (Meisenheimer) or with acetone (Janovsky).

OCH3 O2N NO2 **1** —MeOK→ MeO OMe O2N NO2 **2** (81%)

O2N NO2 **3** —Me2CO, KOH→ H O2N NO2 O **4** (10%)

1	Meisenheimer, J.	*Leibigs Ann.*	**1902**	*323*	205
2	Fendler, J.H.	*J. Org. Chem.*	**1967**	*82*	2507
3	Jones, P.R.	*J. Org. Chem.*	**1986**	*51*	3016
4	Straw, M.J.	*Chem. Rev.*	**1970**	*70*	667
5	Janovsky, I.V.	*Chem. Ber.*	**1886**	*19*	2155
6	Michiya Kimura	*Chem. Pharm. Bull.*	**1968**	*16*	634

2,4-Dinitroanisole-methoxide complex (2).[2] To 2,4 dinitroanisole **1** (2.47 g, 12.5 mmol) in dioxane (3 mL) was added KOMe (0.84 g, 12 mmol) in MeOH (2.4 mL). The red crystals were filtered under N_2, washed with benzene and anh. Et_2O and dried to afford 2.27 g of **2** (81%), mp 130-132°C.

Janovsky complex (4).[7] m-Dintrobenzene **3** (1.00 g, 6 mmol) in acetone (10 mL) was shaken vigorously with KOH pellets for 15 min. The purple mixture was filtered, the KOH pellets washed with acetone and to the combined solutions was added benzene (20 mL). The black-violet precipitate was recrystallized from benzene-acetone to give 135 mg of **4** (10%), mp 250°C.

MELDRUM´S Acid

A cyclic malonate derivative (acidic methylene) used in place of malonate in alkylations or reactions with aldehydes.

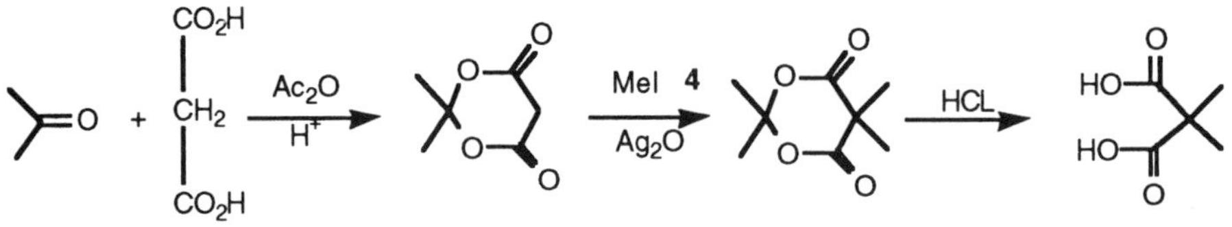

1	Meldrum, A.N.	*J. Chem. Soc.*	**1908**	*93*	598
2	Davidson, D.	*J. Am. Chem. Soc.*	**1948**	*70*	3426
3	Chau, C.C.	*Synthesis*	**1984**		224
4	Dhimane, K.	*Tetrahedron Lett.*	**1987**	*28*	885
5	Ping, L.	*Org.Prep.Proced.Int.*	**1992**	*24*	185

Meldrum's acid (3).[2] To malonic acid **1** (52 g, 0.5 mol) in acetic anhydride (60 mL, 0.6 mol) was added conc H_2SO_4 while stirring. Acetone **2** (40 mL, 0.55 mol) was added. Cooling for 24 h, filtration, washing the solid with ice water and recrystallization from acetone-water gave **3** (35 g, 49%), mp 94-95°C (dec.)

Dimethylmalonic acid (6) To a suspension of silver oxide (26 g, 0.5 mol) in MeCN (50 mL) and MeI **4** (20 mL, 0.33 mol) was added **3** (14.4 g, 0.1 mol) in MeCN (50 mL) over 1 h. After stirring and 24 h standing, filtration and evaporation of the solvent, the oily residue was treated with water. Filtration gave dimethyl Meldrum acid **5**, 12.6 g (73%); vacuum distillation gave 11.8 g, bp 127-129°/33 mm, mp 59.5°C. To a 2M HCl solution was added **5** (3.44 g, 24 mmol) and the mixture was boiled. Cooling and filtration afforded **6**, 2.1 g (80%), mp 192-193°C (dec).

MENCKE - LASZLO Nitration of Phenols

Ortho nitration of phenols by metal nitrates. Also oxidation of mercaptans to disulfides.

OH —$Cu(NO_3)_2$→ OH, NO_2

HO (1) —"Clayfen" $Fe(NO_3)_3$→ O_2N, HO (2 (55%))

1	Mencke, J.B.	*Rec. Trav. Chim. Pays Bass*	**1925**	*44*	141
2	Laszlo, P.	*Tetrahedron Lett.*	**1982**	*23*	5035
3	Laszlo, P.	*Tetrahedron Lett.*	**1983**	*24*	3101
4	Laszlo, P.	*J. Org. Chem.*	**1983**	*48*	4771
5	Laszlo, P.	*Pure Appl. Chem.*	**1990**	*63*	2027

2-Nitroestrone (2).[4] "Clayfen" - was prepared from a mixture of $Fe(NO_3)_3.H_2O$ (45 g, 0.11 mol), K-10-bentonite clay and Me_2CO (750 mL). Freshly prepared "Clayfen" (2.0 g) was added to estrone **1** (0.54 g, 2 mmol) and PhMe (150 mL). The suspension was stirred overnight at 20° and filtered under vacuum. The yellow filtrate was purified by column chromatography on silica gel (n-hexane:EtOAc 8:2). Evaporation of the corresponding fraction afforded 0.347 g of **2** (55%), mp 178-180°C.

MENZER Benzopyran synthesis

Benzopyranone synthesis from phenols and β-ketoesters or unsaturated acids.

OH, HO, OH **1** + EtO_2C, O, Ph **2** → 240 - 250°C, 45 h → OH O, HO, O, Ph **3** (33%)

OH, HO, OH + H_3C, H_3C C=CH--CO_2H → 180° → OH O, HO, O (20%)

1	Menzer, Ch.	*C.R.*	**1952** *232* 1488
2	Lacey, R.N.	*J. Chem. Soc.*	**1954** 859
3	Mercier, Ch.	*C.R. Serie C.*	**1973** *273* 1053

2-Phenyl-4(4H)-one-5,7-dihydroxy-1-benzopyran (3).[1] A mixture of phloroglucinol **1** (2.77 g, 22 mmol) and ethyl benzoylacetate **2** (7.65 g, 40 mmol) was heated to 240-250°C for 1.5 h. The cooled mixture was extracted with 5% NaOH and the aqueous solution was extracted with Et_2O. Acidification afforded a brown product which after sublimation at 250-300°C/0.01 mm and recrystallization from EtOH gave 2.2 g of **3** (33.5%), mp 278°C.

MERRIFIELD Peptide synthesis

Growing of a polypeptide chain by attachment of the first aminoacid to an insoluble polymer and attaching successively aminoacids. At the end cleavage of the peptide from the polymer.

Polystyrene-$C_6H_3(NO_2)$-CH_2Cl **2** —(Cbz-L-valine **3**, TEA)→ [P]-CH_2O-C(=O)--CH(iPr)-NH-Cbz **4**

—(HBr, HOAc)→ [P]-CH_2-O-C(=O)--CH(iPr)-NH_2 **5** —(Cbz-glycine **6**, DCC)→

[P]-CH_2-O-C(=O)--CH(iPr)-NH-C(=O)-CH_2-NH-Cbz **7** —(1. HBr -HOAc; 2. NaOH)→ HO-C(=O)-CH(iPr)-NH-C(=O)-CH_2-NH_2 **8**

1	Merrifield, R.B.	*J. Am. Chem. Soc.*	**1963**	85	2149
2	Merrifield, R.B.	*J. Org. Chem.*	**1964**	*29*	3100
3	Dorman, L.C.	*J. Org. Chem.*	**1969**	*34*	158
4	Wünsch, E.	*Angew. Chem. Int. Ed.*	**1971**	*10*	786

L-Valyl polymer (4). **3** (15.5 g, 61.7 mmol), TEA (8.65 mL, 61.7 mmol) and EtOAc (350 mL) was refluxed with polymer **2** (40 g) for 48 h, AcOH (12 mL) and TEA (28 mL) was added and refluxing for 4 h. Filtration and washing (EtOAc, EtOH, H_2O, MeOH), yielded 46.9 g of **4**.

Polymer (5). **4** (10 g, 5.6 mmol valine) and 30% HBr in AcOH (30 mL) was shaken for 5 h at 25°C, diluted with AcOH (50 mL), filtered, washed (AcOH, EtOH, DMF), to afford **5**.

Polymer (7). **5** (10 g) was treated with Cbz-glycine **6** (2.34 g, 11.2 mmol) in DMF (20 mL). After 20 min shaking, DCC (2.31 g, 11.2 mmol) in DMF (4.6 mL) was added and shaken for 18 h at 25°C. After filtration, more **6** (1.17 g) and DCC (1.16 g) was added and all shaken for 2 h. Filtration, shaking with AcOH (3 mL) and TEA (1 mL) and washing (DMF,EtOH and AcOH) gave **7**.

Glycyl-L-valine (8). **7** was decarbobenzoxylated with HBr-AcOH and shaken with 2N NaOH (4.5 mL) in EtOH (40.5 mL) for 1 h at 25°C. **8** was purified by ion-exchange chromatography.

M E Y E R S Asymmetric synthesis

Chiral oxazoles in asymmetric synthesis of carboxylic acids, aldehydes, chiral dihydronaphthalenes.

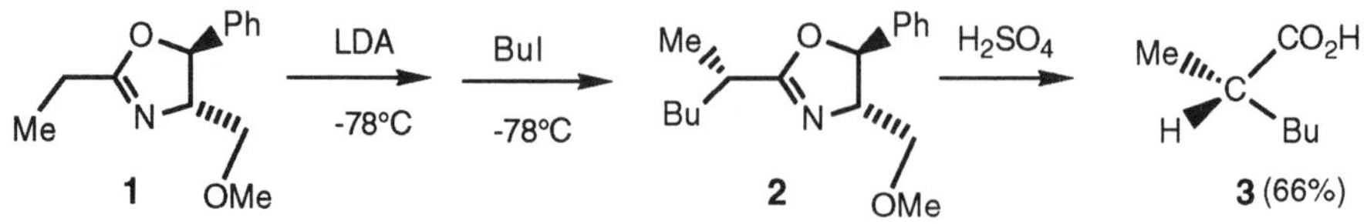

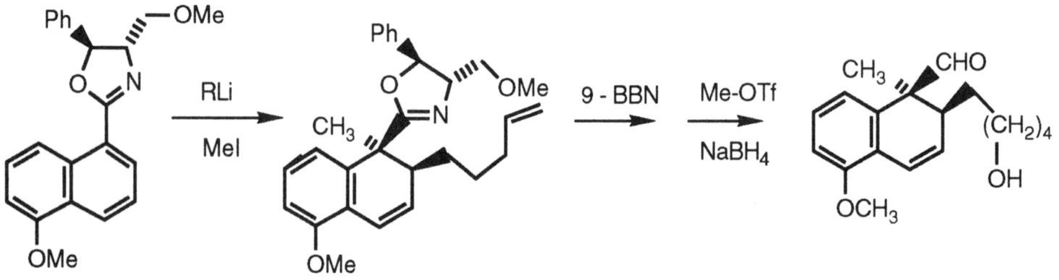

1	Meyers, A.I.	*J. Am. Chem. Soc.*	**1974**	*96*	268
2	Meyers, A.I.	*Acc. Chem. Res.*	**1978**	*11*	375
3	Meyers, A.I.	*J. Am. Chem. Soc.*	**1976**	*98*	567
4	Meyers, A.I.	*J. Org. Chem.*	**1980**	*45*	2785
5	Meyers, A.I.	*J. Org. Chem.*	**1987**	*52*	4592
6	Meyers, A.I.	*J. Am. Chem. Soc.*	**1988**	*110*	4611,7854
7	Meyers, A.I.	*Tetrahedron*	**1989**	*45*	6949

(S)-(+)-2-Methylhexanoic acid (3).[3] (4S,5S)-**1** (15.4 g 70 mmol) in THF (160 mL) under N_2 at -78°C, was treated with LDA (from 9.8 mL of iPr_2NH and 2.2 M n-BuLi (33 mL) in THF 75 mL) over 20 min. After 20 min the mixture was cooled to -98°C and BuI (14.7 g, 80 mmol) in THF (20 mL) was added over 20 min. After 2 h at -98°C the mixture was warmed to 20°C, poured into brine and extracted with Et_2O. Bulb to bulb distillation afforded pure **2** $[\alpha]_{589}^{24}$ = -32.2°. The crude oxazoline **2** (17.2 g) was refluxed for 3.5 h in 4N H_2SO_4. Extraction with Et_2O (3.75 mL), washing with 5% K_2CO_3 (3x100 mL), acidification (pH=1) of the aqueous extract with 12 M HCl and extraction with Et_2O, gave on distillation 5.8 g of **3** (66%), $[\alpha]_{589}^{24}$ = + 14.5°.

MEYER - SCHUSTER Propargyl alcohol rearrangement

Acid catalyzed rearrangement of acetylenic alcohols into α,β-unsaturated carbonyl derivatives.

$$\text{Ph}\underset{\text{OH}}{\underset{|}{\text{CH}}}\text{-C}\equiv\text{CH}\ (\mathbf{1}) \xrightarrow[115°]{H_2SO_4} \text{Ph-CH=CH-CH=O}\ \mathbf{2}\ (33\%)$$

$$\text{Ph}_2\underset{\text{OH}}{\underset{|}{\text{C}}}\text{-C}\equiv\text{C-}\underset{\text{Me}}{\underset{|}{\text{CH}}}\text{-NMe}_2 \xrightarrow{H_2SO_4} \text{Ph-}\underset{\text{Ph}}{\underset{|}{\text{C}}}\text{=CH-}\underset{\text{O}}{\underset{\|}{\text{C}}}\text{-}\underset{\text{Me}}{\underset{|}{\text{CH}}}\text{-NMe}_2$$

1 Meyer, K.H., Schuster, K. *Chem. Ber.* **1922** *55* 819

2 McGregor, W.S. *J. Am. Chem. Soc.* **1948** *72* 183

3 Swaminathan, S. *Chem. Rev.* **1971** *71* 429

4 Huggill, H.P.W.. *J. Chem. Soc.* **1950** 335

Cinnamaldehyde (2).[2] Phenyl ethynyl carbinol **1** (5.4 g, 40 mmol) and water (5 mL) was distilled with steam into 28% H_2SO_4 (25 mL) heated at 115°C. The crude oily product 3.1 g (57%) was separated as the phenylhydrazone by treatment with phenyl hydrazine hydrochloride (17.5 g, 0.12 mol) and NaOAc (10 g) in water (150 mL) to afford 3.0 g of crude **2** phenylhydrazone (33%), mp 154-157°C. Two recrystallizations from EtOH raised the mp to 168-169°C.

MICHAEL Addition

Base promoted 1,4-additions of nucleophiles (usually C) to α,β-unsaturated esters, ketones, nitriles, sulfones, nitro-compounds; often stereoselective addition.

H
O
1
+ Me_2CuLi
2
0°
H
O
Me
3(82%)

COOMe
+
OMe
O
KNH_2, -78°
chiral crown ether
O
O
MeO
OMe
Ph
Me
(80%) 83% ee

OLi
Ph
(98% Z)
+
O
t Bu
Ph
-78°
O Ph O
Ph
t Bu
(87%) 98% anti

1	Komnenos, A.	*Liebigs Ann.*	**1883**	*218*	145
2	Michael, A.	*J. prakt. Chem.*	**1887**	*35(2)*	348
3	Heathcock, C.H.	*Tetrahedron Lett.*	**1986**	*27*	6169
4	Yamaguchi, M.	*Tetrahedron Lett.*	**1984**	*25*	5661
5	Enders, D.	*Tetrahedron*	**1986**	*42*	2235
6	Piers, E.	*Can. J. Chem.*	**1969**	*47*	137
7	Seebach, D.	*Helv. Chim.*	**1985**	*68*	1592
8	Bergman, E.D.	*Org. React.*	**1959**	*10*	179
9	Heathcock, C.H.	*Topics in Stereochem.*	**1991**	*20*	87

Eremophil-11-en-3-one (3).[6] To a slurry of CuI (6.0 g, 31.5 mmol) in ether (120 mL) at 0° (N_2) was injected 1.59 M MeLi in ether (39.4 mL; 62.5 mmol). To the resulting **2** was added **1** (2.0 g, 10 mmol) in 80 mL ether over 20 min. After 2 h at 0° the mixture was poured into 1.2N HCl (800 mL) and extracted with ether. Evaporation gave 1.9 g of **3**, bp 123-124/0.03 mm; semicarbazone mp 191.5-193.5°.

MICHAELIS - BECKER - NYLEN Phosphonylation

Nucleophilic attack of lithium dialkylphosphonates on pyridium salts to produce pyridine phosphates.

Pyridine N-oxide → (Me_2SO_4) → N-methoxypyridinium $MeSO_4^-$ **1** → (LiP(O)(OEt)$_2$ **2**) → 2-H-2-$P(O)(OEt)_2$-N-MeO-dihydropyridine → pyridine-2-$P(O)(OEt)_2$ **3** (67%)

1	Michaelis, A., Becker	*Chem. Ber.*	**1897**	*30*	1003
1	Michaelis, A.,	*Chem. Ber.*	**1898**	*31*	1048
2	Nylen, T.	*Chem. Ber.*	**1924**	*57*	1023
3	Gordon, M.	*J. Org. Chem.*	**1966**	*31*	333
4	Redmore, D.	*J. Org. Chem.*	**1970**	*35*	4114
5	Kemm, K.M.	*J. Org. Chem.*	**1981**	*46*	5188

Diethyl pyridine-2-phosphonate (3).[4] n-Butylithium (23% in hexane) (63 mL, 0.15 mol) was added dropwise to diethyl phosphonate (25.0 g, 0.18 mol) at -20 to -30°C over 2 h. To the resulting **2** was added N-methoxypyridinium methosulfonate **1** (from pyridine N-oxide 14.3 g, 0.15 mol and dimethyl sulfate 18.9 g, 0.15 mol) in diethyl phosphonate (40 mL) over 1 h at -15°C. The mixture was stirred at room temperature overnight and water was added (100 mL). After extraction with $CHCl_3$ (3x75 mL), the organic layer was extracted with 4N HCl, basified and reextracted to yield 22.9 g of **3** (67%), bp 105-112°C (0.08 mm). Redistillation gave pure **3**, bp 96-97°C (0.03 mm), picrate mp 86-87°C.

MIESCHER Degradation

Three carbon degradation of a carboxylic acid side chain (see Barbier-Wieland).

1 → PhMgBr; Ac_2O → 2 → NBS → 1. $PhNMe_2$, 2. Ac_2O → 3 → 1. CrO_3, 2. NaOH → 4

1	Miescher, K.	*Helv. Chim. Acta.*	**1944**	*27*	1815
2	Spring, F.S.	*J. Chem. Soc.*	**1950**		3355
3	Wettstein, A.	*Experientia*	**1954**		407

3 α-Oxy-12-β-acetoxypregnan-20-one (4).[1] To diphenylcholene **2** (100 g, 0.14 mol) in boiling CCl_4 (500 mL) was added NBS (40 g, 0.23 mol). After 10 min reflux the mixture was filtered, the filtrate treated with dimethylaniline (100 g, 0.23 mol), the CCl_4 distilled, the residue diluted with ether and extracted with dil HCl. The residue after evaporation of the solvent (100 g) was crystallized from acetone-water. Reacetylation (Ac_2)/Py) gave diene **3** (28 g). To this product in AcOH (100 mL) was added CrO_3 (20 g, 0.2 mol) in water (20 mL) and AcOH (100 mL) dropwise over 1 hr at 0°C. After 15 h stirring at 20°C, MeOH (50 mL) was added and the solvent removed in vacuum. The residue was treated with aq NaOH and filtered to obtain crude **4** (24.5 g) purified with Girard reagent.

MILAS Olefin Hydroxylation

Hydroxylation of a double bond to a 1,2-diol with hydrogen peroxide and OsO_4 as catalyst.

OsO_4 / H_2O_2

1

HO OH

2a (23%)

2b (4β, 5β) (28%)

1 Milas, W.A. *J. Am. Chem. Soc.* **1936** *58* 1302

2 Milas, W.A. *J. Am. Chem. Soc.* **1959** *81* 3114

4α-5α-Dihydroxycholestan-3-one (2a) and **4β-5β-dihydroxycoprostan-3-one (2b).**[2] To a solution of cholestenone **1** (4 g, 10 mmol) in ethyl ether (150 mL) was added a solution of OsO_4 (0.2 g) in ethyl ether (10 mL) and hydroxgen peroxide 30% (5 mL). After 24 h at 20°C the solvent was removed under vacuum for 24 h. If the residue darkened, a few mL of hydrogen peroxide were added and the operation repeated. The residue was dissolved in hot EtOH (50 mL) and cooled to give **2a**, 1.00 g (23%), mp206-208°C. From the filtrate by dilution with water separated **2b**, 1.20 g (28%), mp 112-112.5°C, after recrystallization from ethanol.

MILLER - SNYDER Aryl Cyanide Synthesis

Synthesis of benzonitriles from aldehydes via oxime ethers. Formation of p-cyanophenol from p-nitrobenzaldoxime and p-nitrobenzonitrile (used as a sometimes recyclable chain carrier).

NOH, NO_2, CN, **2**, DMSO, Δ, NO_2, **1**, CN, O, N, H, NO_2, CN, OH, + **2**, **3** (83%)

HC=O, O, O, **4**, NH_2OH, 70°, KOH, **2**, C≡N, O, O, **5**, + **3**

1	Miller, M.J., Loudon, G.M.	*J. Org. Chem.*	**1975**	*40*	126
2	Snyder, M.R.	*J. Org. Chem.*	**1974**	*39*	3343
3	Snyder, M.R.	*J. Org. Chem.*	**1975**	*40*	2879

p-Cyanophenol (3).[3] To a heated (70°C) mixture of anh. Na_2CO_3 (4.24 g, 40 mmol) in DMSO (25 mL) was added a solution of p-nitrobenzaldoxime **1** (3.32 g, 20 mmol) and p-nitrobenzonitrile **2** (0.15 g, 1 mmol) in DMSO (35 mL). The orange mixture was heated to 114°C for 10 h and was poured into ice water (150 mL) containing 32% HCl (5 mL). Extraction with Et_2O and reextraction from Et_2O with 5% NaOH followed by acidification and recrystallization afforded 1.97 g of **3** (83%), mp 103-107°C, recrystallized from PhH, mp 111-112°C.

M I N I S C I Aromatic amination

Iron catalyzed free radical amination of aromatics or free radical carbamylation of protonated heterocycles.

Me_2NCl (**1**) + benzene (**2**) $\xrightarrow{H_2SO_4 - FeSO_4}$ $C_6H_5NMe_2$ **3** (76%)

quinoxaline (**4**) + H_2NCHO (**5**) $\xrightarrow[H_2O_2]{H_2SO_4 - FeSO_4}$ quinoxaline-2-$CONH_2$ **6** (82%)

1	Minisci, F.	*Tetrahedron Lett.*	**1965**		433
2	Minisci, F.	*Chem. Ind. (Milano)*	**1966**	*48*	716
3	Minisci, F.	*Tetrahedron Lett.*	**1970**		15
4	Minisci, F.	*Synthesis*	**1973**		1
5	Heinisch, G.	*Synthesis*	**1988**		119
6	Minisci, F.	*Tetrahedron*	**1985**	*41*	4157

N,N-Dimethylaniline (3).[2] To N-chlorodimethylamine **1** (4.3 g, 0.054 mol), HOAc (50 mL), PhH (30 mL) and H_2SO_4 (83 mL) was added with stirring finely powdered $FeSO_4$. The temperature rose from 15°C to 27°C within 5 min. After 15 min the mixture was poured into ice made basic (NaOH) and extracted with PhH. Distillation gave 5.0 g of **3** (76%), bp 193-194°C.

Quinoxaline-2-carboxamide (6).[3] Quinoxaline **4** (13.0 g, 0.1 mol) and 98% H_2SO_4 (5.5 mL) in **5** (100 mL) was treated simultaneously with 34% H_2O_2 (15.0 mL, 0.15 mol) and $FeSO_4 \cdot 7H_2O$ (41.7 g, 0.15 mol) with efficient stirring at 10-15°C for 15 min. Excess **5** was distilled and the residue extracted with $CHCl_3$. Evaporation afforded 14.2 g of **6** (82%), mp 200°C.

MISLOW - BRAVERMAN - EVANS Rearrangement

Reversible 2,3- sigmatropic rearrangement of allylic sulfoxides to allyl sulfenates which are cleaved by phosphites to allylic alcohols.

1	Mislow, K.	*J. Am. Chem. Soc.*	**1966**	*88*	3138
2	Braverman, S.	*Chem. Commun.*	**1967**		270
3	Evans, D.A.	*J. Am. Chem. Soc.*	**1971**	*93*	4956
4	Evans, D.A.	*Acc. Chem. Res.*	**1974**	*7*	147
5	Grieco, P.A.	*J. Chem. Soc. Chem. Commun.*	**1972**		702
6	Grieco, P.A.	*J. Org. Chem.*	**1973**	*38*	2245

(+)-(E)-Nuciferole (3).[6] To **1** (195 mg, 1 mmol) in THF (10 mL) at -50°C under N_2, was added dropwise 1.66 M BuLi in hexane (0.65 mL, 1.08 mmol). **2** (548 mg, 15 mmol) in THF (1 mL) was added dropwise over 10 min. The mixture was stirred at -50°C for 1 h and at 25°C for 2 h, poured into brine and extracted with Et_2O/hexane (3:1). After distillation the residue wasdissolved in MeOH (1.5 mL) and added to benzenethiol (660 mg, 5.4 mmol) in MeOH (20 mL). 1.66 M BuLi (0.78 mL) was added under N_2 , and all was heated for 7 h at 65°C, then kept for 10 h at 25°C. Quenching with water, extraction (Et_2O), evaporation and purification by preparative TLC (silicagel Et_2O/hexane 2:1) afforded 127 mg of **3** (58%).

MITSUNOBU Displacement

Inter and intramolecular nucleophilic displacement of alcohols with inversion by means of diethyl azodicarboxylate (DEAD)-triphenylphosphine and a nucleophile. Also dehydration, esterification of alcohols or alkylation of phenols,

CO_2Me HO OH OH **1** — Ph_3P, DEAD → CO_2Me HO HO O–PPh_3 + → CO_2Me HO O **2** (77%)

Ph OH — Ph_3P, HN_3, DEAD, 20° → Ph N_3 (52%)[7]

1	Mitsunobu, O.	*Bull. Chem. Soc. Jpn.*	**1967**	*40*	2380
2	Miller, M.J.	*J. Am. Chem. Soc.*	**1980**	*102*	7026
3	Evans, S.A.	*J. Org. Chem.*	**1988**	*53*	2300
4	McGovan, D.A.	*J. Org. Chem.*	**1981**	*46*	2381
5	Mitsunobu, O.	*Synthesis*	**1981**		1
6	Crich, D.	*J. Org. Chem.*	**1988**	*54*	257
7	Hassner, A.	*J. Org. Chem.*	**1990**	*55*	2243

(-)Methyl cis-3-hydroxy-4,5-oxycyclohex-1-enecarboxylate (2).[4] To (-)methyl shikimate **1** (220 mg, 1.06 mmol) and triphenylphosphine (557 mg, 2.12 mmol) in THF at 0°C, under N_2 was added with stirring (DEAD) (370 mg, 2.12 mmol). After 30 min at 0°C and 1 h at 20°, it was vacuum distilled (Kugelrohr) at 165°C (0.1 mm) and taken up in Et_2O. Cooling gave bis(carbethoxy)hydrazine (10 mg mp 133°C). The filtrate was concentrated and chromatographed (preparative TLC-silica gel Et_2O) to afford 140 mg of **2** (77%) on standing; recrystallized from Et_2O-petroleum ether, mp 81-82°C, α_D^{25} = 55.4°.

MUKAIYAMA Stereoselective aldol

Stereoselective aldol condensation via tin (II) enolates (e.g., **2**).

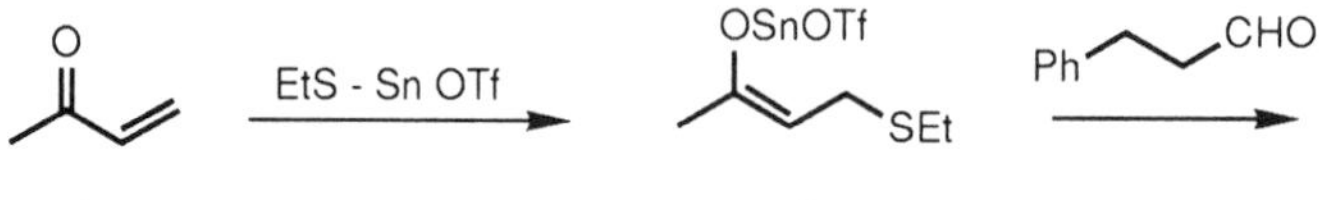

1	Mukaiyama, T.	*Chem. Lett.*	**1982**		353
2	Mukaiyama, T.	*J. Am. Chem. Soc.*	**1973**	*95*	967
3	Mukaiyama, T.	*Chem. Lett.*	**1986**		187
4	Mukaiyama, T.	*Org.React.*	**1982**	*28*	187

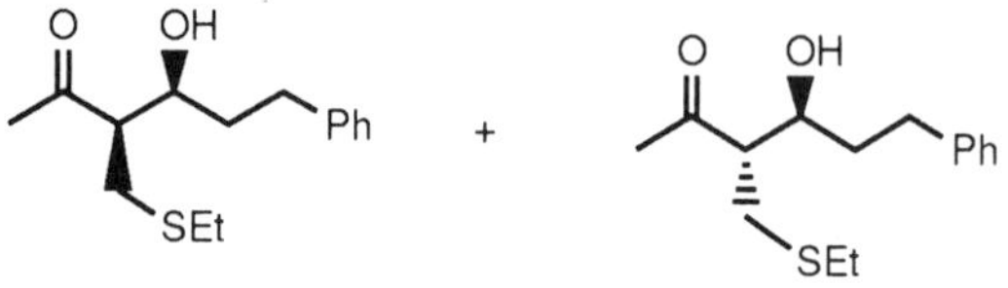

Syn 3-(Ethylthiomethyl)-4-hydroxy-6-phenyl-2-hexanone (4) and **anti (5).**[3] To ethanethiol (10.0 mg, 0.17 mmol) in THF (2 mL) was added 1.54 M n-butyllithium in hexane (0.11 mL) at 0°C under Ar. Stannous triflate (69.0 mg, 0.17 mmol) was added and after 20 min the mixture was cooled to -45°C. Methyl vinyl ketone **1** (118 mg, 1.98 mmol) in THF (1.5 mL) was added followed by 3-phenylpropanal **3** (350 mg, 2.61 mmol) in THF (1.5 mL). After 12 h, aq. citric acid was added and the organic material extracted with CH_2Cl_2. The residue after evaporation was dissolved in MeOH and treated with citric acid. After 30 min stirring, the mixture was quenched with pH 7 phosphate buffer, extracted with CH_2Cl_2, the solvent evaporated and the residue chromatographed to afford 336 mg of **4** (75%), syn:anti (90:10).

MURAHASHI Allylic Alkylation

Allylic alkylation of allyl alcohols in the presence of copper iodide and phosphinimines

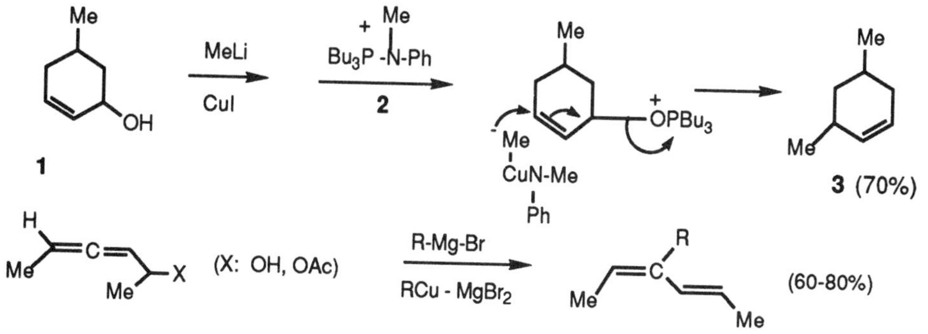

1	Murahashi, S.I.	*J. Am. Chem. Soc.*	**1977**	*99*	2361
2	Lovisalles, J.	*J. Organomet. Chem.*	**1977**	*136*	103
3	Murahashi, S.I.	*J. Am. Chem. Soc.*	**1978**	*100*	4610
4	Trost, B.M.	*J. Org. Chem.*	**1980**	*45*	4256
5	Goering, N.L.	*J. Org. Chem.*	**1981**	*46*	2144
6	Fan, C., Cazes, B.	*Tetrahedron Lett.*	**1988**	*29*	1701

(Methylphenylamino)tri-n-butylphosphonium iodide (2).[5] Phenyl azide and Bu_3 P in Et_2O afforded the phosphinimine which with excess of MeI gave **2**, (from EtOAc).

3,5-Dimethylcyclohexene (3). Cis-5-methyl-2-cyclohexenol **1** (1.12 g, 10 mmol) was treated with MeLi (in Et_2O)(6.7 mL, 1.4 M) and all was added to a suspension of CuI (1.90 g,10 mmol) in THF (20 mL). After 30 min the homogeneous solution was chilled to -78°C and 1.49 M solution MeLi (6.7 mL) in Et_2O was added followed by **2** (4.35 g, 10 mmol) in THF (40 mL). After 3 h stirring at 20°, the mixture was quenched by sat. NH_4Cl (30 mL). The Et_2O layer was washed and evaporated to give 763 mg of **3** (70%) purified on a GC-capillary column .

NAZAROV Cyclopentenone synthesis

Acid catalyzed cyclization of dienones to cyclopentenones.

CO_2Et — $SnCl_4$, 24 h, 20°C → CO_2 Et

1 → **2** (30%)

$SiMe_3$, Me — $FeCl_3$ → H, H, Me + H, H, Me

4 (78%) (22%)

1 Nazarov, J.N.	*Bull. acad. sci (USSR)*	**1946**		633
2 Marino, J.P.	*J. Org. Chem.*	**1981**	*46*	3696
3 Denmark, S.E.	*Tetrahedron*	**1986**	*42*	2821
4 Peel, M.L.	*Tetrahedron Lett.*	**1986**	*27*	5947

2-(Carbethoxy)-3-methyl-3,4,5,6-tetrahydro-2H-pentalen-1-one (2).[2] A solution of ethyl 2-ethylidene-3-(1-cyclopenten-1-yl)-3-oxo-propanoate **1** (210 mg, 1 mmol) in CH_2Cl_2 (10 mL) was stirred with $SnCl_4$ (780 mg, 3 mmol) for 24 h at 20°C. After quenching with water (50 mL), the organic layer was washed ($NaHCO_3$), dried ($MgSO_4$) and purified by preparative TLC to afford 62.5 mg of **2** (30%).

N E B E R Rearrangement

Rearrangement of N,N-dimethylhydrazone or tosylate derivatives of oxime to azirines and from there to α-amino ketones.

=N—N⁺ Me3
I⁻
NaH
=N
HCl
O
NH2
HO
1
HO
2 (42%)

Cl O NOTs
NH–C–CH2–C–NH2
Cl Cl
3
MeONa
MeOH, 25°
Cl O
NH–C
N
Cl Cl NH2
4 (89%)

1	Neber, P.W.	*Liebigs Ann.*	**1926** *449*	109
2	Neber, P.W.	*Liebigs Ann.*	**1936** *526*	277
3	Morow, D.H.	*J. Org. Chem.*	**1965** *30*	579
4	Hyatt, J.A.	*J. Org. Chem.*	**1981** *46*	3953
5	O'Brine, C.	*Chem. Rev.*	**1964** *64*	81
6	Yamura, Y.	*Synthesis*	**1973**	215

17β-Amino-3β-hydroxy-17α-preg-5-en-20-one (2).[2] To 2'-(3β-hydroxypregn-5-en-20-ylide)-1',1',1'-trimethyl hydrazonium iodide **1** (15.0 g, 38.6 mmol) in DMSO (100 mL) and anh. EtOH (1000 mL) was added 51% NaH in mineral oil (7.5 g). After 2 h reflux the mixture was concentrated in vacuum to 150 mL and poured into water (1000 mL). Extraction with PhH:Et_2O (1:1), washing the extract and evaporation gave an oil which was heated with 6N HCl (10 mL) and EtOH (250 mL) for 30 min at 95°C. The mixture was poured into water (1000 mL) and filtered to give 5.4 g of **2** (42%).

NEBER - BOSSET Oxindole-cinnoline synthesis

Synthesis of N-aminooxindols or of cinnolines.

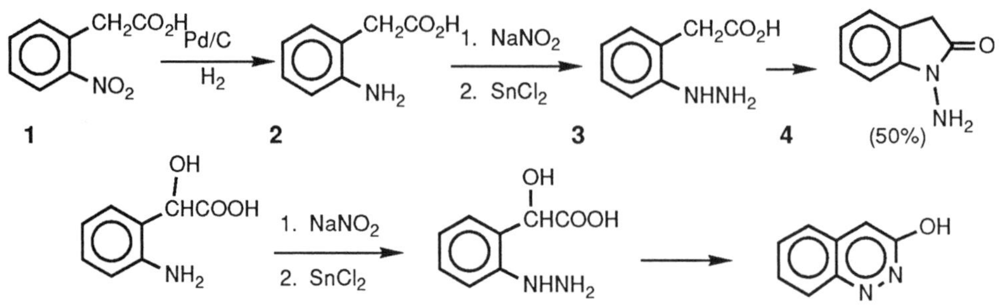

1	Neber, P.W.	*Liebigs Ann.*	**1929** *471*	113
2	Bosset, G.	*C.Z. (Ph.D. Thesis)*	**1920** *II*	3015
3	Baumgarten, H.F.	*J. Am. Chem. Soc.*	**1960** *82*	3977

1-Aminooxindole (4).[3] A solution of o-nitrophenylacetic acid **1** (20 g, 0.11 mol) as the Na salt was hydrogenated in the presence of 10% Pd on C (0.5 g). After removing the catalyst by filtration through Celite, the o-aminophenylacetic acid **2** was diazotized with $NaNO_2$ (7.7 g, 0.11 mol) and after cooling conc. HCl (84 mL) was added. The diazonium salt of **2** was treated with $SnCl_2.2H_2O$ (75 g, 0.33 mol) in HCl (75 mL). The mixture was stirred for 30 min and maintained for 24 h at 0°C. Treatment of tin salt with H_2S, removal of the SnS, concentration of the filtrate and sublimation of the residue in vacuum afforded 7.48-8.46 g of **4** (46-52%), mp 125-126.5°C.

NEF Reaction

Conversion of nitroalkanes to carbonyl compounds by acidification of nitronates.

1. NaOH
2. 2 N HCl
Δ
NO_2
O
O
1 **2** **3** (46%)

1	Nef, J.U.	*Liebigs Ann.*	**1894**	*280*	263
2	Weinstein, B.	*J. Org. Chem.*	**1962**	*27*	4049
3	Langrene, M.	*C.R. (C)*	**1974**	*284*	153(3)
4	Noland, W.E.	*Chem. Rev.*	**1955**	*55*	137
5	Miyakoshy, T.	*Synthesis*	**1986**		766
6	McMurry, J.	*Acc. Chem. Res.*	**1974**	*7*	281
7	Pinnick,, H.W.	*Org. React.*	**1990**	*38*	655

1,2,3,4,5,6,7,8,9,10,11,12-Dodecahydro-9-ketophenantrene (3).[2] A cold solution of NaOH (5 g, 0.125 mol) in EtOH (100 mL) was added slowly to a chilled stirred solution of 1,2,3,4,5,6,7,8,9,10,11,12-dodecahydro-9-nitrophenantrene **1** (10.9 g, 0.0424 mol) in EtOH (100 mL). The mixture was maintained at 0-2°C under N_2 for 30 min and then poured into cold 2N HCl (300 mL). The solution was stirred for 2 h at 5°C and allowed to warm to 20°C over a day. The organic material was extracted with Et_2O (3x300 mL), the extract dried and the solvent removed to give a brown residue 9.5 g (mixture of **2** and **3**). Heating isomerised **2** to **3**. Distillation of 1 g crude product gave 0.46 g of **3** (46%), bp 70°C/0.12 mm; DNPH, mp 231-231.5°C (EtOH).

NENITZESCU Indole Synthesis

Indole synthesis from quinones and amino crotonates.

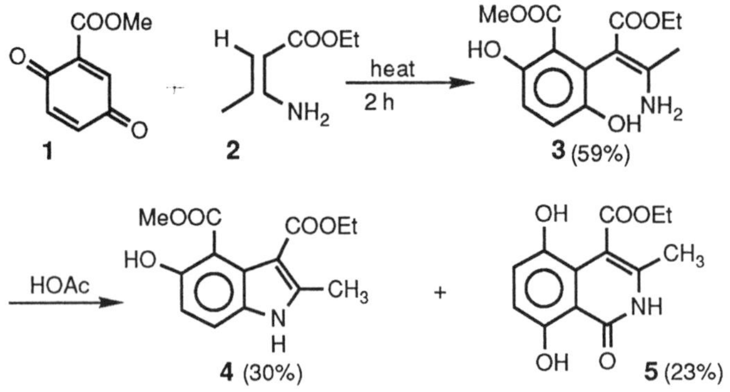

1	Nenitzescu, C.D.	*Chem. Ber.*	**1925**	*58*	1063
2	Nenitzescu, C.D.	*Bull. Soc. Chim. Rom.*	**1929**	*11*	37
3	Allen, G.R.	*J. Org. Chem.*	**1968**	*33*	198
4	Bernici, J.L.	*J. Org. Chem.*	**1981**	*46*	4197
5	Rapderey, T.	*Austr. J. Chem.*	**1984**	*37*	1263
6	Allen, G.R. Jr.	*Org. React.*	**1973**	*20*	337

Ethyl 4-Carbomethoxy-5-hydroxy-2-methylindole-3-carboxylate (4) and ethyl 5,8-dihydroxy-3-methylisocarbostyril-4-carboxylate (5).[3] **1** (850 mg, 5 mmol) and **2** (645 mg, 5 mmol) in EtOH (10 mL) were refluxed for 2 h. After removal of the solvent, the residue was triturated with Et_2O and filtered to give 871 mg of **3** (59%), mp 132.5-134.5°C. **3** (1.96 g, 6.65 mmol) and **2** (200 mg, 1.2 mmol) in AcOH (30 mL) was heated on a steam bath for 15 h and evaporated. The residue was dissolved in EtOAc, treated with charcoal, filtered and evaporated. The residue was triturated with CH_2Cl_2 (25 mL) to give 383 mg of **5** (23%), mp 253-254°C. The CH_2Cl_2 filtrate was chromatographed on magnesia-silica and recrystallized to afford 550 mg of **4** (30%), mp 141-143°C.

NIEMENTOWSKI Quinazolone Synthesis

Quinazolone synthesis from anthranilic acid.

COOH, NH_2 + H_2N–(O = CH) —120°→ quinazolone (O, N, N, H)

1 **2** **3** (90%)

1	v. Niementowski, S.	*J. Prakt. Chem.*	**1895**	*51*	564(2)
2	Eddicott, M.M.	*J. Am. Chem. Soc.*	**1946**	*68*	1299
3	Pater, R.	*J. Heterocycl. Chem.*	**1971**	*8*	699

O'DONNELL Amino Acid Synthesis

Amino acid synthesis by alkylation of glycine derivatives.

H_2C - $COOC_2H_5$ / N = $C(C_6H_5)_2$ + p-O_2N - C_6H_4 - CH_2Cl —1. K_2CO_3 2. HCl→ p-O_2N - C_6H_4 - CH_2 - CH-COO^- / NH_3^+

1 **2** **3** (80%)

1	O'Donnell, M.J.	*Tetrahedron Lett.*	**1982**	*23*	4259
2	O'Donnell, M.J.	*Tetrahedron Lett*	**1984**	*25*	3651
3	O'Donnell, M.J.	*Synthesis*	**1984**		313
4	O'Donnell, M.J.	*Chem. Commun*	**1985**		1168

p-Nitrophenylalanine (3).[3] **1** (1.34 g, 5 mmol), (from glycine and Ph_2CO), **2** (0.86 g, 5 mmol) and K_2CO_3 (2.0 g, 15 mmol) in MeCN (10 mL) were refluxed with stirring for 1 h. Filtration, evaporation, work up with 1N HCl (3 h), then with conc HCl (6.5 mmol,6 h reflux), evaporation and work up with MeOH (100 mL) and propylene glycol (5 mL) gave 0.85 g of **3** (80%).

NOYORI Chiral homogeneous hydrogenation

Homogeneous chiral hydrogenation of unsaturated alcohols, or carboxylic acids, enamides, ketones in the presence of BINAP Ru or Rh complex **8** as catalyst.

1. Mg
2. Ph_2 POCl
THF/PhMe

1(S)-(+) CSA
4

1 **3** (R)**5** (68%)

Cl_3SiH **6**
Et_3N/PhMe

Ru $(OAc)_2$

(R) **7** (R) **8**

8 + **9** $\xrightarrow[100\ atm]{H_2/30°C}$ **10** (96%; 99% e,e)

$\xrightarrow[H_2]{8}$

1	Noyori, R.	*J. Am. Chem. Soc.*	**1980**	*102*	7932
2	Noyori, R.	*J. Org. Chem.*	**1986**	*51*	629
3	Noyori, R.	*J. Am. Chem. Soc.*	**1986**	*108*	7117
4	Noyori, R.	*J. Am. Chem. Soc.*	**1987**	*109*	5858
5	Noyori, R.	*J. Am. Chem. Soc.*	**1989**	*111*	9134
6	Noyori, R.	*Chem. Soc. Rev.*	**1989**	*18*	187
7	Noyori, R.	*Science*	**1990**	*248*	1194
6	Otsuka, S.	*Synthesis*	**1991**		668
9	Noyori, R.	*Acc. Chem. Res.*	**1990**	*23*	345

cont. on next page

(R)-(+)-2,2'-Bis(diphenylphosphino)-1,1'-binaphthyl (BINAP) (7).[2] To Mg (2.62 g, 0.108 g-at) under N_2 was added iodine (50 mg), THF (40 mL), 1,2-dibromoethane (0.51 mL). 2.2'-Dibromo-1,1'-dinaphthyl **1** (20 g, 46.4 mmol) in PhMe (360 mL) was added dropwise over a period of 4 h at 50-75°C. After 2 h stirring at 75°C the mixutre was cooled to 0°C and diphenyl phosphinyl chloride **2** (23.2 g, 98 mmol) in PhMe (23 mL) was added over 30 min. The mixture was heated to 60°C for 3 h, cooled, quenched with water (60 mL), stirred at 60°C for 10 min and the organic layer concentrated to 60 mL. After 24 h at 20°C, the product was filtered, stirred with heptane (45 mL) and PhMe (5 mL), filtered and dried to afford 27.5 g of (±) **3** (91%), mp 295-298°C (pure 304-305°C). (±) **3** (65.4 g, 0.1 mol), (1S)-(+)-camphorsulfonic acid monohydrate **4** (25 g, 0.1 mol) and EtOAc (270 mL) were heated to reflux and HOAc (90 mL) was added to get a clear solution. Gradually cooling to 2-3°C, filtration and washing (EtOAc) gave 35.3 g of a 1:1:1 complex of **3**:**4**:AcOH. The complex was suspended in PhMe (390 mL), treated with water (30 mL) at 60°C and cooled. The organic layer was concentrated to 50 mL and treated with hexane (50 mL). Filtration and drying gave 22.2 g of (R)-(+) **5** (68%), mp 262-263°C, α_D^{24} = +399° (c 0.5 PhH). (R) **5** (50 g, 76.4 mmol), xylene (500 mL), Et_3N (32.4 g, 320 mmol) and trichlorosilane (41.4 g, 304 mmol) under Ar were heated 1 h at 100°C, 1 h at 120°C and 5 h at reflux 30% NaOH (135 mL) was added under stirring at 60°C, the organic layer was concentrated and the residue treated with MeOH (200 mL) to give 47.5 g of (R)-BINAP **7** (95%), mp 241-242°C, α_D^{24} = -228° (c 0.679 PhH).

Methyl (R)-3-hydroxybutanoate (10).[4] To $Ru(OCOCH_3)_2$ and (R)-**7** (806 mg, 0.975 mmol) in CH_2Cl_2 (210 mL) was added 1.42 N HCl in 90% MeOH (1.41 mL, 2 mmol). After 2.5 h stirring at 20°C the solvent was evaporated to give (R)-BINAP **8** (722 mg) stored under Ar. Catalyst **8** can also be prepared from $Ru(COD)Cl_2$, BINAP and TEA.[3] A solution of methyl-3-oxobutanoate **9** (100 g, 0.862 mol) in MeOH (100 mL) was treated with catalyst **8** (341 mg, 0.429 mmol) and hydrogenated at 100 atm and 30°C. Vacuum distillation afforded 97.5 g of **10** (96%), bp 40°C/2 mm, α_D^{24} = 24.2° (neat), 99% ee.

OLEKSYSZYN Aminophosphonic acid synthesis

Synthesis of 1-aminoalkanephosphonic and 1-aminoalkanephosphinic acids from ketones or aldehydes, chlorophosphines and carbamates.

Cyclohexanone (**1**) + PCl_3 (**2**) + $Ph-CH_2-O-CO-NH_2$ (**3**) $\xrightarrow[\Delta]{AcOH}$ 1-amino-cyclohexyl-$P(O)(OH)_2$ **4** (58%)

$MeCH=O$ + $Et-PCl_2$ + $Ph-CH_2\,O-CO-NH_2$ $\longrightarrow$ $Me-CH(NH_2)-P(O)(Et)(OH)$ (51%)

1 Oleksyszyn, J. *Synthesis* **1978** 479
2 Soroka, M. *Liebigs Ann.* **1990** 331

1-Aminocyclohexylphosphonic acid (4).[1] Cyclohexanone **1** (7.35 g, 75 mmol) was added at 20°C to a stirred mixture of benzyl carbamate **3** (7.55 g, 50 mmol) and PCl_3 **2** (6.87 g, 50 mmol) in AcOH (10 mL). The mixture was refluxed for 40 min, treated with 4 M HCl (50 mL) and again refluxed for 0.5 h. After cooling, the organic layer was removed and the aqueous solution was refluxed with charcoal. After filtration and evaporation in vacuum, the residue was dissolved in MeOH (25-40 mL). The methanolic solution was treated with propene oxide until pH 6-7 is reached. The precipitate was filtered, washed with Me_2CO.and recrystallized from MeOH-water to give 7.74 g of **4** (58%), mp 264-265°C.

OLOFSON Reagent

The use of vinyl chloroformate for N-dealkylation of tertiary amines, protection of amino groups, protection of hydroxyl groups or formation of 2-ketoimidazoles. Synthesis of vinyl carbonates by means of fluoro or chloroformates.

1 + Cl–C(=O)–O–CH=CH2 **2** —(20°C)→ **3** (90%) —(HCl)→ NHHCl **4** (100%)

$$Me\text{-}CH(=O) + F\text{-}CO_2Et + KF \xrightarrow[\text{Me CN, 55°}]{\text{18-Crown 6}} CH_2{=}CH\text{-}O\text{-}C(=O)\text{-}OEt$$

5 (89%)

1	Olofson, R.A.	*Tetrahedron Lett.*	**1977**		1567
2	Olofson, R.A.	*Tetrahedron Lett.*	**1977**		1570
3	Pratt, P.F.	*Tetrahedron Lett.*	**1981**	*22*	2431
4	Cooley, J.H.	*Synthesis*	**1989**		1
5	Olofson, R.A.	*J. Org. Chem.*	**1990**	*55*	1
6	Olofson, R.A.	*Pure Appl. Chem.*	**1988**	*60*	1715

Piperidine hydrochloride (4).[2] From a mixture of N-ethylpiperidine **1** (1.13 g, 10 mmol) and vinyl chloroformate **2** (1.38 g, 13 mmol) in dichloroethane at 20° are obtained 1.575 g of carbamate **3** (90%), bp 63-65°C/0.4 torr. Treatment with excess HCl gave **4** in quantitative yield.

OPPENAUER Oxidation

A mild oxidation of alcohols to ketones, using metal (Al, K) alkoxides and a ketone.

HO

H_3CO

N

N

K t-BuO

Ph_2CO

O

H_3CO

N

N

1

2 (95-98%)

1	Oppenauer, R.V.	*Rec. Trav. Chim.*	**1937**	*56*	137
2	Woodward, R.B.	*J. Am. Chem. Soc.*	**1945**	*67*	1425
3	Byrne, B.	*Tetrahedron Lett.*	**1987**	*28*	769
4	Bersin, T.	*Angew. Chem.*	**1940**	*53*	266
5	Djerassi, C.	*Org. React.*	**1951**	*6*	207

Quininone (2).[2] Potassium metal (10 g, 0.25 at g) was dissolved in t-BuOH (200 mL). The excess of t-BuOH was removed in vacuum until a dry mobile powder was obtained. A suspension of t-BuOK and quinine **1** (32.4 g, 0.1 mol), benzophenone (91.0 g, 0.5 mol) in PhH (500 mL) was refluxed for 15-18 h under N_2. The cooled mixture was poured into ice and extracted with 10% HCl until the HCl was colorless. The acid extract was washed with Et_2O and dripped with stirring into NH_4OH and ice. Extraction with Et_2O, washing of the extract with brine and evaporation of the extract afforded a total of 30-32 g of **2** (95-98%), mp 106-108°C.

OPPOLZER Chiral reagent

Asymmetric Diels-Alder reaction, ene reaction, hydrogenation, halogenation by means of chiral olefin, such as camphor derivative **1** or **2**.

$SO_2N(iC_3H_7)_2$

1 **2**

2 + **3** — Et_2AlCl, -78° →

4 (80%) 99.6% ee

1	Oppolzer, W.	*Helv. Chim. Acta*	**1983**	*66*	2358
2	Oppolzer, W.	*Helv. Chim. Acta*	**1984**	*67*	1397
3	Oppolzer, W.	*Angew. Chem. Int. Ed.*	**1984**	*23*	876
4	Oppolzer, W.	*Tetrahedron Lett.*	**1986**	*27*	831
5	Oppolzer, W.	*Pure & Appl. Chem.*	**1990**	*62*	1241

Diels-Alder adduct (4). A 1 M solution of Et_2AlCl in CH_2Cl_2 (70 ml) was added at -78°C to a mixture of *N*-crotonyl sultam **2** (10 g, 35 mmol) and cyclopentadiene (5 ml, 77 mmol). Stirring the solution for 18 h at -78°C, subsequent addition of sat. aqueous $NaHCO_3$ (100 ml), stirring of the two layers for 15 min, extraction (CH_2Cl_2), drying ($MgSO_4$), evaporation and crystallization (MeOH) afforded cycloadduct **4** (9.8 g, 80%), mp 184-186°C (*endo/exo* > 99.5 : 0.5; diastereomeric excess = 99.6%).

ORTON Haloaniline Rearrangement

Rearrangement of N-haloanilides to o- or p-haloanilides

BrNCOMe, Me, Me —Δ→ HN-COMe, Me, Me, Br

1 2

1	Orton, K.J.	*J. Chem. Soc.*	**1909**	*95*	1456
2	Orton, K.J.	*J. Chem. Soc.*	**1911**	*99*	1185
3	Richardson, M.	*J. Chem. Soc.*	**1929**		1873
4	Dewar, M.J.S.	*J. Chem. Soc.*	**1955**		1845
5	Haberfield, P.	J. Am. Chem. Soc.	***1965***	87	5502

4-Bromo-2,6-dimethylacetanilide (2).[3] N-Bromo-2,6-dimethylacetanilide **1** (1.0 g, 4.11 mmol) was added to boiling water. The resulting solid, after recrystallization from MeOH gave **2**, mp 194-195°C.

OSTROMISLENSKY Butadiene Synthesis

Catalytic butadiene synthesis from ethanol and acetaldehyde.

$$CH_3\text{-}CHO + CH_3\text{-}CH_2OH \xrightarrow[350°C]{Ti_2O_3/SiO_2} CH_2{=}CH\text{-}CH{=}CH_2$$

2 **1** **3** (35%)

1	Ostromislensky J.	*J. Russ. Phys. Chem. Soc.*	**1915**	*47*	1494
2	Jones, N.E.	*J. Am. Chem. Soc.*	**1949**	*71*	1822
3	Egloff, E.	*Chem. Rev.*	**1945**	*36*	73

O V E R M A N Pyrrolidine synthesis

Carbon-carbon bond formation via tandem Mannich condensation, cationic aza-Cope rearrangement of homoallylamines, leading to pyrolidines.

Ph CHO, Δ

1. R = H
2. R = Me

3 (87%)

1	Overman, L.E.	*J. Am. Chem. Soc.*	**1979**	*101*	1310
2	Overman, L.E.	*Tetrahedron Lett.*	**1979**		4041
3	Overman, L.E.	*J. Am. Chem. Soc.*	**1983**	*105*	6629
4	Padwa, A.	*J. Org. Chem.*	**1990**	*55*	4801

2-Methoxy-2-methyl-N-propyl-3-butenamine (2).[3] A solution of **1** (5.74 g, 40 mmol), obtained from isoprene and NBS-H_2O followed by propylamine, in THF (80 mL), was added at 4°C in 1 h to a suspension of KH (1.6 g, 40 mmol) in THF. After 30 min stirring, MeI (5.68 g, 40 mmol) was added dropwise at 4°C. The mixture was stirred for 24 h at 20°C, filtered and the filtrate distilled to afford 5.46 g of **2** (86%), bp 69-70°C, 14 mm. **2** (2.28 g, 14.4 mmol) in EtOH (5 mL) at 0°C was treated with HBF_4 (1.57 mL, 14.4 mmol of a 9.2 M aqueous solution). Concentration and recrystallization from PhH gave 3.18 g of **2**.HBF_4 (90%), mp 168-170°C.

3-Acetyl-5-phenyl-1-propylpyrrolidine (3). **2**.HBF_4 (735 mg, 3 mmol), benzaldehyde (350 mg, 3.3 mmol) and dry PhH (5 mL) was heated at reflux for 5 h. To the cooled mixture 1 N NaOH (3 mL) was added, and the aqueous layer was extracted with Et_2O. Evaporation and distillation gave 500 mg of **3** (87%) as a colorless oil; 4:3 mixture of epimers.

PAAL - KNORR Pyrrole Synthesis

Pyrrole synthesis from 1,4-butanedione and amines.

H_2N OH + O O → N OH

1 **2** **3** (73%)

O O H $(NH_4)_2\ CO_3$ DMF → H N

4 **5** (40%)

1	Paal, C.	*Chem. Ber.*	**1885**	*18*	367
2	Knorr, L.	*Chem. Ber.*	**1885**	*18*	299
3	Buu-Hoi Ng, P.	*J. Org. Chem.*	**1955**	*20*	639
4	Wasserman, H.H.	*Tetrahedron*	**1976**	*32*	1863

β-(2,5-Dimethyl-1-pyrryl)ethanol (3).[3] To redistilled 2-aminoethanol **1** (61.0 g, 1 mol) was added in small portions acetonylacetone **2** (114 g, 1 mol); an exothermic reaction immediately took place, with formation of the product. Vacuum distillation yielded 120 g of **3** (73.6%), bp 156-157°C (35 mm) which solidified on cooling, mp 50°C.

PADWA Annulation

Pyrrolines and pyrroles by (4+1) annulation of 2,3-bis(phenylsulfonyl)-1,3-butadiene and amines.

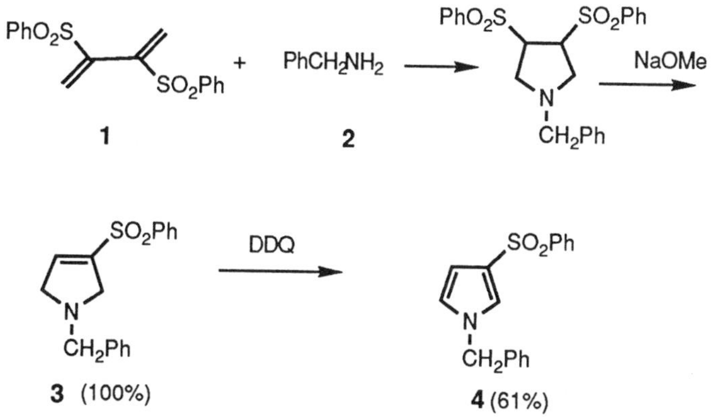

1	Padwa, A.	*Tetrahedron Lett.*	**1988**		2417
2	Padwa, A.	*Tetrahedron Lett.*	**1989**		3259
3	Padwa, A.	*J. Org. Chem.*	**1989**	*54*	810, 2862
4	Padwa, A.	*J. Org. Chem.*	**1990**	*55*	4801
5	Padwa, A.	*Org. Prep. Proc. Int.*	**1991**	*23*	545

N-Benzyl-3-(phenylsulfonyl)-3-pyrroline (3).[4] A solution of 2,3-bis(phenylsulfonyl)-1,3-butadiene **1** (1.22 g, 3 mmol) and benzylamine **2** (0.321 g, 3 mmol) in a 1:1 mixture MeOH:CH_2Cl_2 (250 mL) was stirred at 20°C for 12 h. A solution of sodium methoxide (3mmol) was added to the clear solution, and after 1 h stirring at 20°C was quenched with NH_4Cl saturated solution (50 mL). Usual work up gave 0.897 g of **3** (100% yield).

N-Benzyl-3-(phenylsulfonyl)pyrrole (4). A solution of **3** (0.897 g, 3 mmol) and DDQ (0.681 g, 3 mmol) in PhH (80 mL) was stirred for 3 h at 20°C. The solution was diluted with CH_2Cl_2 (150 mL), washed with 5% $NaHCO_3$ solution and concentrated. Chromatography (silica gel - CH_2Cl_2) afforded 0.579 g of **4** (65%).

PAQUETTE Olefin Synthesis

Desulfonation of sulfones to alkenes (alternative to Ramberg-Backlund).

BuLi
$LiAlH_4$ Δ

1 2 (17%)

1. BuLi, MeI
2. BuLi, LAH
3. 100°

1	Paquette, L.A.	J. Am. Chem. Soc.	**1976**	*98*	4936
2	Paquette, L.A.	J. Org. Chem.	**1981**	*46*	4021
3	Paquette, L.A.	J. Am. Chem. Soc.	**1975**	*97*	3538
4	Paquette, L.A.	Org. Syn.	**1977**	*57*	53

1,4,9,10-Tetrahydro-5,6-benzo-4a,10a-ethenophenantrene (2)[2] To 1,4,9,10-tetrahydro-5,6-benzo-4a,10a-methanothiomethanophenantrene-12,12-dioxide **1** (1 g, 3.08 mmol) in dioxane (40 mL) under N_2 at 0°C was added 1.2 M n-butyl-lithium (2.6 mL) in hexane. The solution (yellow, then brown) was added to a refluxing slurry of $LiAlH_4$ (0.9 g, 24 mmol) in dioxane (25 mL) and was heated under reflux for 23 h, cooled and treated with sat. Na_2SO_4 solution (5 mL). The solid was filtered and rinsed with hexane. The aqueous layer was discarded and the combined organic phases were washed (brine) and evaporated. The residue was chromatographed (silica gel-petroleum ether) to give 134 mg of **2** (17%) (colorless oil).

PARHAM Cyclic ether synthesis

Benzoheterocycle synthesis from dihalides.

1	Parham, W.E.	*J. Org. Chem.*	**1976**	*41*	1184
2	Brewer, P.D.	*Tetrahedron Lett.*	**1977**		4573
3	Bradsher, C.K.	*J. Org. Chem.*	**1978**	*43*	3800
4	Bradsher, C.K.	*J. Org. Chem.*	**1981**	*46*	1384, 4600

7-Methyl-2,3,4,5-tetrahydro-1-benzoxepine (4).[4] 1-Bromo-4-(2-bromo-4-methylphenoxy)butane **3** (9.66 g, 30 mmol) in THF (200 mL) and hexane (50 mL) at -100°C was treated with butyllithium in hexane (33 mmol). After 30 min at -100°C, the mixture was allowed to warm to 25°C over a period of 15 h and was maintained at this temperature for an additional 6 h. Aqueous work up and distillation gave 3.53 g of **4** (73%), bp 91-97°C (3.5-3.7 torr).

PARNES Geminal dimethylation

Gem dimethylation of cyclohexane derivatives (vicinal dihalocyclohexanes or methylcyclohexane) with tetramethylsilane (TMS) and AlX_3.

1	Parnes, Z.N.	*Chem. Commun.*	**1980**	*16*	748
2	Parnes, Z.N.	*Zh. Org. Khim.*	**1981**	*17*	1357
3	Parnes, Z.N.	*J. Org. Chem. USSR (Engl.)*	**1988**	*24*	291
4	Parnes, Z.N.	*Dokl. Akad. Nauk. SSSR*	**1991**	*317*	405

1,1-Dimethylcyclohexane (3).[3] a) To a cooled solution of cis 1,2-dichloro cyclohexane **1** (58.3 mg, 0.38 mmol) in CH_2Cl_2 (3 mL) was added TMS **2** (201 mg, 2.3 mmol) and $AlBr_3$ (610.2 mg, 2.3 mmol). After 30 min at 20°C the mixture was poured into water. GC analysis indicated a 37% yield of **3**.

b) To a mixture of methylcyclohexane **6** (4.9 g, 50 mmol), t-BuCl (6.0 g, 65 mmol) and TMS **2** (22 g, 250 mmol) in CH_2Cl_2 (30 mL) cooled at -78°C was added $AlCl_3$ (10 g, 75 mmol). After 30 min stirring at 20°C the mixture was poured into water (30 mL) and extracted with CH_2Cl_2. The organic layer after distillation afforded 5.3 g of **3** (80%), bp 119-120°C, n_D^{20} = 1.4296.

PASSERINI Condensation

Synthesis of α-hydroxycarboxamides from an isocyanide and an aldehyde or ketone.

$$\mathbf{1}\ (\text{pyridine-2-CHO}) + \mathbf{2}\ (t\text{-BuNC}) \xrightarrow[-5^{\circ}C]{F_3CCOOH} \mathbf{3}\ (32\%)$$

1	Passerini, M.	*Gazz. Chim. Ital.*	**1921**	*51*	126
2	Passerini, M.	*Gazz. Chim. Ital.*	**1925**	*55*	726
3	Baecker, J.	*J. Am. Chem. Soc.*	**1948**	*70*	3712
4	Kaiser, C.	*J. Med. Chem.*	**1977**	*20*	1258
5	Lumna, W.C.	*J. Org. Chem.*	**1981**	*46*	3668
6	Uggi, J.	*Angew. Chem.*	**1962**	*74*	9
7	Eckert, H.	*Synthesis*	**1977**		332

N-Tert butyl-2-hydroxy-2-(2-pyridyl)ethanamide (3).[5] To a mixture of pyridine-2-carboxaldehyde **1** (30 g, 0.28 mol) and tert butyl isocyanide **2** (10.5 g, 0.12 mol) in chloroform (100 mL) cooled at -5°C, was added dropwise with stirring trifluoroacetic acid (30 g, 0.26 mol). The mixture was warmed to 20°C, stirred with 1N NaOH (300 mL) for 2 hr and the organic layer separated. The $CHCl_3$ solution was extracted with 1.2 N HCl, the acidic solution was basified and extracted with ether. After removal of unreacted aldehyde (with $NaHSO_3$) and evaporation of the solvent, the residue (23 g) deposited 8 g of **3** (32%), mp 105-109°C, mp 114-115°C from EtOAc.

PATERNO - BÜCHI 2+2 Cycloaddition

Photochemical 2+2 cyclization of carbonyls and olefins to oxetanes.

hυ, PhH

1 → **2** (85%)

1 Paterno, E.	*Gazz. Chim. Ital.*	**1909**	*39*	341
2 Büchi, G.	*J. Am. Chem. Soc.*	**1954**	*76*	4327
3 Lange, G.C.	*Tetrahedron Lett.*	**1971**	*12*	715
4 Carless, H.A.J.	*Tetrahedron Lett.*	**1987**	*28*	5933

PAYNE Rearrangement

Rearrangement of epoxy alcohols.

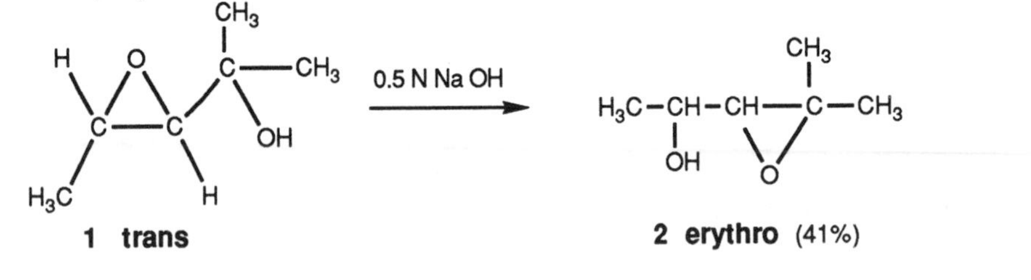

1 trans **2 erythro** (41%)

1 Payne, G.B.	*J. Org. Chem.*	**1962**	*27*	3818
2 Swindell, C.S.	*J. Org. Chem.*	**1990**	*55*	3

Erythro-4-Methyl-3,4-epoxypentan-2-ol (2).[1] trans-2-Methyl-3,4-epoxypentan-2-ol **1** (32.8 g, 0.28 mol) was treated with 0.5 N NaOH (150 mL) at 5°C. The solution was kept at 20°C for 1 h. Aft saturation with $(NH_4)_2SO_4$ (100 g), it was extracted with $CHCl_3$ (3.50 mL) and the washed extract concentrated (max temp 80-85°C). Distillation afforded 13.6 g of **2** (41%), bp 80-81°C, n_D^{25} = 1.424

von PECHMANN Diazo-Olefin Cycloaddition

3+2 dipolar cycloaddition of diazo compounds to olefins. Formation of 2-pyrazolines.

$B(OBu)_2$ + $N\equiv\overset{+}{N}-\overset{-}{C}H-CO_2Et$ →(Δ) [EtO_2C, $B(OBu)_2$ pyrazoline] → [pyrazoline, CO_2Et]

1 2 3 (44%)

1	Pechmann, H. v.,	*Chem.Ber.*	**1898**	*31*	2950
2	Matteson, D.S.	*J. Org. Chem.*	**1962**	*27*	4293
3	Huisgen, R.	*Angew. Chem.*	**1964**	*75*	616

5-Carbetoxy-2-pyrazoline (3).[2] Dibutyl ethyleneboronate **1** (9.2 g, 50 mmol) and **2** (9.2 g, 50 mmol) was kept at 20-30°C under N_2 then heated on a water bath for 12 days. After cooling to -15°C the upper layer (tributyl borate) was removed. The lower layer was washed with ligroin and treated with EtOH (15 mL). Distillation gave 3.16 g (44%) of **3**, bp 65-70°C/0.1 mm.

von PECHMANN- DUISBERG Coumarin Synthesis

Coumarin synthesis from phenols and ethyl acetoacetate.

1 + 2 →(H_2SO_4, 0° → 20°) 3 (83%)

1	Pechmann, H.v., Duisberg, C.	*Chem. Ber.*	**1883**	*16*	2119
2	Kaufman, K.D.	*J. Org. Chem.*	**1967**	*32*	504
3	Miyano, M.	*J. Org. Chem.*	**1972**	*37*	259
4	Sethna, S.	*Org. React.*	**1953**	*7*	2

PEDERSEN Crown Ethers

Crown ether formation and its use in substitutions, oxidations, etc.

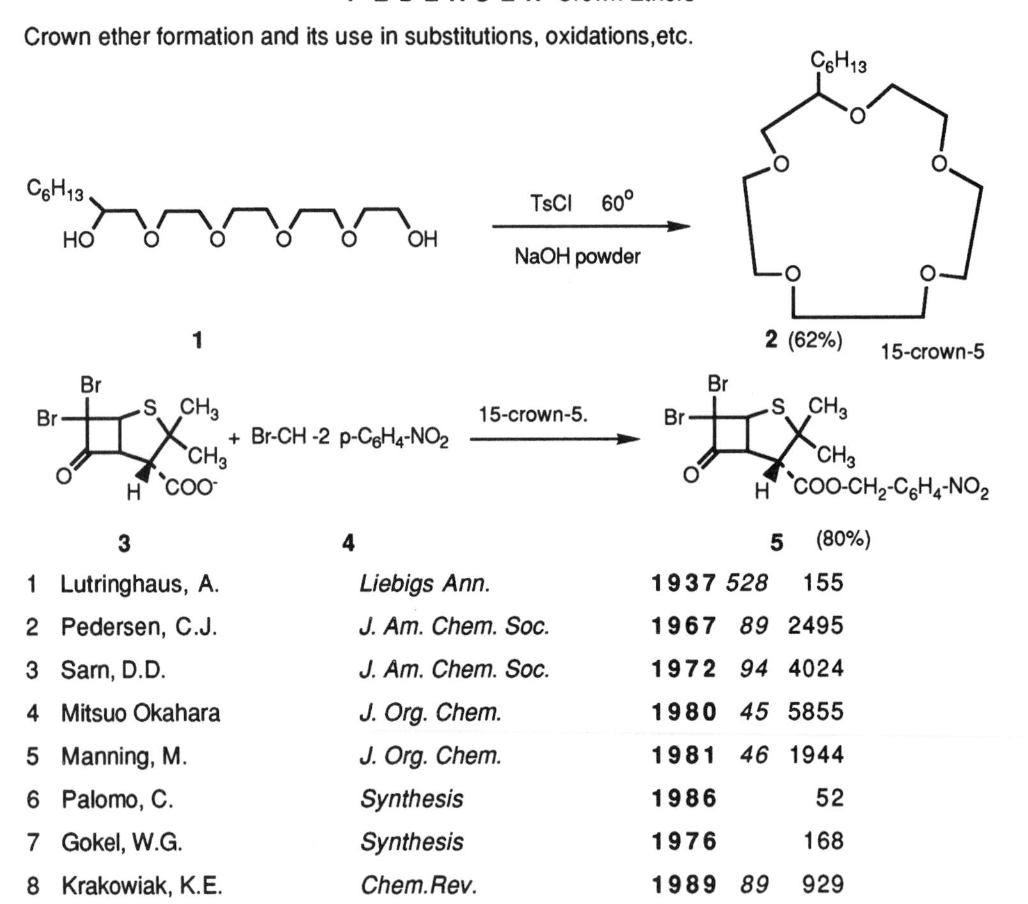

1	Lutringhaus, A.	*Liebigs Ann.*	**1937**	*528*	155
2	Pedersen, C.J.	*J. Am. Chem. Soc.*	**1967**	*89*	2495
3	Sarn, D.D.	*J. Am. Chem. Soc.*	**1972**	*94*	4024
4	Mitsuo Okahara	*J. Org. Chem.*	**1980**	*45*	5855
5	Manning, M.	*J. Org. Chem.*	**1981**	*46*	1944
6	Palomo, C.	*Synthesis*	**1986**		52
7	Gokel, W.G.	*Synthesis*	**1976**		168
8	Krakowiak, K.E.	*Chem.Rev.*	**1989**	*89*	929

4-Nitrobenzyl 6,6-dibromopenicillinate (5).[6] To sodium 6,6-dibromopenicillinate **3** (11.4 g, 30 mmol) and 15-crown-5 **2** [4] (1.5 mL) in MeCN (60 mL) is added 4-nitrobenzyl bromide **4** (6.05 g, 28 mmol) and stirring at 20° is continued for 24 h. After addition of CH_2Cl_2 (50 mL) and washing with water (3x30 mL), the organic solution is dried and evaporated in vacuum to give **5**, recrystallized from EtOH, 11.5 g of **5** (80%), mp 122-124°C.

P E R K I N Carboxylic Acid (Ester) Synthesis

Synthesis of cycloalkane carboxylic acids from α,ω-dihaloalkanes and diethyl sodiummalonate.

Br(CH2)5Br **2** + $CH_2(COOEt)_2$ **1** —EtONa→ cyclohexane-1,1-$(COOC_2H_5)_2$ **3** (34%) —HCl→ cyclohexane-COOH

1 Perkin, W.H.	*Chem. Ber.*	**1883**	*16*	1793
2 Perkin, W.H.	*J. Chem. Soc.*	**1888**	*53*	202
3 Dox, A.W.	*J. Am. Chem. Soc.*	**1921**	*43*	1366
4 Rice, L.M.	*J. Org. Chem.*	**1961**	*26*	54

Ethyl cyclohexane-1,1-dicarboxylate (3).[4] To NaOEt from Na (13.1 g, 0.569 at) in anh. EtOH (300 mL) was added dropwise **1** (45.7 g, 0.285 mol), followed by 1,5-dibromopentane **2** (65.6 g, 0.285 mol). The mixture was refluxed for 2 h, the solvent distilled and water added. After extraction (Et_2O), the extract was distilled to afford 20 g of **3** (34%), bp 105-110°C (5 mm).

P E R K I N Coumarin Rearrangement

Rearrangement of coumarins to benzofurans.

1 (7-hydroxy-3-chlorocoumarin-4-carboxylic acid: HO, O, O, Cl, COOH) —5% KOH, 95% 1h→ **2** (78%) (HO, O, COOH, COOH)

1 Perkin, W.H.	*J. Chem. Soc.*	**1870**	*23*	368
2 Holton, G.W.	*J. Chem. Soc.*	**1949**		2949
3 Johnson, I.R.	*Org. React.*	**1942**	*1*	210

PERKOW Vinyl Phosphate Synthesis

Reaction of α-haloketones with trialkylphosphite to give ketophosphonate or vinylphosphate.

Br-CH$_2$-C(=O)-C(=O)-CH$_2$-Br (**1**) + 2 P(OEt)$_3$ (**2**) —(10°, 2 h)→ H$_2$C=C(O-P(O)(OEt)$_2$)-C(O-P(O)(OEt)$_2$)=CH$_2$ **3** (98%)

(MeO)$_3$P + O=C$_6$H$_8$(X)(Me) ⟶ (MeO)$_2$P(=O)-O-C$_6$H$_8$(Me)

1	Perkow, W.	*Naturwissenschaften*	**1952**	*39*	353
2	Perkow, W.	*Chem. Ber.*	**1954**	*87*	755
3	Borowitz, I.J.	*J. Org. Chem.*	**1971**	*36*	3282
4	Hennig, M.L.	*J. Org. Chem.*	**1973**	*38*	3434
5	Mitsonobu, S.	*J. Org. Chem.*	**1981**	*46*	4030
6	Lichtenthaler, F.W.	*Chem. Rev.*	**1961**	*61*	607

1,3-Butadiene-2,3-diol bis (diethyl phosphate) (3).[4] To a stirred solution of 1,4-dibromo-2,3-butadione **1** (12.2 g, 50 mmol) in Et_2O (50 mL) was added dropwise triethylphosphite **2** (16.6 g, 100 mmol). The reaction mixture was stirred below 10°C for 2 h. The solvent was removed and the residue 17.2 g **3** (98%) was identified by IR, NMR and MS.

PETERSON Olefination

Synthesis of alkenes from α-silyl carbanions and carbonyl compounds. In cases where separation of β-silylalcohol diastereomers (e.g. **4**) can be achieved, pure Z or E olefins can be isolated.

Li + $SiPh_3$ (**1**) → $SiPh_3$, Li (**2**)

2 + Ph CHO (**3**) → $SiPh_3$, Ph, O^- (**4**) → Ph (**5**) (46%) E + Z

Bu_3Sn, S Ph, Si Me_3 —BuLi, THF / Ph CHO, - 78°→ Bu_3 Sn, Ph S, Ph, H (63%)

1 Peterson, D.J. *J. Organomet. Chem.* **1968** *33* 780

2 Peterson, D.J. *J. Org. Chem.* **1967** *32* 1717

3 Peterson, D.J. *J. Am. Chem. Soc.* **1975** *97* 1464

4 Chan, T.H. *J. Org. Chem.* **1974** *39* 3264

5 Mikami, K. *Tetrahedron Lett.* **1986** *27* 4198

6 Ager, D.J. *Org. React.* **1990** *38* 1

1-Phenylheptene (4).[4] To stirred n-BuLi in Et_2O (2.2 mL, 5 mmol), was added dropwise triphenylvinylsilane **1** (1.43 g, 5 mmol) in Et_2O (50 mL) over 1.75 h. Benzaldehyde **3** (0.53 g, 5 mmol) was added over 5 min and the mixture was stirred and refluxed for 30 h then poured into 10% NH_4Cl (50 mL). The aqueous layer was extracted with Et_2O. Evaporation of the solvent afforded 2.2 g of a pale yellow oil, which after treatment with n-pentane gave triphenylsilanol (0.6 g), mp 156-157.5°C. The pentane solution after evaporation and distillation afforded 0.4 g of **4** (46%), bp 46°C/0.01 mm or 94°C/3 mm as a mixture of E:Z isomers (1:1).

P F A U - P L A T T N E R Cyclopropane synthesis

Diazoalkane insertion into olefins with formation of cyclopropanes or ring enlargement of aromatics to cycloheptatrienes; see also formation of pyrazolines (von Pechman).

N_2CHCO_2Et, 130° → EtO_2C– ; NaOH, Δ → HO_2C–

1 **2** (52%)

1	Pfau, A.S., Plattner, P.A.	*Helv. Chim. Acta.*	**1939**	*22*	202
2	Pfau, A.S., Plattner, P.A.	*Helv. Chim. Acta.*	**1942**	*25*	590
3	Huisgen, R.	*Angew. Chem.*	**1964**	*75*	616
4	Gordon, M.	*Chem. Rev.*	**1952**	*50*	141
5	Hafner, K.	*Angew. Chem.*	**1958**	*70*	419

Azulene (2)[2] A mixture of 2-isopropyl-4,7-dimethylindane **1** (200 g, 1.91 mol) and ethyl diazoacetate (50 g, 0.5 mol) was heated for 1 h at 130°C. Vacuum distillation and recovery of **1** (160 g) gave a brown residue which was heated with 40% NaOH (40 mL) and EtOH (200 mL). The unreacted ester was extracted with Et_2O and the aqueous solution was acidified to obtain crude **2**, which after distillation afforded 24 g of **2**(52%), bp 160-185°C/ 2mm.

PFITZINGER Quinoline synthesis

Quinoline-4-carboxylic acids from isatin and α-methylene carbonyl compounds.

KOH, 100°C

1 + 2 → 3 (48%)

1	Pfitzinger, W.	*J. Prakt. Chem.*	**1886**	*33*	100(2)
2	Pritzinger, W.	*J. Prakt. Chem.*	**1888**	*38*	582(2)
3	Borsche, D.	*Liebigs Ann.*	**1910**	*377*	70
4	Henze, H.R.	*J. Am. Chem. Soc.*	**1948**	*70*	2622
5	Buu Hoi, Ng. P.	*J. Chem. Soc.*	**1949**		2882
6	Buu Hoi, Ng. P.	*Bull. Soc. Chim. Fr.*	**1966**		2765

2-Phenyl-3-propoxycinconic acid (3)[4] To a solution of 33% KOH (25 mL) were added isatin **1** (21.05 g, 0.15 mol) and α-propoxyacetophenone **2** (26.7 g, 0.15 mol). Enough EtOH was added to render the mixture homogeneous and this was heated to 100°C for 48 h. After decolorization with charcoal the filtrate was acidified with 50% AcOH and the product **3**, mp 216°C (40%), was recrystallized from dilute EtOH or Me_2CO.

PFITZNER- MOFFATT Oxidation

Oxidation of alcohols to ketones or aldehydes by means of DCC-DMSO.

$p\text{-}O_2N\text{-}C_6H_4\text{-}CH_2\text{-}OH$ (1) —DCC / DMSO→ $p\text{-}O_2N\text{-}C_6H_4\text{-}CHO$ (2) (92%)

3 —DCC/DMSO / H_3PO_4→ 4 (90%)

1	Pfitzner, K.E., Moffatt, J.G.	*J. Am. Chem. Soc.*	**1963**	*85*	3027
2	Pfitzner, K.E., Moffatt, J.G.	*J. Am. Chem. Soc.*	**1965**	*87*	5661,5670
3	Johnson, C.R.	*Tetrahedron Lett.*	**1965**	*25*	2101
4	Schobert, R.	*Synthesis*	**1987**		741
5	Epstein, W.W.	*Chem. Rev.*	**1967**	*67*	247

p-Nitrobenzaldehyde (2).[1] To a solution of p-nitrobenzyl alcohol **1** (0.135 g, 1 mmol) in DMSO was added dicyclohexylcarbodiimide (DCC) (0.618 g, 3 mmol). TLC indicated quantitative oxidation. **2** was isolated as the DNPH derivative in 92% yield, mp 316-317°C.

4-Androsten-3,17-dione (4).[2] A mixture of testosterone **3** (29 mg, 0.1 mmol) and DCC (62 mg, 0.3 mmol) in DMSO (0.5 mL) was stirred with H_3PO_4 (4.8 mg, 0.05 mmol). After 2 h an additional amount of DCC (20 mg, 0.1 mmol) was added. Stirring was continued for a few hours then the solvent was evaporated (45°C) and the residue chromatographed by preparative TLC (silica gel, $CHCl_3$:EtOAc 4:1) to give 28 mg of **4** (98%).

PICTET - HUBERT - GAMS Isoquinoline Synthesis

Isoquinolines from phenethylamides, phenanthridine from o-acylaminobiaryl with $POCl_3$ - $SnCl_4$.[6]

CH_3O, CH_3O, NH, O, OCH_3, OCH_3 **1** —$POCl_3$→ CH_3O, CH_3O, N, OCH_3, OCH_3 **2** (87%)

1 Pictet, A., Hubert *Chem. Ber.* **1896** *29* 1182

2 Pictet, A., Gams, A. *Chem. Ber.* **1909** *42* 2943

3 Falck, J.R. *J. Org. Chem.* **1981** *46* 3742

4 Whaley, M.W. *Org. React.* **1951** *6* 151

5 Boyer, J.H. *Synthesis* **1978** 205

Papaverine (2).[3] $POCl_3$ (423 mg, 3 mmol), was added dropwise to refluxing **1** (423 mg, 3 mmol) in MeCN. After 1 h the mixture was poured into ice water, washed (Et_2O), basified (2N NH_4OH), extracted with EtOAc and chromatographed to give 371 mg of **2** (87%).

PICTET - SPENGLER Isoquinoline Synthesis

Isoquinoline synthesis of phenethylamines and pyruvic acid derivatives.

HO, HO, NH_2 HCl **3** + Ph, COOH, OH **4** —80°, $NaHCO_3$→ HO, HO, NH, HOOC, Ph **5** (92%)

6 Pictet, A., Spengler, F. *Chem. Ber.* **1911** *44* 2030

7 Hudlicky, T. *J. Org. Chem.* **1981** *46* 1738

8 Bates, H.A. *J. Org. Chem.* **1986** *51* 3061

9 Govindachari, T.R. *Org. React.* **1951** *6* 151

10 Valentine,D. *Synth.* **1978** 329

PILLOTY - ROBINSON Indole Synthesis

Indole (carbazole) synthesis from azines, see also Fischer.

N-N 1 → Ac_2O / TsOH → 2 (63%) N $COCH_3$

1	Pilloty, C.	*Chem. Ber.*	**1910**	*43*	489
2	Robinson, R.	*J. Chem. Soc.*	**1918**	*113*	639
3	Posvic, H.	*J. Org. Chem.*	**1974**	*39*	2575
4	Robinson, B.	*Chem. Rev.*	**1969**	*69*	227

N-Acetyloctahydrocarbazole-1,2,3,4,5,6,7,8 (2).[3] Cyclohexylketazine **1** (4.8 g, 25 mmol) and TsOH (0.5 g) in Ac_2O (25 mL) were refluxed for 1 h. Evaporation of the solvent in vacuum and crystallization of the residue from MeOH:H_2O afforded 3.7 g of **2** (64%), mp 72-74°C.

PINNER Imino Ether Synthesis

Synthesis of imino ethers, amidines and ortho esters from nitriles.

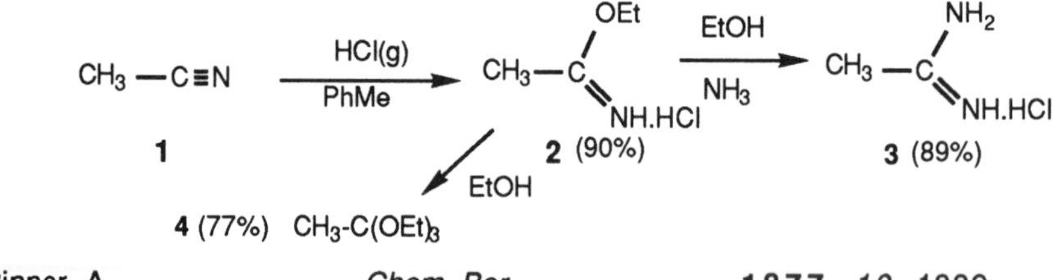

1	Pinner, A.	*Chem. Ber.*	**1877**	*10*	1889
2	Rapoport, H.	*J. Org. Chem.*	**1981**	*46*	2465
3	Roger, R.	*Chem. Rev.*	**1961**	*61*	179
4	Pompaert, J.	*Synthesis*	**1972**		622

P I R K L E Resolution

1-(1-Naphthyl)ethyl isocyanate **2** for chromatographic resolution of alcohols, hydroxy esters thiols via diastereomeric derivatives.

Me
H—C-NH_2 —$COCl_2$→ Me
H-C-NCO + CF_3
H-C-OH —80°→

(R)-**1** (R)- **2** (96%) **3**

CF_3 O H Me
H-C-O-C—N-C-H —1. chromat. 2. EtONa→ CF_3
H-C-OH

(RR)**5a** (75%) + (RS)**5b** (R)(-) **6** (95%)

1	Pirkle, W.H.	*J. Org. Chem.*	**1974**	*39*	3904
2	Pirkle, W.H.	*J. Org. Chem.*	**1979**	*44*	2169
3	Irie, H.	*J. Chem. Soc. Perkin Tr. I*	**1989**		1209

(R)-(-)(1-Naphthyl)-2,2,2-trifluoroethanol (6).[1] (R)-(+)-1-(1-Naphthyl)ethylamine **1** (17.13 g, 0.1 mol) in PhH (200 mL) was treated with dry HCl to give the **1**. HCl. The mixture was diluted with PhMe (100 mL), phosgene was bubbled in and after a few min it was heated to reflux for 4 h. Distillation gave 19.06 g of R-(-)- **2** (96.3%), bp 106-108°C (0.16 mm), α_D^{24} = -50.5°. Racemic **3** (6.20 g, 0.27 mol) and (R)-(-) **2** (5.34 g, 0.27 mol) were mixed and heated to 80°C for 65 h, until the NCO band at 2260 cm^{-1} disappeared. Chromatography on alumina (PhH) gave a major fraction, 4.34 g of **5a** (75%), mp 139.7-140.6°C (PhH).

5a (4.32 g, 10 mmol) was added to NaOEt (2.5 g Na and 30 mL EtOH) and refluxed for 30 min. Evaporation, neutralization, extraction with CH_2Cl_2 and chromatography of the residue (silica CH_2Cl_2) gave 2.17 g of **6** (95.7%). Distillation afforded **6** mp 52-53°C, α_D^{25} = +25.8°.

POLONOVSKY N-oxide Rearrangement

Conversion of heterocyclic N-oxides to α-acetoxyheterocycles.

$$\mathbf{1} \xrightarrow[7h,\ \Delta]{Ac_2O} \mathbf{2}\ (59\%)$$

Pyridine → pyridine N-oxide $\xrightarrow{Ac_2O}$ 2-acetoxypyridine (OAc)

1	Polonovsky, M.	*Bull. Soc. Chim. Fr.*	**1927**	*41*	1190
2	Bell, S.C.	*J. Org. Chem.*	**1962**	*27*	1691
3	Sternbach, L.H.	*J. Org. Chem.*	**1965**	*30*	3576
4	Lalonde, R.T.	*J. Am. Chem. Soc.*	**1971**	*93*	2501
5	Gierson, D.	*Org. React.*	**1990**	*39*	85

3-Acetoxy-1-acetyl-7-chloro-2,3-dihydro-5-phenyl-1H-benzo-1,4-diazepine (2).[2] A solution of 7-chloro-2,3-dihydro-1-acetyl-5-phenyl-1H-1,4-benzodiazepine-4-oxide **1** (7.0 g, 22 mmol) in Ac_2O (60 mL) was refluxed for 7 h. The solvent was removed in vacuum and the residue was triturated with Et_2O. The crystalline product after recrystallization from hexane CH_2Cl_2 gave 4.6 g of **2** (59%), mp 177-179°C.

POMERANZ - FRITSCH - SCHLITTER - MULLER

Isoquinoline Synthesis

Isoquinoline synthesis from aromatic aldehydes and an aminoacetal (Pomeranz-Fritsch) or from phenethylamines and glyoxal acetal (Schlitter-Muller).

MeO, CHO, OMe **1** + $CH(OEt)_2$, $H_2N\text{-}CH_2$ **2** —Δ→ MeO, MeO, $(EtO)_2CH$, CH_2, C=N, H **3** (99%) —H_2SO_4→ MeO, N, OMe **4** (8%)

Ph, NH_2 + $CH(OEt)_2$, CH, O —95°, 1 h→ $CH(OEt)_2$, PH, N **7** (85%) → N **8** (45%)

1	Pomeranz, C.	*Monatsh.*	**1893**	*14*	116
2	Fritsch, P.	*Chem. Ber.*	**1893**	*26*	419
3	White, J.D.	*J. Org. Chem.*	**1967**	*32*	2689
4	Popp, F.D.	*Chem. Rev.*	**1958**	*58*	328
5	Gensler, W.J.	*Org. React.*	**1951**	*6*	192
6	Schlitter, E., Mueller, I.	*Helv.Chim.Acta*	**1948**	*31*	914, 1119
7	Birch, A.J.	*J.Chem.Soc. Perkin Tr.*	**1974**		2185

6,8-Dimethoxyisoquinoline (4).[4] 2,4-Dimethoxybenzaldehyde **1** (3.6 g, 22 mmol) and 2-aminoacetaldehyde diethylacetal **2** (3.48 g, 26 mmol) in PhH (50 mL) was refluxed for 2 h. The solvent was distilled in vacuum to give 6.14 g of **3** (99%). Crude **3** (2.84 g, 10 mmol) was added to cooled 75% H_2SO_4 (20 mL) and stirring was continued for 3 h at 0°C and for 3 h at 20°. Quenching with water, extraction with Et_2O, chromatography on neutral alumina (hexane-PhH) and recrystallization from hexane-PhH gave 0.142 g of **4** (7.5%), mp 125-126°C.

POSNER Trioxane synthesis

Reaction of triethylsilyl hydrotrioxide with electron-rich olefins to give dioxetanes that react intramolecularly with a keto group in the presence of t-butyldimethyl silyl triflateto afford 1,2,4-trioxanes; also oxydative cleavage of alkenes. Also used in cleavage of olefins.

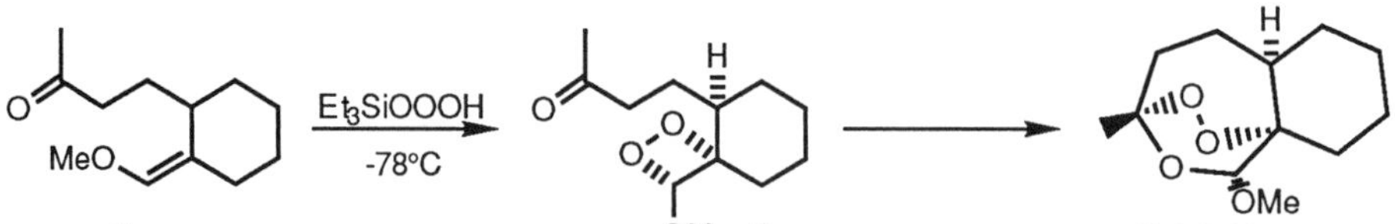

$Et_3SiOOOH$, -78°C; 1 → 2 → 3 (58%)

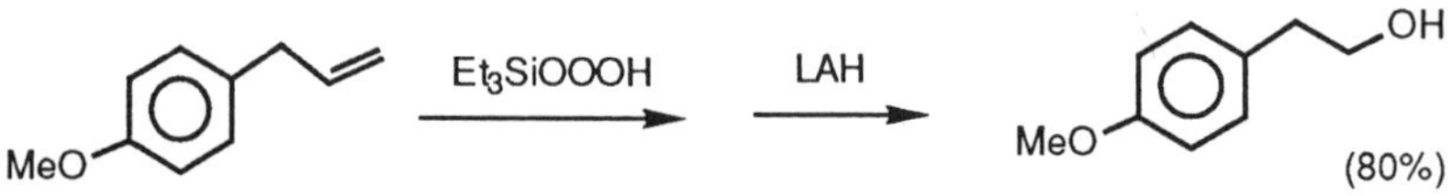

$Et_3SiOOOH$; LAH (80%)

1 Corey, E.J. *J. Am. Chem. Soc.* **1986** *108* 2472]

2 Posner, G.H. *J. Am. Chem. Soc.* **1987** *109* 278

3 Posner, G.H. *J. Org. Chem.* **1989** *54* 3252

4 Posner, G.H. *Tetrahedron Lett.* **1991** *32* 4235

Trioxane (3).[4] Triethylsilyl hydrotrioxide was prepared, in situ, by rapid addition of triethylsilane (0.52 mL, 3.2 mmol) into ozone-saturated methylene chloride (80 ml) at -78°C. To this freshly-prepared solution of $Et_3SiOOOH$ (80 ml, 3.2 mmol) at -78°C was immediately added methoxy vinyl ether **1** (63.7 mg, 0.32 mmol) in CH_2Cl_2 (3 mL) over 10 sec under N_2. After 30 min stirring was added pre-cooled (-78°C) TBDMSOTf (0.080 mL, 0.35 mmol) in CH_2Cl_2 (1 mL) over 1 min. After 5-15 min stirring at -78°C, TEA (0.68 mL, 4.9 mmol) was added, the mixture was kept at -20°C for 1 h then warmed to 20°C. Chromatography (EtOAc:hexane 2:98) gave 42 mg of **3** (58%), mp 68-69°C (from hexane).

P R E V O S T - W O O D W A R D Olefin Hydroxylation

Difunctionalization of alkenes with iodine and silver (or sodium) carboxylates.

PhCOOAg

I_2 Δ

PhCOO

1

2 (28%)

I_2 HOAc

90° 3h

OAc

KOH

24h

OH

OH

3

4 (65%)

1	Prevost, C.	*C.R.*	**1933**	*196*	1129
2	Smissman, E.E.	*J. Org. Chem.*	**1970**	*35*	3532
3	Johnson, R.G.	*Chem. Rev.*	**1956**	*56*	261
4	Wilson, C.V.	*Org. React.*	**1957**	*9*	350
5	Woodward, R.B.	*J. Am. Chem. Soc.*	**1958**	*80*	209
6	Lwowski, W.	*Angew. Chem.*	**1958**	*70*	490

4(a)-Isopropyl-6-benzoyloxymethyl-5H,6H-furo[2,3-d]-$\Delta^{1,7}$-2,4-(3H)-pyrimidinedione (2).[2] A suspension of silver benzoate (11.50 g, 50 mmol) in PhH (200 mL) was treated with I_2 (5.35 g, 25 mmol) in PhH (100 mL). After 15 min stirring, 5-isopropyl-5-allylbarbituric acid **1** (5.25 g, 25 mmol) in hot PhH (200 mL) was added and the mixture was refluxed for 2 h. Cooling, filtration, concentration in vacuum and chromatography of the residue on silica gel ($CHCl_3$) afforded 2.28 g of **2** (28%), mp 170-172°C (Me_2CO - petroleum ether).

PRINS - KRIEWITZ Hydroxymethylation

Acid catalyzed hydroxymethylation of alkenes. 1,3-Dioxane synthesis.

2 + $[CH_2O]_n$ 1 —(H_2SO_4, 70°)→ 3 (22%) + 4 (2%)

5 + 1 → 6 (45%)

1	Kriewitz, O.	*Chem. Ber.*	**1899**	*32*	57
2	Prins, H.	*Chem. Weeckblad*	**1919**	*16*	64,1072
3	Dolby, L.J.	*J. Org. Chem.*	**1963**	*28*	1456
4	Adam, D.R.	*Synthesis*	**1977**		661
5	Andersen, N.A.	*J. Org. Chem.*	**1985**	*50*	4144
6	Arundale, R.	*Chem. Rev.*	**1952**	*51*	505

3-Oxabicyclo[4,3,0]nonane (3); trans-6-hydroxy-cis-3-oxabicyclo[3,3,1]-nonane (4).[3] Paraldehyde **1** (1.83 g, 61 mmol) in AcOH (5 mL) and one drop of H_2SO_4 , was heated under stirring to 50°C. 3-Hydroxymethylcyclohexene **2** (5.0 g, 44 mmol) in AcOH (5 mL) was added dropwise and the mixture was heated for 2 h at 70°. The cooled mixture was diluted with water, neutralized with Na_2CO_3 and extracted with $CHCl_3$. Fractional distillation gave 6.11 g of **3** and **4** in a ratio of 22:2 (GC) as well as other products.

4-Chloromethyl-1,3-dioxane (6).[4] Allyl chloride **5** (600 g, 8 mol) was added over a period of 1.5 h to an ice cooled stirred mixture of **1** (300 g, 10 mol) and sulfuric acid (180 mL). After 30 min the mixture was poured into ice water, the organic layer was washed with 5% K_2CO_3, dried and distilled to give 300 g of **3** (45%), bp 83-86°C (22 mm).

P S C H O R R Arylation

Formation of polycyclics from a diazonium salt. Intramolecular Cu catalyzed arylation of diazonium salts (see Gomberg - Bachmann).

CO_2H $NaNO_2$/HCl Cu CO_2H

NH_2

1 **2** (40%)

O $NaNO_2$/HCl Cu O

NH_2

1	Pschorr, R.	*Chem. Ber.*	**1900**	*33*	1810
2	Hey, D.H.	*J. Chem. Soc.*	**1949**		3162
3	Kupchan, S.N.	*J. Org. Chem.*	**1973**	*56*	405
4	Leake, P.N.	*Chem. Rev.*	**1956**	*56*	27
5	De Tar, D.L.F.	*Org. React.*	**1957**	*9*	409
6	Gokel, G.W.	*Tetrahedron Lett.*	**1980**	*21*	4141

Phenanthren-9-carboxylic acid (2).[3] A solution of o-aminophenylcinnamic acid **1** (1.45 g, 6 mmol) in HCl (3.3 mL) and water (100 mL) was diazotized with $NaNO_2$ (0.7 g, 10 mmol) in water (40 mL). To the diazonium salt was added copper bronze (1.0 g), the mixture was heated on a water bath to complete the reaction. The white solid was filtered, dissolved in ammonia, filtered from insoluble material and the filtrate was acidified. Crystallization from AcOH gave 0.5 g of **2** (40%), mp 250-252°C.

PUMMERER Sulfoxide Rearrangement

Rearrangement of a sulfoxide to an α-acetoxysulfide.

$$(EtO)_2P(O)-CH_2-S(O)-C_6H_5 \xrightarrow[MeSO_3H\ \Delta]{Ac_2O} (EtO)_2P(O)-CH(OCOCH_3)-S-C_6H_5$$

1 → 2

Propyl p-tolyl sulfoxide $\xrightarrow[F_3C\text{-}(CO)_2O]{Ac_2O}$ 1-acetoxypropyl p-tolyl sulfide (OAc)

1	Pummerer, R.	*Chem. Ber.*	**1910**	*43*	1401
2	Johnson, C.R.	*J. Am. Chem. Soc.*	**1969**	*91*	682
3	Dinizo, St. D.	*Synthesis*	**1977**		181
4	Ishibashi, H.	*J. Chem. Res.*	**1987**		296
5	Takahashi, T.	*J. Chem. Soc. Perkin I*	**1991**		1667
6	De Lucchi, O.	*Org. React.*	**1991**	*40*	157

α-Acyloxy-α-(diethoxyphosphinyl)methyl phenyl sulfide (2).[3] To a solution of α-(diethoxyphosphinyl) methyl phenyl sulfoxide **1** (275 mg, 1 mmol) and Ac_2O (204 mg, 2 mmol) in dry CH_2Cl_2, was added methanesulfonic acid (24 mg, 0.25 mmol). The solution was refluxed for 3 h, cooled, diluted with 5% Na_2CO_3 (25 mL) and extracted with Et_2O. Purification by preparative TLC (silica gel F 254, eluent Et_2O) gave 225 mg of **2** (72%), n_D^{20} = 1.5103.

RAMBERG - BACKLUND Olefin Synthesis

Conversion of dialkyl sulfones to alkenes by rearrangement of α-halosulfones with base (via thiaranedioxides).

1. NCS
2. RCO_3H

KOtBu

1 2 3

1	Ramberg, L., Backlund, B.	*Ark. Kem. Mineral Geol.*	**1940**	*13A*	50
2	Paquette, L.	*J. Org. Chem.*	**1981**	*46*	4021
3	Opitz, G.	*Angew. Chem.*	**1963**	*77*	411
4	Paquette, L.	*Org. React.*	**1977**	*25*	1

1,4,9,10-Tetrahydro-5,6-benzo-4a,10a-ethenophenantrene (3).[3] To 1,4,9,10-tetrahydro-5,6-benzo-4a,10a-methanothiomethanophenantrene **1** (14.6 g, 50 mmol) were added N-chlorosuccinimide (NCS) (6.72 g, 50.5 mmol) and dry CCl_4. The mixture was refluxed under N_2 for 29 h, cooled, filtered, and evaporated to give a mixture of isomeric α-chlorosulfides. To this product in $CHCl_3$ (200 mL) at -23°C was added dropwise 0.624N ethereal monoperphthalic acid (163 mL). After 10 h at 20°, work up gave an isomeric mixture of α-chloro sulfone **2**.

Sulfone **2** dissolved in dioxane (250 mL) was treated with t-BuOK (35.1 g, 0.313 mol) under N_2 at 0°C, then heated to reflux for 20 h. Dilution with water and chromatography on silica gel (hexane) gave 5.13 g of **3** (40%) as a yellow oil.

RAPP - STOERMER Benzofuran synthesis

Benzofuran synthesis from salicylaldehydes and α-haloketones

Cl CHO OH **1** + Br O **2** OMe —KOH→ Cl CHO O O OMe **3**

Pyr HCl

Cl O C=O OH **4** (67%)

1	Rapp, E.	*Gazz. Chim. Ital.*	**1895** *25 II*	285
2	Stoermer, R.	*Liebigs Ann.*	**1900** *312*	331
3	Buu Hoi, Ng. P.	*J. Chem. Soc.*	**1957**	2593

5-Chloro-2-(4-hydroxybenzyol)benzofuran (4).[3] A mixture of 5-chlorosalicylaldehyde **1** (15.0 g, 0.096 mol), α-bromo-p-methoxyacetophenone **2** (22.0 g, 0.096 mol), KOH (5.3 g, 0.096 mol) in EtOH (150 mL) was heated to reflux to give 10 g of crude **3** (35%).

The crude **3** (10 g, 0.0325 mol) was heated in pyridine hydrochloride for 30 min. The cooled mixture was stirred with water, the product was filtered and purified via its sodium salt (NaOH then precipitation with HOAc). Recrystallization from aqueous EtOH afforded 6 g of **4** (67%), mp 238°C.

R E E T Z Ti alkylation

Chemo and diastereoselective addition of alkyltitanium reagents to aldehydes.

Me
Ph—CH—C(=O)H
Me_2TiCl_2

1 2 (82%) 80:20 3

1	Reetz, M.T.	*Angew. Chem. Int. Ed.*	**1980**	*19*	1011
2	Reetz, M.T.	*Chem. Ind.*	**1981**		541
3	Reetz, M.T.	*Tetrahedron Lett.*	**1981**		4691
4	Reetz, M.T.	*Angew. Chem. Int. Ed.*	**1982**	*21*	96
5	Posner, G.H.	*Tetrahedron*	**1984**	*40*	1401
6	Reetz, M.T.	*Angew. Chem. Int. Ed.*	**1984**	*23*	566

Threo and erythro 2-phenylbutan-3-ol (2) and **(3).**[1] A cooled solution (-50°C) of 2-phenylpropionaldehyde **1** (2.7 g, 20 mmol) in dry CH_2Cl_2 (50 mL) is added dropwise within 20 min to a mixture of dimethyltitanium dichloride (prepared from dimethyl zinc and $TiCl_4$ in CH_2Cl_2) (2.98 g, 20 mmol) in CH_2Cl_2 (100 mL). After 1 h the mixture is poured into ice water and treated with dil. HCl until the precipitate dissolved. The organic phase is separated and the aqueous phase extracted 3 times with CH_2Cl_2. The combined organic phases are dried ($MgSO_4$) and the solvent removed to give **2** and **3** in a ratio of 80:20 (GC). By distillation (kugelrohr) one obtains 2.45 g of **2** and **3** (82%) (80°C 1 torr).

REFORMATSKY - BLAISE Zinc alkylation

Synthesis of β-hydroxyesters from carbonyl derivatives and α-haloesters via a zinc reagent (Reformatsky). Synthesis of β-ketoesters from nitriles and α-haloesters via a zinc reagent (Blaise).

$$C_6H_5\text{-}CHO\ (\mathbf{1}) + Br\text{-}CH_2COOEt\ (\mathbf{2}) \xrightarrow[B(OEt)_3]{Zn} C_6H_5\text{-}CH(OH)\text{-}CH_2\text{-}COOEt\ \ \mathbf{3}\ (95\%)$$

$$Ph\,CH(OSiMe_3)\text{-}CN\ (\mathbf{4}) \xrightarrow[Zn\ ultrasound]{Br\text{-}CHF\text{-}COOEt} Ph\,CH(OSiMe_3)\,C(=O)\text{-}CHF\text{-}COOEt\ (\mathbf{5}) \xrightarrow{H_2SO_4} \mathbf{6}\ (62\%)$$

1	Reformatsky, S.N.	*Chem. Ber.*	**1887**	*20*	1210
2	Rathke, M.W.	*J. Org. Chem.*	**1970**	*35*	3966
3	Rathke, M.W.	*Org. React.*	**1975**	*22*	423
4	Blaise, E.	*C.R.*	**1901**	*132*	478
5	Kitazume, T.	*Synthesis*	**1986**		855
6	Kishi, Y.	*J. Org. Chem.*	**1983**	*48*	3833
7	Toda, F.	*J. Org. Chem.*	**1991**	*56*	4333
8	Fuerster, A.	*Synthesis*	**1989**		571

Ethyl 3-phenyl-3-hydroxypropionate (3).[2] To zinc powder (6.54 g, 100 m at.) in THF (20 mL) under N_2 was added **1** (10.6 g, 100 ml) in THF (25 mL) and $B(OEt)_3$ (25 mL). Under stirring **2** (121.1 g, 100 mmol) was added at once and after 12 h. conc. NH_4Cl (25 mL) and glycerol were added. Extraction (Et_2O) and distillation gave 18.5 g of **3** (95%), bp 105°C/0.3 mm.

α-Phenyl-γ-fluorotetronic acid (6).[5] To Zn powder (3.92 g, 60 m at.) and cyanohydrin **4** (5.1 g, 25 mmol) in THF (5 mL) was added α-fluorobromo acetate **5** (3.7 g, 20 mmol) in THF (5 mL). All was irradiated in an ultrasonic bath cleaner (32 KHz, 35 W) for 2 h, quenched in ice (300 g) and H_2SO_4 (20 mL) and kept for 24 h at 20°C. Extraction with EtOAc, drying, evaporation of the solvent and recrystallization afforded 2.4 g of **6** (62%), mp 102-105°C.

R E G I T Z Diazo transfer

Synthesis of diazo compounds from active methylenes with tosyl azide (diazo transfer).

$$\mathbf{1} \xrightarrow[\text{DBU}]{\text{Ts}N_3\ \mathbf{2}} \mathbf{3}\ (100\%)$$

$$NC-CH_2-CN \xrightarrow{p-TsN_3} NC-\bar{C}(N_2^+)-CN \xrightarrow{Et_3N} NC-\bar{C}(N{=}N-N^+Et_3)-CN$$

$$(R-SO_2)_2CH_2 \xrightarrow{pTsN_3} (R\ SO_2)_2C^- -N_2^+$$

1	Regitz, M.	*Angew. Chem. Int. Ed.*	**1967**	*6*	733
2	Regitz, M.	*Chem. Ber.*	**1964**	*97*	1482
3	Ledon, H.	*Synthesis*	**1972**		351
4	Ledon, H.	*Synthesis*	**1974**		347
5	Koteswar, Rao Y.	*Indian J. Chem.*	**1986**	*25b*	735

5-Diazo-6,7-dihydro-2,6,6-trimethyl-4(5H)-benzofuranone (3).[5] To 6,7-dihydro-5-formyl-2,6,6-trimethyl-4(5H)-benzofuranone **1** (774 mg, 3.76 mmol) in CH_2Cl_2 (5 mL), was added p-toluenesulfonyl azide **2** (740 mg, 3.76 mmol) in CH_2Cl_2 (5 mL), followed by 1,8-diazobicyclo [5,4,0]-undec-7-ene (DBU) (576 mg, 5.64 mmol). The mixture was stirred for 15 min at 30°C, poured into 10% KOH (20 mL), the organic layer was separated, washed (HCl, water) dried ($MgSO_4$) and evaporated to give 765 mg of **3** (100%), mp 102-103°C, (from hexane); decomposed slowly at rt.

REIMER - TIEMANN Phenol Formylation

Formylation of phenols with $CHCl_3$-base (dichlorocarbene).

1	Reimer, K.	*Chem. Ber.*	**1876**	*9*	824
2	Reimer, K., Tiemann, F.	*Chem. Ber.*	**1876**	*9*	1285
3	Bird, C.	*Chem. & Ind. (London)*	**1983**	*21*	827
4	Newmann, R.	*Synthesis*	**1986**		569
5	Wynberg, H.	*Chem. Rev.*	**1960**	*60*	169
6	Wynberg, H.	*Org. React.*	**1982**	*28*	1

2- and **4-Hydroxybenzaldehyde (2)** and **(3).**[4] To a mixture of sodium phenolate **1** (12.2 g, 0.2 mol), polyethylene glycol (PEG 400), $CHCl_3$ (71.7 g, 0.6 mol) and PhMe (50 mL) was added 50% KOH (80 mL) during a period of 30 min. Stirring was continued for an additional 30 min. The solvent was removed and the residue acidified with 2N H_2SO_4 (300 mL). After steam distillation and recrystallization, 9.76 g of **2** and **3** (40%) was obtained; mp of **3**: 118-119°C.

REISSERT - GROSHEINTZ - FISCHER Cyanoamine Reaction.

Synthesis of aldehydes or alkaloids from acid chlorides via 1-cyano-2-acylisoquinoline or 2-cyanoquinoline intermediates.

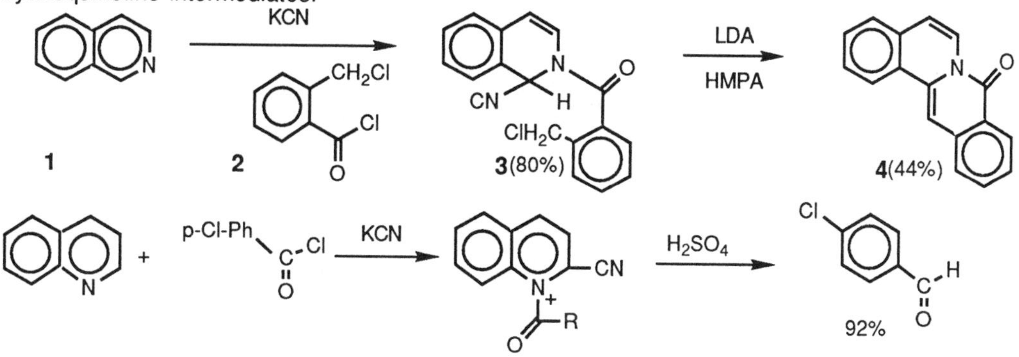

1	Reissert, A.	*Chem. Ber.*	**1905**	*38*	1608, 3415
2	Grosheintz,J.M, Fischer,H.O.L	*J. Am. Chem. Soc.*	**1941**	*63*	2021
3	McEwen, W.E.	*J. Org. Chem.*	**1981**	*46*	2476
4	Mosettig, E.	*Chem. Rev.*	**1955**	*55*	511
5	Popp, F.D.	*Synthesis*	**1970**		591
6	Popp, F.D.	*Bull. Soc. Chim. Belg.*	**1981**	*90*	609

5,6,13,14-Didehydro-9-oxoberberine (4).[3] **1** (2.5 g, 20 mmol), KCN (6.5 g, 100 mmol) CH_2Cl_2 (30 mL), water (16 mL) and benzyltriethylammonium chloride (176 mg, 0.77 mmol) were stirred at 0°C and **2** (7.34 g, 39 mmol) in CH_2Cl_2 (10 mL) was added over 50 min. Extraction with CH_2Cl_2 and evaporation gave 5.6 g of crude **3** (93%). Chromatography afforded 4.81 g of **3** (80%), mp 136-137°C. LDA (from 134 mmol BuLi in 25 mL THF) was stirred for 15 min at 20° and added over 40 min to **3** (2.0 g, 65 mmol) in THF (125 mL) and HMPA (50 mL) at -78°C. After 26 h at 20° the mixture was poured into ice water (1500 mL) and filtered to obtain crude **4** (1.68 g), 350 mg of **4** from MeOH and from the mother liquor total of 710 mg (44%), mp 144-146°C.

R E P P E Acetylene Reactions

Ni or Ti catalyzed tetramerization of trimerization of acetylene and reactions with alcohols, amines, carboxylic acids, thiols.

$$HC \equiv CH \xrightarrow[Ca_2C]{Ni(CN)_2} \text{cyclooctatetraene} \qquad 1 \xrightarrow[TiCl_4]{AlEt_3} \text{benzene}$$

1 2 3

$$HC \equiv CH + CH_2O \longrightarrow HO-CH_2-CH=CH-CH_2-OH$$

$$HC \equiv CH + HOR \text{ (or } HNR_2\text{, } HO-C(=O)-R) \longrightarrow H_2C=CH-OR$$

1	Reppe, W.	*Liebigs Ann.*	**1948**	*560*	1-104
2	Reppe, W.	*Experientia*	**1949**	*5*	93-108
3	Reppe, W.	*Liebigs Ann.*	**1953**	*582*	1-133
4	Reppe, W.	*Liebigs Ann.*	**1955**	*596*	11-20
5	Lutz, E.F.	*J. Am. Chem. Soc.*	**1961**	*83*	2552

Cyclooctatetraene (2).[2] A cooled (0-10°C) solution of $NiCl_2$ in EtOH was treated with 10% ethanolic HCN. After 12 h at 0° the $Ni(CN)_2$ catalyst was filtered and washed. To $Ni(CN)_2$ (20 g) and calcium carbide (50 g) in THF (2000 mL) under N_2 at 5 atm. acetylene was introduced at 15-20 atm and the mixture was heated to 30-60°C while acetylene was introduced from time to time. After removal of the catalyst, distillation afforded 320-400 g of **2**, bp 141-142°C.

Benzene (3).[5] $AlEt_3$ (3.4 g, 30 mmol) in n-heptane (200 mL) was treated with $TiCl_4$ (1.9 g, 10 mmol) in n-heptane (5 mL) under N_2 with efficient stirring. Acetylene **1** was bubbled through a solution of catalyst (30 mL/min). The temperature of the mixture rose slowly from 22°C to 30°C. After 5100 mL of **1** was passed through, the catalyst was removed and **3** was distilled (50%).

von RICHTER Aromatic Carboxylation

Reaction of m- and p-nitrohalobenzenes with CN^- leading to o- and m-halobenzoic acids with loss of the NO_2 group.

1	Richter, v.W.	*Chem. Ber.*	**1871**	*4*	21
2	Richter, v.W.	*Chem. Ber.*	**1875**	*8*	1418
3	Bunnett,J.E.	*J. Org. Chem.*	**1950**	*15*	481
4	Bunnett, J.E.	*J. Org. Chem.*	**1956**	*21*	944
5	Ibne Rasa, K. M.	*J. Org. Chem.*	**1963**	*28*	3240
6	Huisgen, R.	*Angew. Chem.*	**1960**	*72*	314

3-Bromobenzoic acid (2).[4] A mixture of 4-nitrobromobenzene **1** (4.0 g, 20 mmol) and KCN (2.6 g, 40 mmol) in anhydrous EtOH was heated to reflux for 1-2 h. After work up there was obtained 0.4 g (10%) of **2**, mp 155-158°C.

RICHTER - WIDMAN - STOERMER Cinnoline Synthesis

Synthesis of cinnolines from substituted anilines via diazonium salts.

1	Richter, v.W.	*Chem. Ber.*	**1883**	*16*	677
2	Scofield, K.	*J. Chem. Soc.*	**1949**		2393
3	Leonard, N.I.	*Chem. Rev.*	**1945**	*37*	270
4	Widman, O.	*Chem. Ber.*	**1884**	*17*	722
5	Stoermer, R.	*Chem. Ber.*	**1909**	*42*	3115
6	Scofield, K.	*J. Chem. Soc.*	**1945**		512
7	Simpson, J.C.E.	*J. Chem. Soc.*	**1947**		808

4-Hydroxy-3-phenylcinnoline (2).[2] 2-Aminotolane **1** (40 mg, 0.2 mmol) in 32% HCl (1 mL) at 0°C was diazotized with $NaNO_2$ (15.2 mg, 0.22 mmol), kept overnight at 20°, and heated to 95°C. The mixture was diluted with water (5 mL) to afford 25 mg of **2** (55%), mp 260-261°C.

4-Hydroxy-6-cyanocinnoline (4).[6] A cooled suspension of **3** (1.0 g, 6.1 mmol) in 2N HCl (15 mL) was treated with 15% $NaNO_2$ (15 mL) under stirring. The mixture was heated on a water bath for 1 h, cooled, filtered and the crude **4** was dissolved in Na_2CO_3 solution, treated with Norrite charcoal. HOAc precipitated **4**. Recrystallization from EtOH afforded **4**, mp 284-285°C.

R I L E Y Selenium Dioxide Oxidation

Oxidation of aldehyde or ketones to 1,2-dicarbonyl compounds with SeO_2 (sometimes oxidation to α,β-unsaturated ketones).

SeO_2

1

2 (20%)

Ph

Ph H

O (70%)

1	Riley, H.L.	*J. Chem. Soc.*	**1932**		1875
2	Schaefer, J.P.	*J. Am. Schem. Soc.*	**1933**	*66*	1668
3	Waitkins, G.R.	*Chem. Rev.*	**1945**	*36*	235
4	Rabjohn, N.	*Org. React.*	**1976**	*24*	263
5	Sharples, K.B.	*J. Am. Chem. Soc.*	**1976**	*98*	300

Cyclohexane-1,2-dione (2).[1] A solution of SeO_2 (56.0 g, 0.5 mol) in EtOH (300 mL) was added dropwise in 2 h to heated (70-80°C) cyclohexanone **1** (50.0 g, 0.5 mol). Heating was continued for another 2 h, the solvent was removed by distillation and Se was separated and washed with Et_2O. Vacuum distillation of the residue gave 40 g of **1** and **2**. Separation of **2** by extraction with 10% KOH from an Et_2O solution, acidification of the aqueous solution with ice-cooled HCl, saturation with NaCl and extraction with Et_2O gave 11 g of **2** (20%), bp 96-97°C (25 mm); bis phenylhydrazone, mp 152°C.

RITTER Amidation

Acid catalyzed reaction of nitriles with alkenes or alcohols to afford amides.

$H_3C-CH(OH)-CH(Br)-CH_3$ (**1** threo) + Me - CN (**2**) $\xrightarrow{H_2SO_4}$ [$H_3C-CH-CH-CH_3$ (Br+ bridged) / Me—C≡N $\rightarrow$ $H_3C-CH(N^+\equiv C-Me)-CH(Br)-CH_3$] $\xrightarrow{H_2O}$ $H_3C-CH(NH-C(=O)Me)-CH(Br)-CH_3$ **3** threo (81%)

OH / OH diol + Me CN $\xrightarrow{H_2SO_4}$ OH / $N^+\equiv C-Me$ $\longrightarrow$ O / N / Me oxazine

1	Ritter, J.J.	*J. Am. Chem. Soc.*	**1948**	*70*	4045
2	Ritter, J.J.	*J. Am. Chem. Soc.*	**1952**	*74*	763
3	Balaban, A.	*J. Org. Chem.*	**1965**	*30*	879
4	Wohl, R.A.	*J. Org. Chem.*	**1973**	*38*	3099
5	Ibatulin, V.G.	*Bull. Acad. Sci. USSR*	**1986**	*35*	356
6	Krimen, L.I.	*Org. React.*	**1969**	*17*	215
7	Meyers, A.I.	*J. Org. Chem.*	**1973**	*38*	36

Threo-2-Acetamido-3-bromobutane (3).[4] To threo- **1** (15.3 g, 0.1 mol) in ice cooled MeCN **2** (12.3 g, 0.3 mol), was added under stirring H_2SO_4 (70.7 g, 0.7 mol) over 30 min. The solution was warmed to 20°, kept for 3 h at 35°C, and poured into ice water (300 mL). Na_2CO_3 (20 g) was added in portions and the solution was stirred for another 5-10 min. Extraction with Et_2O, drying and evaporation of the solvent gave 15.7 g, of **3** (81%).

ROBINSON Annulation

Fusion of six-membered rings by reaction of cyclanones with vinyl ketones (base or acid catalyzed), a tandem Michael addition - aldol condensation.

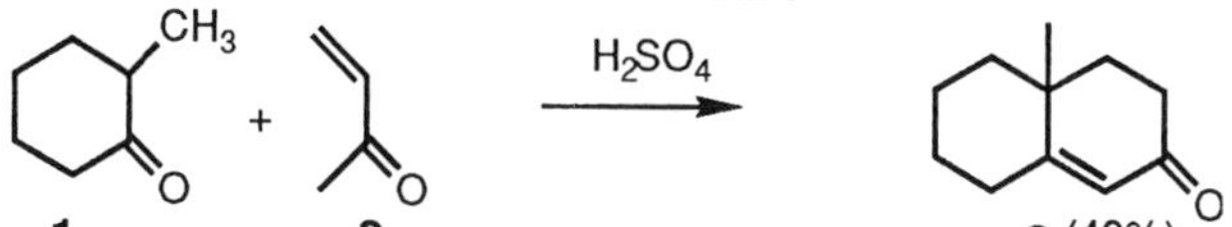

1 + 2 —(H_2SO_4)→ 3 (49%)

1	Robinson, R.	*J. Chem. Soc.*	**1937**		53
2	Gawley, R.E.	*Synthesis*	**1976**		777
3	Huffman, J.W.	*J. Org. Chem.*	**1985**	*50*	4255
4	Brewster, J.C.	*Org. React.*	**1953**	*7*	113

10-Methyl-2-oxo-$\Delta^{1(9)}$-octalin (3).[2] **1** (45.0 g, 0.5 mol), **2** (36.0 g, 0.5mol) and H_2SO_4 (0.3 mL) in PhH (100 mL), were refluxed for 16 h. Dilution with hexane (100 mL), washing with 5% KOH, drying, evaporation of the solvent and distillation gave 32.8 g or **3** (49%), bp 112-115°C (5 mm).

ROBINSON - ALLAN - KOSTANECKI Chromone Synthesis

Synthesis of chromones or coumarines from o-acyloxy aromatic ketones.

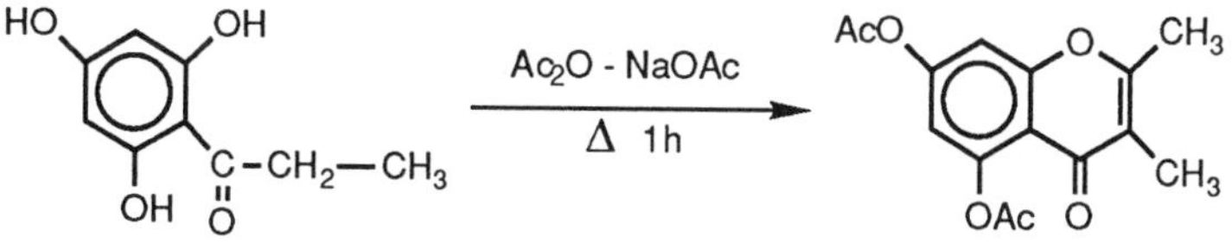

4 —(Ac_2O - NaOAc, Δ 1h)→ 5 (77%)

5	Kostanecki, S.	*Chem. Ber.*	**1901**	*34*	102
6	Robinson, R., Allan, J.	*J. Chem. Soc.*	**1924**	*125*	2192
7	Szell, Th.	*J. Chem. Soc. (C) Org.*	**1967**		2041
8	Hauser, C.R.	*Org. React.*	**1955**	*8*	91

ROBINSON - FOULDS Quinoline synthesis

Cyclization of 2-aminostyrene derivatives to quinolines.

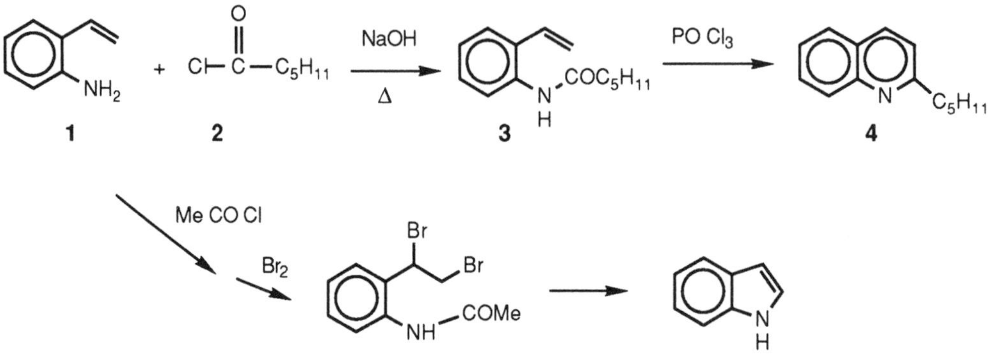

1 Robinson, R., Foulds, R.P. *J. Chem. Soc.* **1914** *105* 1968

2 Taylor, T.W.J. *J. Chem. Soc.* **1936** 181

2-n-Hexanamidostyrene (3).[2] 2-Aminostyrene **1** (0.5 g, 3 mmol) in Me_2CO (6 mL) was treated with a few drops of conc. NaOH and n-hexanoyl chloride **2** (0.6 g, 4 mmol). After 30 min reflux, the mixture was poured into water and the precipitated oil solidified on standing. Recrystallization from petroleum ether gave **3**, mp 61°C.

2-n-Amylquinoline (4). 3 was heated with 8 times its weight of phosphoryl chloride for 5 min. The mixture was poured into water and after filtration, the filtrate was made alkaline and extracted with Et_2O. The picrate of **4** melted at 104-105°C.

ROBINSON - GABRIEL Oxazole Synthesis

Oxazole synthesis from amides of α-aminoketones.

$$C_6H_5\text{-}CO\text{-}CH(Me)\text{-}NH\text{-}CO\text{-}C_6H_5 \xrightarrow{H_2SO_4} \text{2,5-diphenyl-4-methyloxazole}$$

1 → **2** (72%)

1	Robinson, R.	*J. Chem. Soc.*	**1909**	*95*	2167
2	Gabriel, S.	*Chem. Ber.*	**1910**	*43*	1283
3	Wassermann, H.H.	*J. Org. Chem.*	**1973**	*38*	2407
4	Krasowtsky, B.M.	*Chem. Heter. Compound*	**1986**	*22*	2291

2,5-Diphenyl-4-methyloxazole (2).[3] α-Benzamidopropiophenone **1** (0.3 g, 1.1 mmol) was added to concentrated H_2SO_4 (3 mL) under stirring. After 10 min copious quantities of water were added until the milky white product was completely precipitated. Filtration and recrystallization from petroleum ether (30-60°C) afforded 0.2 g of **2** (72%), mp 80-81°C.

ROELEN Carbonylation

Cobalt catalyzed addition of CO-H_2 to olefins. Synthesis of aldehydes.

$$\text{1-pentene} + CO + H_2 \xrightarrow[230\ \text{atm}]{Co\ 150°} \text{hexanal}$$

1 → **2** (75%)

1	Roelen, O.	*U.S. Patent*	**1943**	2.327.066	
2	Roelen, O.	*Angew. Chem.*	**1948**	*60*	62
3	Adkins, H.	*J. Am. Chem.Soc.*	**1948**	*70*	383
4	Kropf, H.	*Angew. Chem. Int. Ed.*	**1966**	*5*	648

ROSENMUND - BRAUN Aromatic Cyanation

Cu catalyzed nucleophilic substitution of aromatic halogen by cyanide (see Ullman-Goldberg).

CuCN 2.5 h, Δ quinoline

1 → **2** (88%)

1	Rosenmund, K.W.	*Chem. Ber.*	**1916**	*52*	1749
2	Braun, J.	*Liebigs Ann.*	**1931**	*488*	111
3	Newmann, M.S.	*J. Org. Chem.*	**1961**	*26*	2525
4	Bunnett, J. F.	*Chem. Rev.*	**1951**	*49*	392

ROSENMUND Arsonylation

Cu catalyzed arsonylation by substitution of aromatic halides. See also Bart-Scheller.

o-HOOC-C_6H_4-Br + AsO_3K_3 $\xrightarrow{Cu}$ o-HOOC-C_6H_4-$AsO(OH)_2$

3 **4** **5** (41%)

5	Rosenmund, K.W.	*Chem. Ber.*	**1921**	*54*	438
6	Hamilton, C.S.	*J. Am. Chem. Soc.*	**1930**	*52*	3284
7	Hamilton, C.S.	*Org. React.*	**1944**	*2*	415

o-Carboxyphenylarsonic acid (5).[8] **3** (20.0 g, 0.1 mol), AsO_3K_3 (50% solution), **4** (63 mL) and 10% KOH (63 mL) in EtOH (20 mL) with copper powder was refluxed at 90-95°C for 12 h, filtered hot, acidified with 20 mL HCl and evaporated. The residue was extracted with MeOH, the solvent evaporated, the residue washed with Et_2O to afford 10.1 g of **5** (41%) (from water).

ROSENMUND - SAITZEW Reduction to Aldehyde

Hydrogenation of acyl chlorides to aldehydes in the presence of poisoned Pd catalyst.

$$\underset{\mathbf{1}}{p\text{-}Cl\text{-}C_6H_4\text{-}COCl} + H_2 \xrightarrow[Pd/BaSO_4]{H_2} \underset{\mathbf{2}\ (77\%)}{p\text{-}Cl\text{-}C_6H_4\text{-}CH{=}O}$$

1	Saitzew, N.	*J. Prakt. Chem.*	**1873**	*114*	1301
2	Rosenmund, K.W.	*Chem. Ber.*	**1918**	*51*	585
3	Brown, H.C.	*J. Am. Chem. Soc.*	**1958**	*80*	5372
4	Burgsthaler, W.	*Synthesis*	**1976**		767
5	Sonntag, A.D.	*Chem. Rev.*	**1953**	52	245
6	Mossettig, E.	*Org. React.*	**1948**	*4*	363

p-Chlorobenzaldehyde (2).[4] To a well stirred suspension of 5% Pd on $BaSO_4$ (5 g) in PhH (750 mL) containing 2,4-dimethylpyridine (16.05 g, 0.15 mol) and equilibrated with H_2, was added **1** (43.75 g, 0.25 mol) in PhH (250 mL). After 1-2 h the catalyst was filtered. Evaporation and recrystallization from Et_2O-petroleum ether gave 20.4 g of **2** (77%), mp 45°C.

ROTHEMUND Porphine Synthesis

Porphine synthesis from pyrrole and aldehydes.

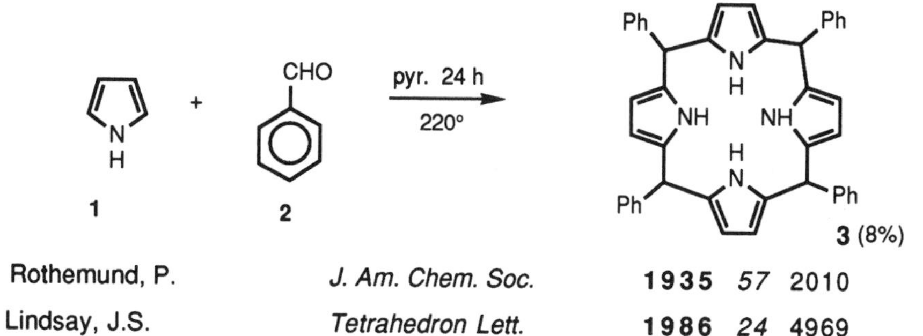

1	Rothemund, P.	*J. Am. Chem. Soc.*	**1935**	*57*	2010
2	Lindsay, J.S.	*Tetrahedron Lett.*	**1986**	*24*	4969

ROZEN Hypofluorite reagents

Acetyl hypofluorite (AcOF) and methyl hypofluorite (MeOF) as fluorinating agents of olefins and aromatics.

$$Ph_2C{=}CH_2\ (\mathbf{3}) + CH_3OF\ (\mathbf{1}) \longrightarrow Ph_2CF{-}CH_2OCH_3\ \mathbf{4}\ (71\%)$$

$$NaOAc \xrightarrow{F_2} AcOF\ (\mathbf{2})$$

$$Ph{-}CH{=}CH{-}COOEt\ (\mathbf{5}) \xrightarrow{AcOF} (Ph{-}\overset{+}{C}H{-}CHF{-}CO_2Et)\ AcO^- \longrightarrow PhCH(OAc){-}CHF{-}CO_2Et\ (\mathbf{6})$$

7 + AcOF (2) ⟶ 8 (61%) + 9 (8%)

1	Rozen, S.	*J. Chem. Commun.*	**1981**		443
2	Rozen, S.	*J. Org. Chem.*	**1984**	*49*	806
3	Rozen, S.	*Synthesis*	**1985**		665
4	Rozen, S.	*J. Org. Chem.*	**1987**	*52*	2588
5	Rozen, S.	*J. Am. Chem. Soc.*	**1991**	*113*	2648
6	Rozen, S.	*Acc. Chem. Soc.*	**1988**	*21*	307

Methyl hypofluorite (1).[5] A flow of 10% F_2 in Ar was passed at a rate of 100 mL/min into 2.5 mL MeOH and 2.5 mL EtCN at -78°C. After 1 h a 2N (5 mmol) solution of **1** was obtained.

1,1-Diphenyl-1-fluoro-2-methoxyethane (4).[5] 1,1-Diphenylethene **3** (250 mg, 1.4 mmol) in CH_2Cl_2 (5 mL) was treated with **1** (4.5 mmol) in MeCN (70 mL) at -40°C. After 40 min the mixture was poured into water, extracted with CH_2Cl_2 and the solvent evaporated. Chromatography of the residue (silica gel Merck EtOAc/petroleum ether) afforded 221 mg of **4** (71%).

1-Methoxy-2-fluoronaphthalene (8).[2] To **7** (3.16 g, 20 mmol) in CH_2Cl_2-$CFCl_3$ was added a solution (-75°C) of **2** (excess of 10-50%). The reaction was quenched with 500 mL water after 80-90% conversion, the organic layer washed with aq. $NaHCO_3$ and the solvent evaporated. Chromatography afforded 2.18 g of **8** (61%) and 0.28 g of **9** (8%).

RUFF - FENTON Degradation

Oxidative degradation of aldoses via α-hydroxy acids to lower chain aldoses.

CHO / H—C—OH / HO—C—H / H—C—OH / H—C—OH / CH_2OH $\xrightarrow[Ca(OH)_2]{Br_2}$ CO_2Ca / H—C—OH / HO—C—H / H—C—OH / H—C—OH / CH_2OH (**1**) $\xrightarrow[Ba(OAc)_2]{Fe_2(SO_4)_3,\ H_2O_2}$ CHO / HO—C—H / H—C—OH / H—C—OH / CH_2OH (**2** (50%))

1	Ruff, O.	*Chem. Ber.*	**1898**	*31*	1573
2	Fenton, O.	*Proc. Chem. Soc.*	**1893**	*9*	113
3	Fletcher, H.G.	*J. Am. Chem. Soc.*	**1950**	*72*	4546

D-Arabinose (2).[3] A mixture of calcium gluconate dihydrate **1** (200 g, 0.43 mol) (obtained by oxidation of D-glucose), $Ba(OAc)_2$ (20 g, 0.08 mol) and $Fe_2(SO_4)_3$ (10 g, 0.025 mol) was stirred in water (2000 mL) until a precipitate appeared. The suspension was filtered and the brown solution was treated with 30% H_2O_2 (129 mL) at 35°C. A second portion of 30% H_2O_2 (120 mL) was added and the temperature was raised to 40°C. After filtration on Norrite and concentration under vacuum, MeOH was added and the precipitate filtered and recrystallized to give 55-60 g of **2** (50%), mp 162-164°C, $[\alpha]_D^{20} = +103°$.

RUPE Rearrangement

Isomerisation of ethynyl carbinols to unsaturated carbonyl compounds.

OH

HCOOH

Δ

1

H_2O

O

CH-CHO

+

2 (49%)

3 (0.8%)

1	Rupe, H.	*Helv. Chim. Acta.*	**1920**	*9*	672
2	Rupe, H.	*Helv. Chim. Acta.*	**1938**	*17*	238
3	Chanley, J.D.	*J. Am. Chem. Soc.*	**1948**	*70*	246

1-Acetyl-1-cyclohexene (2) and **Cyclohexylideneacetaldehyde (3).** A mixture of 1-ethynyl-1-cyclohexanol **1** (65.0 g, 0.5 mol) and 90% HCOOH (400 mL) was refluxed for 45 min. The cooled mixture was poured into water (2000 mL) and extracted with petroleum ether. The organic layer was washed with 10% NaOH, the solvent evaporated and the residue was carefully fractionated. One obtains 32 g of **2** (49%), bp 111°C (49 mm), n_D^{20} = 1.5020, semicarbazone, mp 188-192°C.

RUZICKA - FUKUSHIMA Rearrangement

Base or acid catalyzed rearrangement of 17α-hydroxy-20-keto steroids to D-homo steroids.

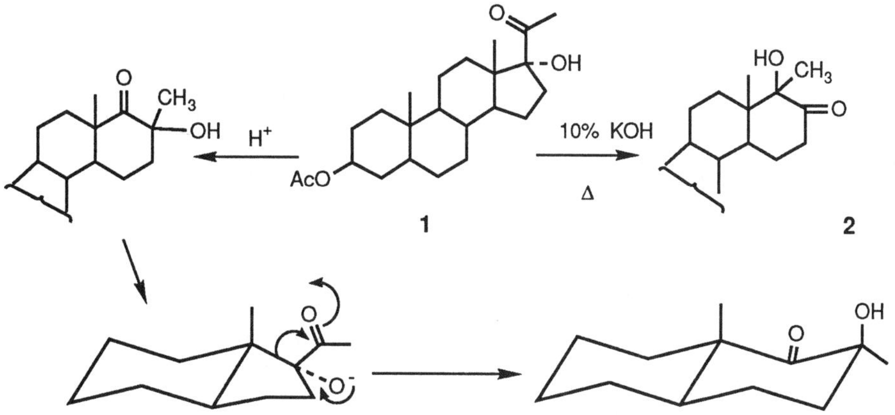

1	Ruzicka, L.	*Helv. Chim. Acta*	**1938**	*21*	1760
2	Fukushima, D.K.	*J. Am. Chem. Soc.*	**1955**	*77*	6585
3	Wendler, N.L.	*J. Am. Chem. Soc.*	**1956**	*78*	5027

3β-17aβ-Dihydroxy-17aα-methyl-D-homoandrostane-17-one (2).[2] A solution of 3β-acetoxy-17α–hydroxyallopregnane-20-one **1** (250 mL, 0.68 mmol) in EtOH (200 mL) and 10% KOH (200 mL) was refluxed for 4 h. The cooled alkaline solution was extracted with EtOAc, the organic layer washed with brine and the solvent removed in vacuo. Chromatography of 223 mg of crude product on silica gel (80 g) and elution with CH_2Cl_2:EtOH afforded 150 mg of **2** (60%). Recrystallization from Me_2CO gave 120 mg of **2**, mp 199-200°C, $[\alpha]_D^{26}$ = -36.5°.

SAKURAI Allylation

Ti mediated inter or intramolecular addition of allylic silanes to α,β-unsaturated ketones or to aldehydes.

$TiCl_4$

-78°

+ Me_3 Si

1 2

H

$CH_2CH=CH_2$

3 (85%)

Ph - CHO + $(iPrO)_2TiCl_2$ + **2**

O-iPr

Ph

4 5 6 (90%)

1	Sakurai, H.	*J. Am. Chem. Soc.*	**1977**	*99*	1673
2	Wilson, S.R.	*J. Am. Chem. Soc.*	**1982**	*104*	1124
3	Schnitzer, H.	*Synthesis*	**1988**		263
4	Sakurai, H.	*Tetrahedron Lett.*	**1976**		1295
5	Sakurai, H.	*J. Org. Chem.*	**1984**	*49*	2808
6	Seebach, D.	*Angew. Chem. Int. Ed.*	**1985**	*24*	765
7	Sakurai, H.	*Pure Appl. Chem.*	**1982**	*54*	1

9-Allyl-2-decalone (3).[1] To Δ ^{1,9} 2-octalone **1** (300 mg, 2 mmol) and $TiCl_4$ (380 mg, 2mmol) in CH_2Cl_2 (5 mL) at -78°C was added a solution of trimethyl allyl silane **2** (159 mg, 2.8 mmol) in CH_2Cl_2 (3 mL); the reaction is exothermic. After stirring at -78°C for 18 h and at -30°C for 5 h, hydrolysis, extraction and distillation gave 353 mg of **3** (85%), bp 120°C (5 mm).

Phenylallylcarbinol isopropyl ether (6).[6] Benzaldehyde **4** (106 mg, 1 mmol) was treated with diisopropoxydichlorotitanium **5** (225 mg, 1.1 mmol) at -75°C and subsequently **2** (136.8 mg, 1.2 mmol) was added . Work up led to isolation of 171 mg of **6** (90%).

SANDMEYER Isatin synthesis

Isatin synthesis from anilines.

Cl_3CHO, H_2NOH, 100°

1 → **2** (75%) + **3**

H_2SO_4, 105°

4 (98%)

1 Sandmeyer, T.	*Helv. Chim. Acta.*	**1919**	*2*	234
2 Sheilley, F.E.	*J. Org. Chem.*	**1956**	*21*	171

4,5-Dichloroisatin (4).[2] A suspension of **1** (10.0 g, 66 mmol) in water (50 mL) and 35% HCl (12 mL) was treated with Na_2SO_4 (66 g), chloral hydrate (10.5 g, 63 mmol) in water (224 mL) and hydroxylamine hydrochloride (13.0 g, 187 mmol) in water (60 mL), and stirred at 100°C for 150 min. The mixture was kept 24 h at 20°C, the solid was filtered and dissolved in 1% NaOH (750 mL). The insoluble **3** was removed and the solution was acidified with 35% HCl (25 mL) to afford 10.8 g of **2** (75%), mp 190-191°C; after recrystallization from PhH or EtOH-H_2O, mp 197°C.

To preheated 98% H_2SO_4 (70 mL) under stirring was added **2** (20.2 g, 87 mmol) during 30 min. The mixture was heated to 105°C for 10 min. and then allowed to cool to 30°C and poured into cold water (400 mL). The precipitate was collected, washed and dried to yield 18.5 g of **4** (98.8%), mp 253-254°C, from EtOH, mp 260.5-261°C.

SANDMEYER - GATTERMANN Aromatic substitution

Substitution of an amine group via the diazonium salt by nucleophiles such as Cl^-, Br^-, I^-, CN^-, $R\text{-}S^-$, HO^-, some via cuprous salt catalysis.

$$\underset{\mathbf{1}}{p\text{-}O_2N\text{-}C_6H_4\text{-}NH_2} \xrightarrow[HCl]{NaNO_2} p\text{-}O_2N\text{-}C_6H_4\text{-}\overset{+}{N}{\equiv}N\ Cl^- \xrightarrow[100^\circ]{CuCl} \underset{\mathbf{2}\ (94\%)}{p\text{-}O_2N\text{--}C_6H_4\text{-}Cl}$$

$$\underset{\mathbf{3}}{C_6H_5NH_2} \xrightarrow[PEG]{NaNO_2} C_6H_5\text{-}\overset{+}{N}{\equiv}N\ Cl^- \xrightarrow[0^\circ]{HI} \underset{\mathbf{4}\ (100\%)}{C_5H_5\text{-}I}$$

1	Sandmeyer, T	*Chem. Ber.*	**1884**	*17*	1633
2	Cattermann, L.	*Chem. Ber.*	**1890**	*23*	1218
3	Hodgson, H.H.	*J. Chem. Soc.*	**1944**		22
4	Suzuki, N.	*J. Chem. Soc. Perkin Tr.*	**1987**		645
5	Hodgson, H.H.	*Chem. Rev.*	**1947**	*40*	251
6	Pfeill, E.	*Angew. Chem.*	**1953**	*65*	155

p-Nitrochlorobenzene (2). A hot solution of p-nitroaniline **1** (2.6 g, 19 mmol) in 32% HCl (50 mL) was cooled and treated with $NaNO_2$ (2.0 g, 29 mmol). After 30 min of stirring Cu_2Cl_2 (2.0 g) and $CuCl_2.H_2O$ (3.4 g) was added. The mixture was boiled for 30 min and steam distilled. From the distillate by extraction there are obtained 3-3.05 g of **2** (94-96%), mp 83-84°C.

Iodobenzene (4). Aniline **3** (9.1 g, 0.1 mol) and 57% HI solution (0.3 mol) was diazotized by addition of $NaNO_2$ (6.9 g, 0.1 mol) in polyethylene glycol 200 (PEG 200, 2.0 g) and CH_2Cl_2 at 0°C. There are obtained 20.4 g of **4** (100%).

SARETT Oxidizing reagent

Oxidization of an alcohol to an aldehyde or ketone with CrO_3-pyridine.

CrO_3 / pyr

1 → **2** (89%)

1	Sarett, L.H.	*J. Am. Chem. Soc.*	**1953**	*75*	422
2	Holum, J.R.	*J. Org. Chem.*	**1961**	*26*	4814
3	Gassmann, P.G.	*J. Org. Chem.*	**1964**	*28*	323

4b-Methyl-7-ethylenedioxy-1,2,3,4,4a,4b,5,6,7,8,10,10aβ-dodecahydrophenanthrene-1,4-dione (2).[1] A solution of 4b-methyl-7-ethylenedioxy-1,2,3,4,4aα,4b,5,6, 7,8,10,10aβ,dodecahydrophenanthrene-4β-ol-1-one **1** (3.12 g, 10 mmol) in pyridine (30 mL) was treated with 3.1 g of CrO_3 in 30 mL of pyridine. The reaction mixture was maintained for 24 h at 20°C, poured into water, extracted three times with PhH:Et_2O (1:1), filtered through Supergel and the solvent evaporated. The residue, after recrystallization from Et_2O gave 2.76 g of **2** (89%), mp 117-120°C.

Reagent. The yellow CrO_3-pyridine reagent is prepared by adding red CrO_3 at 15-20° to 10 parts of pyridine in portions with stirring. The complex is moderately soluble in pyridine and used as a suspension to which a 10% solution of the alcohol in pyridine is added. Note: if pyridine is added to CrO_3 the mixture usually inflamed.

SCHMIDT Rearrangement

The use of HN_3 in the conversion of carboxylic acids to amides, of aldehydes into intriles, or of ketones into tetrazoles or amides (with retention).

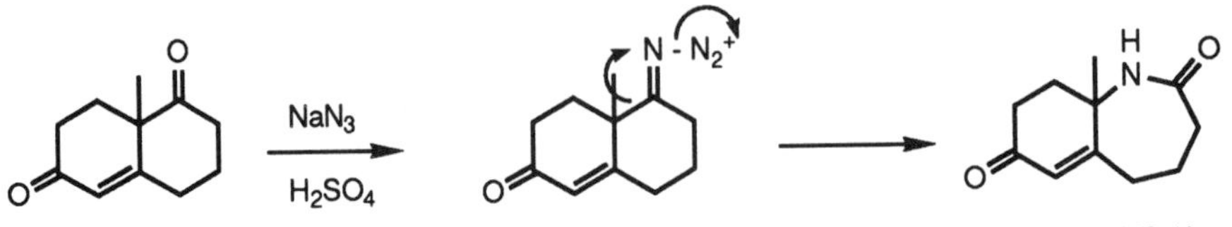

1	Schmidt, K.F.	*Z. Angew. Chem.*	**1923**	*36*	511
2	Greco, C.V.	*Tetrahedron*	**1970**	*26*	4329
3	Bach, R.D.	*J. Org. Chem.*	**1982**	*47*	239
4	Pavlov, P.A.	*Chem. Heter. Compd.*	**1986**	*22*	140
5	Hassner, A.	*J. Org. Chem.*	**1988**	*53*	22
6	Applequist, D.E.	*Chem. Rev.*	**1954**	*54*	1084
7	Wolff, H.	*Org. React.*	**1964**	*3*	307

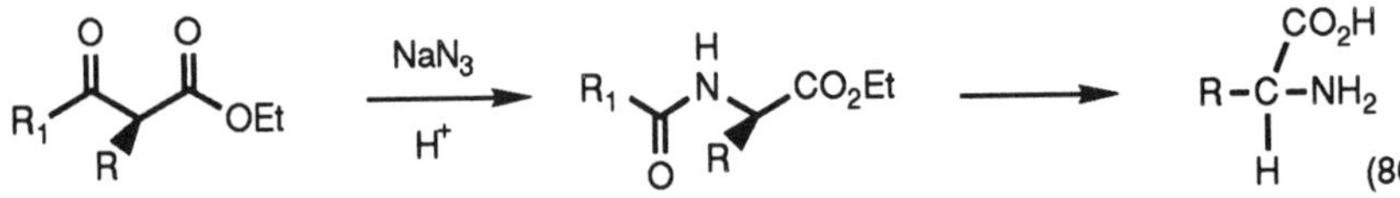

Lactam (2).[2] Cooled 9-methyl-$\Delta^{5,10}$-octalin-1,6-dione **1** (2 g, 11.2 mmol) in 98% H_2SO_4 (40 mL) was treated with NaN_3 (1.46 g, 22.4 mmol) at 5°C for 45 min, stirred for another 60 min at 5°C, and poured onto ice (200 g). Extraction with CH_2Cl_2, washing (5% KOH, water) drying ($MgSO_4$) and evaporation afforded 1.7 g of **2** (74%), mp 221-231°C. Recrystallization (EtOAc) gave 1 g of **2** (46%), mp 234-236°C.

SCHMITZ Diaziridine Synthesis

Diaziridine synthesis from chloramine, ammonia and (excess) aldehyde. In the presence of excess aldehyde formation of bicyclic triazolidines takes place.

C_5H_{11}-CHO + NH_2Cl + NH_3 ⟶ [C_5H_{11}-diaziridine (NH, NH) **2**]

1

2 —(C_5H_{11}-CHO, NH_3)⟶ **3** (87%)

1	Schmitz, E.	*Angew. Chem.*	**1959**	*71*	127
2	Schmitz, E.	*Chem. Ber.*	**1962**	*95*	680
3	Nilsen, A.T.	*J. Org. Chem.*	**1976**	*41*	3221
4	Schmitz, E.	*Chem. Ber.*	**1967**	*100*	142

cis (trans)-2,4,6-tri-(n-pentyl)-1,3,5-triazabicyclo[3.1.0]hexane (3).[3] A solution of t-butyl hypochlorite (2.71 g, 26 mmol) in t-BuOH (3 mL) was added at -35°C over 5 min to a stirred 1O N methanolic ammonia solution (25 mL), followed by hexanal **1** (5.0 g, 50 mmol). The mixture was stirred for 2.5 h at 20°C, the solvent was removed in vacuum and the residue was extracted with boiling hexane to afford 4.25 g (87%) of a mixture of cis and trans **3** in a ratio of 3.3:6.7 (exo). The less soluble fraction from hexane gave 0.67 g of **3** trans-exo (13%), mp 51-52°C; **3** cis-exo, mp 50-54°C.

SCHOLL Polyaromatic synthesis

Preparation of condensed polynuclear aromatics by Friedel-Crafts catalysts.

$AlCl_3$, $SnCl_4$

1 → 2 (66%)

1	Scholl, R.	*Chem. Ber.*	**1910**	*43*	2201
2	Scholl, R.	*Monatsh.*	**1912**	*33*	1
3	Nenitzescu, C.D.	*Chem. Ber.*	**1958**	*91*	2109
4	Vingiello, F.A.	*J. Org. Chem.*	**1971**	*36*	2053
5	Allen, C.F.H.	*Chem. Rev.*	**1959**	*59*	987

Dibenzo-(a.1)-pyrene (2).[4] To a refluxing mixture of $AlCl_3$ (0.6 g, 4.5 mmol) and $SnCl_4$ (0.5 g, 1.9 mmol) in PhH (25 mL), was added a hot solution of 1-phenyl-benz(a)antracene **1** (30 mg, 0.1 mmol) in PhH (25 mL). After 5 h reflux the mixture was poured into 10% HCl (500 mL), the organic layer was separated and the solvent evaporated in vacuum. To the residue was added 95% EtOH (5 mL) to obtain 0.2 g of **2** (66%), mp 160-163°C; recrystallized from PhH:EtOH, mp 162-163°C.

SCHÖLLKOPF Amino acid synthesis

Asymmetric synthesis of amino acids from dihydropyrazines.

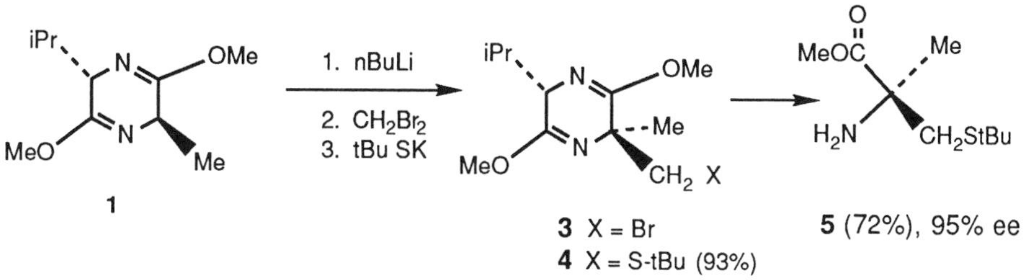

1

1. nBuLi
2. CH_2Br_2
3. tBu SK

3 X = Br
4 X = S-tBu (93%)

5 (72%), 95% ee

1 Schöllkopf, U *Synthesis* **1981** 969
2 Schöllkopf, U. *Liebig's Ann.* **1982** 1925
3 Schöllkopf, U. *Synthesis* **1983** 37
4. Schöllkopf, U. *Synthesis* **1985** 1052

(3S,6R) pyrazine (4).[3] To **1** (2.77 g, 14 mmol) in THF (25 mL) at -70°C was added 1.8N BuLi in hexane (8.3 mL, 15 mmol) followed after 15 min by CH_2Br_2 (26.1 g, 0.15 mol) in THF (15 mL). After stirring 30 h at -70°C, work up afforded 3.2 g of **3** (79%), bp 760-80°C (0.1 torr).
To t-butylmercaptan (0.32 g, 3.5 mmol) in DMSO (10 mL), was added KOtBu (0.37 g, 3.3 mmol) and after 5 min a solution of **3** (0.87 g, 3 mmol) in DMSO (2 mL). Stirring was continued for 5 h at 70°C and the solution was mixed with ether (30 mL) and shaken with water (10 mL). After drying and distillation 0.837 g of **4** (93%), bp 80-90°C (0.1 torr) was isolated.

(R)-S-t-Butyl-α-methylcystein methyl ester (5).[3] A suspension of **4** (0.75 g, 2.5 mmol) in 0.25 N HCl (20 mL) was stirred at 20°C for 5 days, washed with Et_2O and evaporated. The residue (**5** HCl and $ValOCH_3$ HCl) was dissolved in water, Et_2O was added, followed by ammonia (to pH=8-10) with shaking. The Et_2O extract was distilled ($ValOCH_3$ forerun) to afford 0.37 g of **5** (72%), bp 60-70°C (0.1 torr), $[\alpha]_D^{20}$ = -16.3° (c 1.0 EtOH), e.e.>95%.

SCHOLTZ Indolizine synthesis

Indolizine synthesis from reaction of pyridinyl ketones with aldehydes in the presence of ammonium acetate.

1	Scholtz, M.	*Chem. Ber.*	**1912**	*45*	734
2	Barow, E.T.	*Chem. Ber.*	**1948**	*42*	638
3	Krohnke, F.	*Chem. Ber.*	**1971**	*104*	1624
4	Uchida, T.	*Synthesis*	**1976**		209

1-(4-Chlorobenzylideneamino)-3-(4-chlorophenyl)-indolizine (3).[4] A mixture of 2-acetylpyridine **1** (2.4 g, 20 mmol), p-chlorobenzaldehyde **2** (5.6 g, 40 mmol) and ammonium acetate (20 g, 0.33 mol) in AcOH (30 mL) was heated for 30 min under stirring on a water bath. To the mixture MeOH (20 mL) was added and stirring was contined for another 15 min. at the same temperature. After 24 h the product was filtered and chromatographed on alumina (neutral activity 1, eluent $CHCl_3$). There are obtained 2.9 g of **3** (39%), mp 167-168°C.

SCHWARTZ Hydrozirconation

Hydrozirconation with $Cp_2Zr(Cl)H$ and Michael addition of the Zr reagent.

THF

$Cp_2Zr(Cl)$–CH=CH–$(CH_2)_5CH_3$

NiAcac$_2$

1 **2** **3**

4 (61%)

1	Wailes, C.P.	*J. Organometal. Chem.*	**1970**	*24*	405
2	Schwartz, J.	*J. Am. Chem. Soc.*	**1974**	*96*	8115
3	Schwartz, J.	*J. Am. Chem. Soc.*	**1980**	*102*	1333
4	Schwartz, J.	*Angew. Chem. Int. Ed.*	**1976**	*15*	333
5	Negishi, Ei-ichi	*Aldrichimica Acta*	**1985**	*18*	31
6	Schwartz, J.	*Chimica Scripta*	**1989**	*29*	411

3-(1-Octen-1-yl)cyclopentanone (4).[3] Chlorobis (η^5-cyclopentadienyl) hydridozirconium **1** (38.68 g, 0.15 mol) in THF (50 mL) under Ar was treated with 1-octyne **2** (23.6 mL, 0.16 mol) at 15-25°C. After 18 h stirring at 20°C, 2-cyclopentenone **3** (10.9 mL, 0.13 mol) was added and the mixture was kept for 10 min in an ice bath. Nickel acetylacetonide (3.34 g, 13 mmol) was added in three portions at 10 min intervals below 40°C. After 2 h stirring at 5°C and 2 h at 20°C the mixture was quenched with 1 N HCl (150 mL) and ice water (200 mL). Extraction with hexane, washing and evaporation of the solvent, afforded 24.79 g of crude product. Chromatography on silica gel (2% EtOAc in hexane and 4% EtOAc in hexane) gave 15.43 g of **4** (61.2%).

SCHWEIZER Allylamine synthesis

Synthesis of E-allylamines from vinylphosphonium salts and aldehydes (via Wittig reaction).

Br⌒⌒OH **2** —(nBu_3 P **1**; Δ, 4d)→ $n\text{-}Bu_3P^+$⌒⌒ Br^- **3**

Ph CHO **4** + phthalimide (HN) **5** —(NaH; **3**)→ N-(3-phenylallyl)phthalimide **6** (80%) —(NH_2NH_2)→ Ph⌒⌒NH_2 **7** (83%)

1	Schweizer, E.E.	*J. Org. Chem.*	**1966**	*31*	467
2	Rouhut, M.M.	*J. Org. Chem.*	**1963**	*28*	2565
3	Evans, D.A.	*J. Am. Chem. Soc.*	**1978**	*100*	1548
4	Meyers, A.I.	*J. Org. Chem.*	**1981**	*46*	3119

(E)-3-Phenylpropenylamine (7).[4] $n\text{-}Bu_3P$ **1** (131 g, 0.65 mol), 1,2-dimethoxyethane (DME) (300 mL) and **2** (72.2 g, 0.68 mol) were refluxed for 4 h. 48% HBr (1 mL) and isopentyl acetate (243 mL, 2.2 mol) were added and reflux continued for 3 d. After evaporation, the residue was treated with DME (400 mL) and Na_2SO_3 (68.9 g, 0.65 mol) and refluxed for 48 h. Filtration, evaporation and treatment with EtOAc gave 100 g of **3** (57%), mp 149.5-150.5°C. NaH (1.5 mmol) was washed (pentane), treated in THF with **4** (0.3 g, 1 mmol), **3** (0.4 g, 1.3 mmol) and phthalimide **5** (0.19 g, 1.3 mmol) and heated at 60°C (TLC $CHCl_3$:Et_2O:hexane 5:1:4). Treatment with 5% citric acid in water and extraction with Et_2O gave 281 mg of **6** (80%), mp 150-151°C.

To **6** (174 mg, 0.6 mol) in anh. EtOH (19 mL) was added 95% hydrazine (60 mL, 1.8 mmol). 4.5 h reflux, acidification to pH=2, heating for 1 h, filtration, dilution of the filtrate, extraction with Et_2O and basification gave 93 mg of **7** (83%), mp 101-102°C, 100% E.

SCHWEIZER Rearrangement

Thermal reaction of "allenyl azines", derived from propargylphosphonium salts with ketenes, isocyanates, CS_2 or phthalic anhydride to form bi- and tricyclic fused pyrazolo heterocycles.

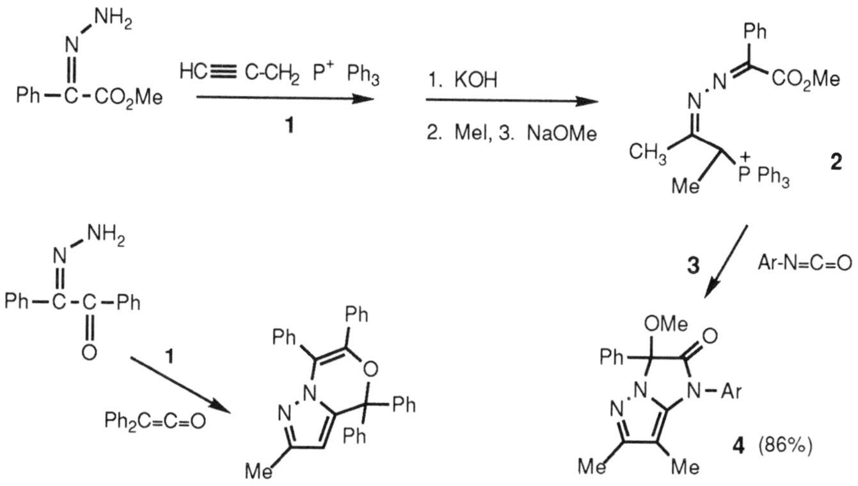

1	Schweizer, E.E.	*J. Org. Chem.*	**1978**	*43*	4328
2	Schweizer, E.E.	*J. Org. Chem.*	**1987**	*52*	1810
3	Schweizer, E.E.	*J. Org. Chem.*	**1990**	*55*	1687
4	Schweizer, E.E.	*J. Org. Chem.*	**1990**	*55*	6363

6,7-Dimethyl-3-methoxy-3-phenyl-1-(2,6-dimethylphenyl)-1H-imidazo(1,2-b) pyrrazol-2(3H)-one (3).[2] Phosphorane **2** (1.48 g, 3 mmol) and 2,6-dimethylphenylisocyanate **3** (0.59 g, 4 mmol) in PhMe (30 mL) was heated under reflux with stirring for 16 h (TLC EtOAc : hexane 1:6). Evaporation and column chromatography (silica gel EtOAc/hexane 1:6) afforded crude **4**. Recrystallization from Et_2O-petroleum ether gave 0.930 g (86%), mp 128-5-130°C.

SEMMLER - WOLFF - SCHROETER Oxime Aromatization

Aromatization of cyclohexanone oximes to anilines.

Ac_2O / AcOH / HCl, refl. 18 h.

1 → 2 (53%)

1	Semmler, W.	*Chem. Ber.*	**1892**	*25*	3352
2	Wolff, L.	*Liebigs Ann.*	**1902**	*322*	351
3	Schroetter, G.	*Chem. Ber.*	**1930**	*63*	1308
4	Cook, J.M.	*J. Org. Chem.*	**1980**	*45*	2585

SIMONIS Benzopyrone synthesis

Benzopyrone synthesis from phenols and β-ketoesters

1 + 2 → P_2O_5, 100°, 2h → 3 (10%)

1	Simonis, H.	*Ber.*	**1913**	*46*	2014
2	Simonis, H.	*Ber.*	**1914**	*47*	2229
3	Lacey, R.N.	*J. Chem. Soc.*	**1954**		854
4	Sethna, S.M.	*Chem. Rev.*	**1945**	*36*	14
5	Sethna, S.M.	*Org. React.*	**1953**	*7*	15

SHARPLESS Asymmetric Epoxidation

Enantioselective epoxidation of allyl alcohols by means of titanium alkoxide, (+) or (-) diethyl tartarate (DET) and t-butyl hydroperoxide (TBHP).

$Ti(OR)_4$; (+)DET

TBHP: CH_2Cl_2 -23°C

1 → **2** (77%) 95% ee

$Ti(OiPr)_4$

DET, TBHP

erythro 98:2 threo

1	Sharpless, K.B.	*J. Am. Chem. Soc.*	**1980**	*102*	5974
2	Sharpless, K.B.	*J. Am. Chem. Soc.*	**1981**	*103*	464, 6237
2	Sharpless, K.B.	*J. Am. Chem. Soc.*	**1987**	*109*	5765
4	Sharpless, K.B.	*Aldrichimica Acta*	**1983**	*16*	67

2(S),3(S)-Epoxygeraniol (2).[1] To CH_2Cl_2 (200 mL) at -23°C was added sequentially under stirring titanium tetraisopropoxide (5.68 g, 5.94 mL, 20 mmol), L(+)DET (4.12 g, 3.43 mL, 20 mmol) and after 5 min geraniol **1** (3.08 g, 3.47 mL, 20 mmol) and 3.67 M of (TBHP) 40 mmol in CH_2Cl_2 . After 18 h at -20°C 10% aqueous tartaric acid (50 mL) was added under stirring and after 30 min the mixture was heated to 20°C and stirred for 1 h. The organic layer was washed, dried and evaporated. The oily residue was diluted with Et_2O (150 mL) washed with 1 N NaOH (60 mL), brine, dried and the solvent evaporated. Chromatography on silica gel afforded 2.6 g of **2** (77%), 95% ee, $[\alpha]_D^{24}$ = - 6.36°C (c 1.5, $CHCl_3$).

S H A R P L E S S Asymmetric Dihydroxylation

Enantioselective syn dihyroxylation of olefins using AD-mix-β formed from phthalazine-dihydroquinidine **1** or AD-mix-α formed from phthalazine-dihydroquinine **2** and OsO_4.

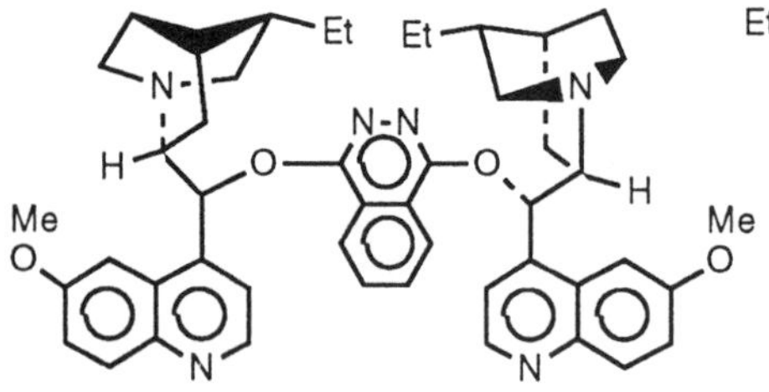

1 $(DHQD)_2$- PHAL 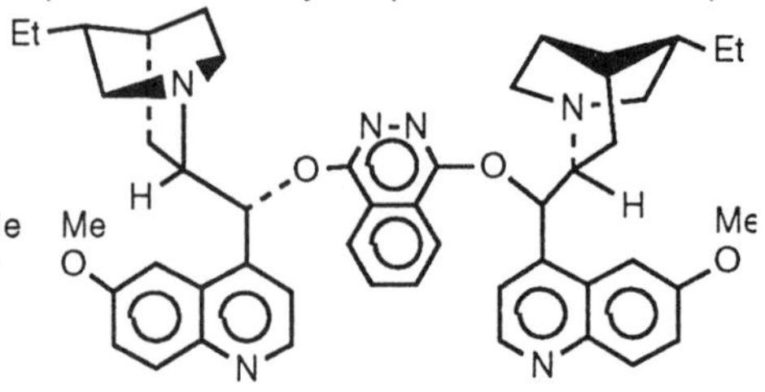 **2** ($(DHQ)_2$ - PHAL

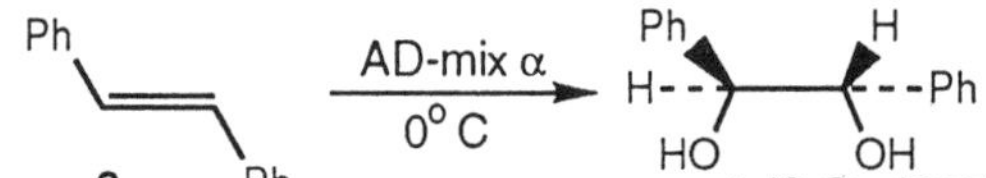

3 → AD-mix α, 0° C → **4** (S,S) (96%; 99%ee)

1	Sharpless, K.B.	*J. Am. Chem. Soc.*	**1988**	*110*	1968
2	Sharpless, K.B.	*Tetrahedron Lett.*	**1990**	*31*	2999, 3817
3	Sharpless, K.B.	*J. Org. Chem.*	**1992**	*57*	2768
4	Soderquist, J. A.	*J. Org. Chem.*	**1992**	*57*	5844
5	Sharpless, K.B.	*Org.Synth.*	**1991**	*70*	47

AD-mix (α or β).[1] A mixture of $K_2OsO_2(OH)_4$ (0.52 g) and $(DHQ)_2$-PHAL **2** (dihydroquinine-phthalazine) or $(DHQD)_2$-PHAL **1** (dihydroquinidine-phthalazine) (5.52 g) was mixed with powdered $K_3Fe(CN)_6$ (700 g) and K_2CO_3 (294 g). The mixture was kept dry, ready for use in the next step. **1,2-Diphenylethanediol (4) (S,S) or (R,R).** AD-mix (α) or AD-mix (β) (1.4 g) was stirred at 20°C into t-BuOH (5 mL) and water (5 mL). Methanesulfonamide (95 mg, 1mmol) was added and the mixture was cooled to 0°C. Under vigorous stirring, trans stilbene **3** (180 mg, 1mmol) was added (monitored by tlc or gc). Na_2SO_3 (1.5 g) was added and after 1 h the mixture was extracted with EtOAc. Washing (2N KOH) and flash chromatography (silica gel EtOAc-hexane) afforded 205.4 mg of **4** (96%), 99%ee, $[\alpha]^{24}_D$ = +90.0° or -90.0° (c 1.2 EtOH).

SHERADSKY Rearrangement

Rearrangement of O-(2-pyridyl)oximes to 2-pyridones via a hetero-Cope rearrangement.

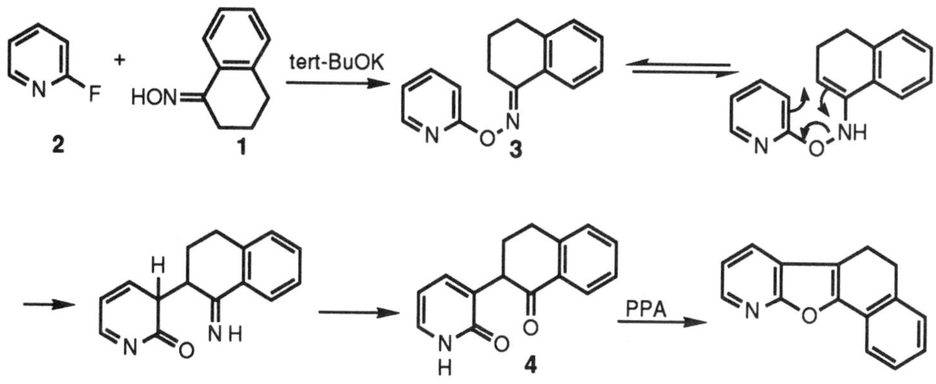

1	Sheradsky, T.	*Tetrahedron Lett.*	**1966**		5225
2	Sheradsky, T.	*Israel J. Chem.*	**1968**	*6*	859
3	Sheradsky, T.	*J. Org. Chem.*	**1971**	*36*	1061
4	Laronze, J.Y.	*Tetrahedron Lett.*	**1989**	*30*	2229

0-(2-Pyridyl)oxime of 1-tetralone (3).[3] 1-Tetralone oxime **1** (3.22 g, 20 mmol) and t-BuOK (2.2 g, 20 mmol) in dry DMSO (40 mL) was stirred under N_2 for 30 min and **2** (1.94 g, 20 mmol) in DMSO (40 mL) was added. Stirring was continued for 1 h at 20°C and 3 h at 90°C. The solution was poured into water (300 mL) and the precipitate crystallized from MeOH to afford 2.38 g of **3** (50%), mp 224°C.

3-(1-Oxo-2-tetralyl)-2-pyridone (4). A solution of **3** (1.0 g, 4.1 mmol) in ethylene glycol (20 mL) was refluxed for 20 h under N_2, and poured into water (100 mL). The preciptate was crystallized from EtOH to yield 0.73 g of **4** (73%), mp 206-207°C.

SHESTAKOV Hydrazine synthesis

Synthesis of α-hydrazino acids from α-amino acids via ureas.

KCNO 60° → **2** (76%); NaClO, NH_2NH_2 → **3** (48%)

1 Shestakov, P. *Z. Angew. Chem.* **1903** *16* 1061

2 Karady, S. *J. Org. Chem.* **1971** *36* 1949

3 Viret, J. *Tetrahedron* **1987** *43* 891

4 Kost, A.N. *Russ. Chem. Rev.* **1964** *33* 159

L-α-(3,4-Dimethoxybenzyl)-α-hydrazinopropionic acid (3)[2] L-α-Amino-α-(3,4-dimethoxybenzyl)propionic acid hydrochloride **1** (44.0 g, 0.16 mol) in water (440 mL) was treated at 5°C with KCNO (77.6 g, 0.96 mol) in small portions. The slurry was heated at 60°C for 4 h and then filtered. The filtrate was acidified with 35% HCl to pH 1,. The product was filtered, washed and dried at 50°C, to give 34.5 g of **2** (74.6%), mp 205-207°C.

To ice-cooled **2** (2.2 g, 7.8 mmol) in 2.5 N KOH (15.6 mL) was added 0.17 N NaOCl (13.7 mL, 9.75 mmol). The solution was heated to 80°C for 1 h, PhMe (45 mL) and hydrazine hydrate (0.8 mL) was added followed by 8 mL of 35% HCl, while the mixture was stirred vigorously. After 30 min stirring at 80°C, the aqueous layer was extracted, the extract was evaporated to dryness and the residue digested with EtOH. The EtOH solution was treated with Et_2NH to pH 6.4 and the product filtered, washed, and dried to give 1 g of **3** (48%), mp 222-224°C, $[\alpha]_D$ = -9° (H_2O).

SIEGRIST Stilbene Synthesis

Synthesis of stilbenes by base catalyzed condensation of reactive toluenes with benzalanilines.

CH_3 + PhN = CH-Ph $\xrightarrow[\text{DMF } 90°]{\text{KOH}}$ Ph

1 **2** **3** (94%)

1	Siegrist, A.E.	*Helv. Chim. Acta.*	**1967**	*50*	906
2	Siegrist, A.E.	*Helv. Chim. Acta.*	**1969**	*52*	1282
3	Newman, M.S.	*J. Org. Chem.*	**1978**	*43*	524

2-(Stilben-4-yl)benzo(b)furan (3).[2] A mixture of 2-(p-tolyl)benzo(b)furan **1** (2.6 g, 12.5 mmol), benzalaniline **2** (2.27 g, 12.5 mmol) and powdered KOH (6.25 g, 0.1 mol) in DMF (100 mL) was stirred and heated to 90°C for 30 min and to 90-95°C for another 60 min. To the cooled mixture was added water (100 mL) and 10% HCl. The product was filtered, washed with water and MeOH, to give after drying 3.5 g of **3** (94%), mp 270-271°C.

SIMMONS - SMITH Cyclopropanation

Cyclopropane formation from alkenes with alkyldiiodides and Zn-Cu (carbenoid addition to double bonds).

CH_2I_2 / Zn - Cu

1 → 2 (48%)

$OSiMe_3$ → CH_2I_2 / Zn - Cu → $SiMe_3$ O

1	Simmons, H.E., Smith, R.,D.	*J. Am. Chem. Soc.*	**1958**	*80*	5323
2	Koch, S.D.	*J. Org. Chem.*	**1961**	*26*	3122
3	Girard, M.	*Synthesis*	**1972**		542
4	Denise, J.M.	*Synthesis*	**1978**		550
5	Mori, A.	*Tetrahedron*	**1986**	*42*	6447
6	Simmons, H.E.	*Org. React.*	**1973**	*20*	1

Zinc-copper couple. Zn dust (120 g, 1.85 at) and CuO (15 g, 0.18 mol) were heated carefully in a stream of H_2 at 425-450°C for 90-100 min, heated for an additional 4 h at the same temperature and cooled under H_2.

cis and trans Tricyclo [7,1,0,0,[4,6]] decane (2).[2] The Zn-Cu couple (10 g) covered with Et_2O (70 mL) was treated with a few drops of CH_2I_2 to start the reaction. A mixture of **1** (10.8 g, 0.1 mol) and CH_2I_2 (145 g, 0.54 mol) in Et_2O (300 mL) was added dropwise during 20 min under stirring and all was refluxed for 36 h. Zn-Cu and CH_2I_2 were removed by washing with $Na_2S_2O_3$, and distillation gave 6.5 g of **2** (48%), mp 38-39°C, bp 42-43°C (3.5 mm).

SKATTEBØL Dihalocyclopropane Rearrangement

Rearrangement of gem-dihalocyclopropanes to allenes or of vinyl dihalocyclopropanes to cyclopentadienes and fulvenes by MeLi.

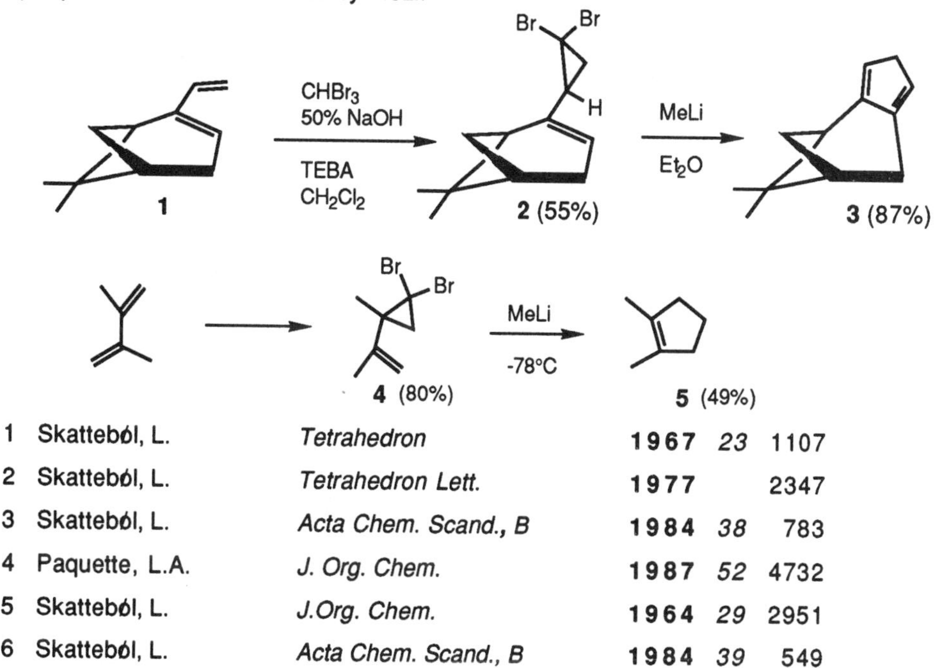

1	Skattebøl, L.	*Tetrahedron*	**1967**	*23*	1107
2	Skattebøl, L.	*Tetrahedron Lett.*	**1977**		2347
3	Skattebøl, L.	*Acta Chem. Scand., B*	**1984**	*38*	783
4	Paquette, L.A.	*J. Org. Chem.*	**1987**	*52*	4732
5	Skattebøl, L.	*J.Org. Chem.*	**1964**	*29*	2951
6	Skattebøl, L.	*Acta Chem. Scand., B*	**1984**	*39*	549

Dibromocyclopropane (2).[4] To a suspension of **1** (29.6 g, 0.42 mol), $CHBr_3$ (26.2 mL, 0.3 mol), triethylbenzylammonium chloride (TEBA), 1.09 g, 4.4 mmol) and EtOH (0.8 mL) in CH_2Cl_2 cooled in an ice bath, 50% NaOH (100 mL) was added dropwise during 10 min. The mixture was stirred for 24 h at 20°C and poured into water (250 mL). Workup afforded 32.5 g of **2** (55%), bp 85-95°C/0.08 Torr.

(1R)-(-)-9,9-Dimethyltricyclo (6.1.1.0$^{2.6}$)deca-2,5-diene (3).[4] **2** (17.6 g, 55 mmol) in dry Et_2O (2000 mL) at 0° was treated with 1.5 M MeLi in Et_2O (220 mmol), stirred for 10 h at 20°C and quenched with ice water (1000 mL). Workup, passing through neutral alumina and bulb to bulb distillation) afforded 7.7 g of **3** (87%), bp 90°C/5 Torr, $[\alpha]_D^{23}$ = -21.9° (c 1.76, EtOH).

SKRAUP Quinoline synthesis

Quinoline synthesis from anilines and acrolein or glycerol.

CH_3O ; NH_2 ; NO_2 + $H_2C{=}CH{-}CHO$ $\xrightarrow[H_3PO_4,\ 100^o]{H_3AsO_4}$ CH_3O ; N ; NO_2

1 **2** **3** (60%)

H_2N ; NH_2 + $CH_2{-}CH{-}CH_2$ (OH OH OH) $\xrightarrow{H_2SO_4}$ N ; N

4 **5** **6** (45%)

1	Skraup, Z.H.	*Chem. Ber.*	**1880**	*13*	2086
2	Yale, H.L.	*J. Am. Chem. Soc.*	**1948**	70	254
3	Wahren, M.	*Tetrahedron*	**1964**	*20*	2773
4	Bergstrom,	*Chem. Rev.*	**1944**	*35*	152
5	Manske, R.H.F.	*Org. React.*	**1953**	*7*	59

6-Methoxy-8-nitroquinoline (3).[2] A mixture of 3-nitro-4-aminoanisole **1** (33.6 g, 0.2 mol), arsenic acid (56.8 g, 0.4 mol) in 85% H_3PO_4 was heated to 100°C and acrolein **2** (15.94 g, 0.284 mol) was added dropwise under stirring. After all of **2** was added (25-30 min), the mixture was maintained for an additional 30 min at the same temperature. The solution was poured into water (800 mL), decolorized with charcoal and made alkaline with NH_4OH. The precipitate after filtration and recrystallization from EtOAc (600 mL) gave 25 g of **3** (60%), mp 157-158°C.

S M I L E S Aromatic rearrangement

Rearrangement by nucleophilic aromatic substitution and aromatic migration from one hetero atom to another (O to N or S to O).

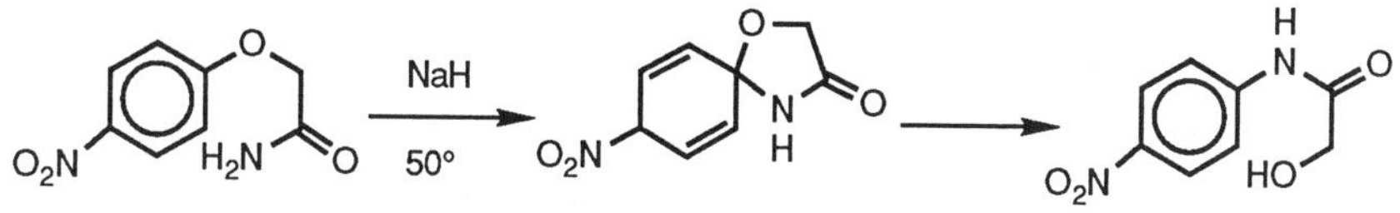

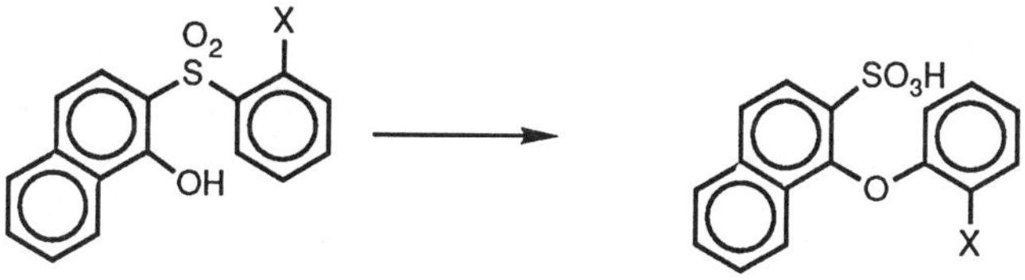

1	Smiles, S.	*J. Chem. Soc.*	**1931**		2364
2	Gillman, N.V.	*J. Org. Chem.*	**1973**	*38*	373
3	Bayles, R.	*Synthesis*	**1977**		31
4	Bunnet, J.	*Chem. Rev.*	**1951**	*49*	362
5	Huisgen, R.	*Angew. Chem.*	**1960**	*72*	314
6	Truce, W.E.	*Org. React.*	**1970**	*18*	100

N-(p-Nitrophenyl)-2-hydroxyacetamide (2).[3] A solution of p-nitrophenoxyacetamide **1** (2.7 g, 13 mmol) in DMF (20 mL) was treated with a 50% suspension of NaH (0.33 g). The mixture was stirred for 1 h at 50°C, water was added and the product recrystallized from EtOAc to afford 1.2 g of **2** (45%), mp 194°C.

SMITH - MIDDLETON - ROZEN Fluorination

Conversion of carbonyls to CF_2 compounds by SF_4 (Smith) or diethylaminosulfur trifluoride (DAST) (Middleton) or by IF on hydrazones (Rozen).

(+) Ph - CH(Et) - CO_2H + SF_4 —(40°, 9 d)→ (+) Ph - CH(Et) - CF_3 (54%)

1 —(N_2H_4, 80°)→ **3** —(IF, 1h, 20°C)→ **4** (65%) + **5** (10%)

1	Berg, M.A.	*Bull. Soc. Chim. Fr.*	**1925**	*37*	637
2	Smith, W.C.	*J. Am. Chem. Soc.*	**1959**	*81*	3165
3	Rozen, S.	*J. Am. Chem. Soc.*	**1987**	*109*	896
4	Middleton, W.J.	*J. Org. Chem.*	**1975**	*40*	574
5	Boswell, G.A.	*Org. React.*	**1974**	*21*	1

Iodine fluoride (IF). A suspension of well-ground iodine (25 g), in $CFCl_3$ (500 mL) was sonicated for 30 min, cooled to -78°C and agitated with a vibromixer. Nitrogen-diluted F_2 (10% v/v) was bubbled through (1.1 equiv.) to give a light brown suspension of IF.

4-tert-butyl-1,1-difluorocyclohexane (4).[3] Ketone **1** (5 g, 33 mmol) in EtOH (15 mL) was added to hydrazine hydrate **2** (10 g) in EtOH (40 mL) and heated to reflux, then diluted with water, extracted with $CHCl_3$, dried ($MgSO_4$) and the solvent evaporated to give 5.5 g of **3** (100%). A solution of **3** (2 g, 11 mmol) in $CHCl_3$ (20 mL) at -78°C was treated with IF (6.42 g, 44 mmol) and the reaction was monitored by GC (5% SE-30 column). There was obtained 1.23 g of **4** (65%), and 10-15% of 2-iodo derivative **5**.

S O M M E L E T Aldehyde synthesis

Aldehyde synthesis from primary alkyl halides with hexamethylene tetramine.

CH_2Br, CO_2Me **1** + $C_6H_{12}N_4$ $\xrightarrow[\Delta]{CHCl_3}$ $\xrightarrow[\Delta]{H_2O}$ CHO, CO_2Me **2** (54%)

Me, HO, CH_2Cl, Me, N + $C_6H_{12}N_4$ ⟶ Me, HO, CHO, Me, N (60%)

1	Sommelet, M.	*C.R.*	**1913**	*157*	852
2	Sommelet, M.	*Bull. Soc. Chim. Fr.*	**1913**	*13*	1085(4)
3	Zaluski, M.C.	*Bull. Soc. Chim. Fr.*	**1970**		1445
4	Angyal, S.J.	*Org. React.*	**1954**	*8*	198

2-Carbomethoxy-3-formylfuran (2).[3] To a suspension of hexamethylene tetramine (82.5 g, 0.588 mol) in $CHCl_3$ (500 mL) was added 2-carbomethoxy-3-bromomethylfuran **1** (120 g, 0.541 mol). After 4 h reflux the complex was filtered and refluxed for 30 min in water. After cooling there was obtained 46 g of **2** (54%), mp 36°C.

S O M M E L E T - H A U S E R Ammonium ylid rearrangement

Rearrangement of quaternary ammonium ylids to amines by aryl transfer.

$Ph\text{-}CH_2\text{-}NH(Me)$ **1** $\xrightarrow{ClCH_2SiMe_3\ \mathbf{2}}$ $Ph\text{-}CH_2\text{-}N(Me)\text{-}CH_2\text{-}SiMe_3$ **3** $\xrightarrow{EtI}$ $PhCH_2\text{-}N^+(Me)(Et)\text{-}CH_2\text{-}SiMe_3\ I^-$ **4**

4 $\xrightarrow{CsF}$ $[C_6H_5\text{-}CH_2\text{-}N^+(Me)(Et)\text{-}\bar{C}H_2]$ $\longrightarrow$ $o\text{-}CH_3\text{-}C_6H_4\text{-}CH_2\text{-}N(Me)\text{-}Et$ **5** + $Ph\text{-}CH_2\text{-}CH_2\text{-}N(Me)\text{-}Et$ **6** (86%, 97:3)

1	Sommelet, M.	*C.R.*	**1937**	*205*	56
2	Hauser, C.R.	*J.Am.Chem.Soc.*	**1951**	*73*	4122
3	Huisgen, R.	*Angew. Chem.*	**1960**	*72*	315
4	Sato, Y.	*J. Org. Chem.*	**1987**	*52*	1844
5	Pine, S.H.	*Org. React.*	**1970**	*18*	404

Ethylmethyl-o-xylylamine (5) and **Ethylmethylphenethylamine (6).**[4] **1** (0.242 g, 2 mmol) and **2** (0.122 g, 1 mmol) in DMSO (1 mL) were heated at 140°C for 3-17 h, poured into water, extracted with Et_2O and the extract evaporated. The residue was treated with Ac_2O (to remove unreacted amine) and after 3 h 10% HCl was added and the mixture extracted with Et_2O. Evaporation afforded **3**, bp 111-116°C (16 mm). **3** (0.2 g, 1 mmol) and Et-I (0.465 g, 3 mmol) in Me_2CO were refluxed for 18 h, cooled and diluted with Et_2O to give crystals of **4** 0.625 g, (94%). **4** (0.724 g, 2 mmol) and CsF (1.5 g, 10 mmol) in HMPA were stirred at 20° for 24 h, poured into 2% $NaHCO_3$ (200 mL) and extracted with Et_2O. Vacuum distillation gave 0.622 g of **5** and **6** (86%) in a ratio of 97:3, bp 155°C (160 mm).

SONN - MULLER Aldehyde synthesis

Aldehyde synthesis from amides or ketoximes, by reduction of imino chlorides.

$$C_6H_5-C(=NOH)-C_6H_5 \ (\mathbf{1}) \xrightarrow{PCl_5} C_6H_5-C(=N-OPCl_4)-C_6H_5 \xrightarrow{rearr.} C_6H_5-C(Cl)=N-C_6H_5 \ (\mathbf{2})$$

$$\xrightarrow[HCl]{SnCl_2} C_6H_5-CH=N-C_6H_5 \xrightarrow{H_2O} C_6H_5-CHO \ \ \mathbf{3}\ (85\%)$$

1	Sonn, A., Muller, E.	*Chem. Ber.*	**1919**	*52*	1929
2	Coleman, C.R.	*J. Am. Chem. Soc.*	**1946**	*68*	2007
3	Ferguson, L.N.	*Chem. Rev.*	**1946**	*38*	244
4	Mossetig, E.	*Org. React.*	**1954**	*8*	240

Benzaldehyde (3).[2] A solution of benzophenone oxime **1** (91.0 g, 0.41 mol) in PhH (200 mL) was treated with PCl_5 (120 g, 0.58 mol). The solvent and the $POCl_3$ formed were removed in vacuum and the residue (**2**) dissolved in CH_2Cl_2 (125 mL) was treated with $SnCl_2$ (300 g, 1.6 mol) in Et_2O (900 mL) and saturated with HCl gas. The yellow product obtained was hydrolized by 4N HCl (1250 mL). Steam distillation and Et_2O extraction followed by evaporation of the solvent afforded 41 g of **3** (85%), bp 87-88°C (15 mm).

S O U L A Phase transfer catalyst

Solid-liquid phase transfer catalyst **2** for aliphatic and aromatic nucleophilic substitution; also synergistic effect with Cu in Ullmann synthesis.

$H_2 + NH_3$, Ni: **1** → **2** (63%)

3 (ONa) + **4** (Cl, NO_2) — **2**, 130° → **5** (92%)

6 (OH) + Br **7** — **2**, K_2CO_3, 110° → **8** (75%)

1	Pederson, C.J.	*J. Am. Chem. Soc.*	**1967**	*89*	7017
2	Lehn, B.	*Tetrahedron Lett.*	**1969**		2885
3	Vogtle, F.	*Angew. Chem. Int. Ed.*	**1974**	*13*	814
4	Soula, G.	*Eur. Patent*	**1978**		5094
5	Soula, G.	*French Patent*	**1979**		16673
6	Soula, G.	*J. Org. Chem.*	**1985**	*50*	3717,3721

Tris (3,6-dioxaoctyl) amine (TDA-2) (2).[6] Raney Nickel (195 g) and **1** (1400 g, 10.4 mol) were heated to 150°C and a stream of NH_3 (51 g) and H_2 (2 g per hour) was passed through at 185°C for 3 h. Then only H_2 was maintained for another 2 h. The catalyst was filtered and the filtrate distilled to give 800 g of **2** (63%), bp 195°C/0.5 mm.

p-Phenoxynitrobenzene (5). **4** (32 g, 0.2 mol), **3** (23 g, 0.2 mol) and **2** (3.2 g, 0.01 mol) in chlorobenzene (100 mL) were heated to 130°C for 9 h. Quenching with water (100 mL), work up and recrystallization from EtOH afforded 39.56 g of **5** (92%).

SPENGLER - PFANNENSTIEL Oxidation

Oxidation of reductive sugars in alkaline solution with molecular O_2.

$Ba(OH)_2$, O_2: **1** → **2**

1 Spengler, O., Pfannenstiel, A. *DR Pat.* 618164

2 Hardegger, E. *Helv. Chim.Acta.* **1952** *35* 618

3-(α-D-Glucosido)-D-arabonic acid (2).[2] A solution of maltose **1** (18.0 g, 53 mmol) in water (200 mL) was added dropwise to a very well stirred solution of $Ba(OH)_2$ cryst. (20 g) in water (150 mL), under a flow of O_2. In 22 h there were absorbed 1250 mL of O_2 (calculated 1250 mL). The mixture was saturated with CO_2 and filtered through Celite and 120 mL of Wofatit KS. Concentration under vacuum afforded 17 g of crude **2** (100%). Separation of **2** was carried out as the brucinate, mp = 152-154°C, $(\alpha]_D$ = 50° (c = 0.5 water).

STAAB Reagent

1,1'-Carbonyldiimidazole **2** an activating reagent for carboxylic acids in formation of esters, amides, peptides, aldehydes and ketones via imidazolides **3**.

1	Staab, H.A.	*Chem. Ber.*	**1956**	*89*	1927
2	Staab, H.A.	*Liebigs Ann.*	**1957**	*609*	75, 83
3	Staab, H.A.	*Liebigs Ann.*	**1962**	*654*	119
4	Komives, T.	*Org. Prep. Proc. Int.*	**1989**	*21*	251
5	Ley, S.V.	*Synlett*	**1990**		255
6	Staab, H.A.	*Angew. Chem. Int. Ed.*	**1962**	*7*	351

Vitamin A aldehyde (5).[3] **4** (1 g, 3.3 mmol) and **2**[2] (490 mg, 3.3 mmol) in Et_2O (50 mL) was refluxed under N_2 for 30 min, treated dropwise at -20°C with 0.41 N $LiAlH_4$ in Et_2O (16.4 mL, 1.68 mmol) and diluted with Et_2O (50 mL). After 30 min the solvent was removed in vacuum, the residue taken up in MeOH and treated at 0°C with 10% H_2SO_4 in MeOH until the precipitate formed redissolved. Addition of 2,4-DNPH (750 mg) in MeOH (10 mL) yielded the dinitrophenylhydrazone of **5** in 56% yield, mp 206-207°C.

S T A U D I N G E R Azide reduction

Conversion of organic azides with phosphines or phosphites to iminophosphoranes (phosphazo compounds) and their hydrolysis to amines.

$$\underset{\mathbf{1}}{NC-N_3} \xrightarrow{P(OMe)_3\ \mathbf{2}} \underset{\mathbf{3}\ (80\%)}{NC-N=P(OMe)_3}$$

$$Ph_3=N-SiMe_3 + Cl_3C-CH=O \longrightarrow Ph_3P=N-\underset{CCl_3}{CH}-OSiMe_3$$

$$\underset{\mathbf{4}}{N_3-CH_2CH_2-COOEt} \xrightarrow[H_2O,\ 20°]{Ph_3P} \underset{\mathbf{5}\ (83\%)}{H_2N-CH_2CH_2-COOEt}$$

1	Staudinger, H.	*Helv. Chim. Acta.*	**1919**	*2*	635
2	Marsh, F.D.	*J. Org. Chem.*	**1972**	*37*	2966
3	Cooper, R.D.G.	*Pure and Appl. Chem.*	**1987**	*59*	485
4	Gololobov, Yu G.	*Tetrahedron*	**1981**	*37*	437
5	Carrie, R.	*Bull. Chem. Soc. Fr.*	**1985**		815

Trimethyl N-cyanophosphoroimidate (3).[2] Cyanogen azide **1** (7.0 g, 0.1 mol) in MeCN (40 mL) was added slowly to trimethyl phosphite **2** (12.4 g, 0.1 mol) in Et_2O (200 mL) with cooling at 15°C. When nitrogen evolution was complete, the volatiles were removed in vacuum and the residue was washed with Et_2O to give 13.1 g of **3** (80%), mp 56.4 - 56.8°C.

Ethyl 3-aminopropanoate (5).[5] To a 1M solution of **4** in THF (from the bromo ester with NaN_3 in DMSO) was added a molar equivalent of Ph_3P, 1.5 equiv of water and a boiling chip (N_2 evolution). After 8 h at 20°C and evaporation, the residue was treated with Et_2O-hexane and Ph_3PO was filtered. This process was repeated and **5** was distilled at 40-45° and 10.5 torr (83%).

STAUDINGER - PFENNINGER Thiirane dioxide synthesis

Thiirane dioxide episulfone synthesis by reaction of diazomethane with sulfenes or SO_2.

$$2\ CH_2N_2 + SO_2 \longrightarrow H_2C\text{—}CH_2\ (\text{cyclic } SO_2)$$

1 **2** **3** (70%)

$$R_1R_2CH\text{-}SO_2Cl \xrightarrow{Et_3N} R_1R_2C{=}SO_2 + CH_2N_2 \longrightarrow R_1R_2C\text{—}C\ (\text{cyclic } SO_2)$$

1	Staudinger, H., Pfenninger, F.	*Chem. Ber.*	**1916**	*42*	1941
2	Hesse, G.	*Chem. Ber.*	**1957**	*90*	1166
3	Opitz, G.	*Z. Naturforschung.*	**1963**	*b18*	775
4	Optiz, G.	*Angew. Chem.*	**1961**	*77*	41
5	Fischer, N.H.	*Synthesis*	**1970**		396

Thiiran-1,1-dioxide (3).[5] A cooled solution of diazomethane **1** (9.6 g, 0.228 mol) in ether (500 mL) at -15°C was saturated with SO_2 dry gas. Evaporation of the solvent left 8.1 g of a residue which after distillation afforded 7.35 g of **3** (70%), bp 64°C (0.3 mm), mp 19°C.

STEGLICH - HASSNER Direct esterification

Direct room temperature esterification of carboxylic acids with alcohols, including tert. alcohols with the help of dicyclohexylcarbodiimide (DCC) and 4-diakylaminopyridine catalysts **3.**

NH Ph + PH CH_2OH → DCC, 25°, 2h (4-dialkylaminopyridine, N(R)R) → NH PH, O CH_2 Ph (80%)

Me CO_2H

3, R : Me or $(CH_2)_4$

OH + (Ph)$_2$O → **3**, 80° → O Ph

1	Steglich, W.	*Angew. Chem. Int. Ed.*	**1969**	*9*	981
2	Steglich, W.	*Angew. Chem. Int. Ed.*	**1978**	*17*	522
3	Hassner, A.	*Tetrahedron*	**1978**	*34*	2069
4	Hassner, A.	*Tetrahedron Lett.*	**1978**		4475
5	Steglich, W.	*Angew. Chem. Int. Ed.*	**1978**	*17*	569

General procedure.[4] A solution of carboxylic acid (10 mmol), N,N-dicyclohexylcarbodiimide (11 mmol), alcohol (11 mmol) and 4-pyrrolidinopyridine **4** (R,R: $(CH_2)_4$) (1 mmol), in Et_2O (25-30 mL) was allowed to stand at 20°C a few minutes to a few days until esterification was complete. N,N-dicyclohexylurea was filtered and the filtrate washed with water (3x50 mL), 5% AcOH (3x50 mL) and again water (3x50 mL), dried ($MgSO_4$) and evaporated in vacuum to afford the ester in 60-100% yield.

STEPHEN Aldehyde synthesis

Aldehyde synthesis from nitriles and $SnCl_2$ - HCl

$$\underset{\mathbf{1}}{\text{R-CN}} \xrightarrow[\text{HCl, } 0^{\circ}]{SnCl_2} \underset{\mathbf{2}}{[\text{RCH=NH}_2]_2^{+} \; SnCl_6^{-2}}$$

$$\mathbf{2} + H_2O \xrightarrow{\text{steam distill}} \underset{\mathbf{3}}{\text{RCHO}}$$

1	Stephen, J.	*J. Chem. Soc.*	**1925**	*127*	1874
2	Stephen, T.W.	*J. Chem. Soc.*	**1956**		4695
3	Ferguson, L.W.	*Chem. Rev.*	**1946**	*38*	243
4	Mosettig, E.	*Org. React.*	**1954**	*8*	246

Aldehydes, general procedure (3).[2] A solution of nitrile **1** (1 mol) in EtOAc saturated with HCl gas at 0°C is added to a solution of $SnCl_2$ (1.1 mol) in EtOAc previously saturated with HCl at 0°C. After several hours at 0°C the aldimine complex $[R\text{-}CHNH_2]_2^{+}$ $SnCl_6^{-2}$ separated as pale yellow prisms. The complex was filtered, washed with Et_2O and dried over KOH under vacuum to afford a pure sample of **2**. Steam distillation of **2** gave aldehyde **3** in the distillate if the aldehyde was volatile. Alternatively, the aldehyde was extracted from the residue of the steam distillate.

STEPHENS - CASTRO Acetylene cyclophane synthesis

Polyacetylene cyclophane synthesis from an iodophenyl copper acetylide.

I
1
Cu^+
NH_4OH
I
Cu
2
pyr
Δ
3 (4.6%)

1	Stephens, R.D., Castro, C.E.	*J. Org. Chem.*	**1963**	*28*	3313
2	Campbell, I.D.	*Chem. Commmun.*	**1966**		87
3	Staab, H.E.	*Chem. Ber.*	**1970**	*103*	1157
4	Staab, H.E.	*Synthesis*	**1974**		424

2.2.2.2.2.2-Metacyclophene-1,9,17,25,33,41-hexayne (3).[4]

m-Iodophenylacetylene **1** (104 g, 0.456 mol) in EtOH (1000 mL) was added under stirring to copper (I) chloride (45.2 g, 0.456 mol) in aqueous ammonia. Filtration, washing (water, EtOH, Et_2O) and drying gave 51 g of **2** (39%). N_2 was bubbled through boiling pyridine (800 mL), **2** (4.5 g, 0.172 mol) was added and the mixture refluxed 24 h. After distillation of 500 mL of pyridine, the solution was poured into water. The precipitate was extracted with boiling Et_2O, PhH and PhMe. From the last solvent one obtains 800 mg of **3** (4.6%), mp 350°C.

STETTER 1,4-Dicarbonyl Synthesis

Michael addition of aromatic or heterocyclic aldehydes (via cyanohydrins) to α,β-unsaturated systems. Also addition of aliphatic aldehydes catalyzed by thiazolium ylids.

Ar CHO **1** + CN **2** —NaCN, DMF→ (HO, CN, Ar, CN) ⟶ Ar, O, CN **3** (80%)

CHO **4** + O **5** —HO, N^{+}, Ph, S **6**, NH_3→ C_6H_{13}, O, O **7** (75%)

1	Stetter, H.	*Angew. Chem.*	**1973**	*85*	89
2	Stetter, H.	*Chem. Ber.*	**1974**	*107*	210
3	Stetter, H.	*Synthesis*	**1975**		379
4	Stetter, H.	*Angew. Chem. Int. Ed.*	**1976**	*15*	639
5	Stetter, H.	*Org. Synth.*	**1985**	*65*	26

4-Oxo-4-phenylbutanenitrile (3).(Ar=Ph).[4] A solution of benzaldehyde **1** (10.6 g, 0.1 mol) in DMF (50 mL) is added over 10 min to stirred NaCN (2.45 g, 0.05 mol) in DMF (50 mL) at 35°C. After 5 min acrylonitrile **2** (4.0 g, 0.075 mol) in DMF (100 mL) is added over 20 min at 35°C. After 3 h stirring and work up, one obtains 9.5 g of **3** (80%), bp 114°C (0.3 torr), mp 70°C.

STIEGLITZ Rearrangement

Rearrangement or ring enlargement of amines via nitrenes.

1 Ph NH$_2$. HCl —KOCl→ Ph NHCl **2** (100%) —NaOMe→ Ph N **3**

$(C_6H_5)_2C(NH_2)-C_6H_5$ **4** —$Pb(OAc)_4$, PhH, Δ, 1h→ $(C_6H_5)_2C=N-C_6H_5$ **5** (85%)

1	Stieglitz, J.	*Chem. Ber.*	**1913**	*46*	2146
2	Stieglitz, J.	*J. Org. Chem.*	**1936**	*1*	31
3	Pinck, L.A.	*J. Am. Chem. Soc.*	**1937**	*59*	8
4	Sisti, A.J.	*J. Org. Chem.*	**1974**	*39*	3932
5	Robertson,	*Chem. Rev.*	**1935**	*16*	417

Phenylphenanthridine (3).[3] Amine hydrochloride **1** (2.72 g, 9.28 mmol) in EtOH (75 mL) was treated with 0.926 N cold solution of KOCl. Colorless crystals appear. The mixture was shaken for 30 min in ice, water was added and the product was filtered and dryed (P_2O_5) to afford 2.7 g of **2** (100%), mp 102°C (hexane). **2** (2.0 g, 6.8 mmol) in anh. pyridine (20 mL) was treated with an excess of $NaOCH_3$ (exothermic). After 20 h the solvent was removed in vacuum, the residue triturated with Et_2O and the extract treated with dry HCl to obtain the hydrochloride of **3**, mp 107-108°C (from petroleum ether), mp 95-100° (from water).

Diphenylmethyleneaniline (5).[4] To a suspension of lead tetraacetate (4.9 g, 10 mmol) in PhH (100 mL) under N_2 was added a solution of triphenylmethylamine **4** (2.6 g, 10 mmol) in PhH (100 mL) dropwise under stirring. The mixture was refluxed for 1 h, cooled, filtered, washed and the solvent evaporated. The residue was crystallized from EtOH to give 2.2 g of **5** (85%), mp 111-112°C.

STILES - SISTI Formylation

Synthesis of aldehydes by formylation of Grignard reagents with p-dimethylaminobenzaldehyde and a diazonium salt.

1	Stiles, M., Sisti, A.	*J. Org. Chem.*	**1960**	*25*	1691
2	Sisti, A.	*J. Org. Chem.*	**1962**	*27*	279
3	Sisti, A.	*J. Chem. Eng. Data*	**1964**	*9*	108

Cyclohexanecarboxaldehyde (4).[2] A solution of sulfanilic acid **3** (100 g, 0.31 mol) in water (200 mL) and Na_2CO_3 (18.4 g, 0.18 mol) was diazotized with HCl (64 mL) and $NaNO_2$ (24.4 g, 0.35 mol) in water (75 mL) at 0-5°C. NaOAc (70 g) in water (200 mL) was added to buffer the solution to pH=6. A solution of p-dimethylaminophenylcyclohexyl carbinol **2** (45.8 g, 0.2 mol) in acetone, obtained from cyclohexylmagnesium bromide **1** and p-dimethylaminobenzaldehyde, (250 mL) was added. The red solution was stirred for 30 min at 0-5°C and for 30 min without cooling. The reaction mixture was diluted with water, extracted with Et_2O and the extract distilled to give 15.45 g of **4** (69%), bp 50-53°C/20 mm.

STILLE Cross Coupling

Coupling of organotin reagents (and Pd catalyst) with aryl or vinyl halides or triflates, acyl chlorides or allyl acetates.

OTf
+ Bu_3Sn
Pd (II)
1 2 3 (91%)

OAc
Me_3SnAr 5
Pd (II) dba_2
MeO
MeO
4 6 (71%)

1 Stille, J.K. *J. Am. Chem. Soc.* **1984** *106* 4630
2 Stille, J.K. *J. Org. Chem.* **1990** *55* 3019
3 Stille, J.K. *Angew. Chem. Int. Ed.* **1986** *25* 508

1-Vinyl-4-tert-butylcyclohexene (3).[1] To LiCl (0.56 g, 13 mmol) and tetrakis (triphenylphosphine) palladium (0) (0.032 g, 0.028 mol, 1.6 mol%) under Ar was added THF (10 mL) followed by a solution of vinyl triflate **1** (0.51 g, 1.8 mmol) and tributylvinyltin **2** (0.56 g, 1.8 mmol) in THF (10 mL). The slurry was heated to reflux for 17 h, cooled to 20°C and diluted with pentane (60 mL). The mixture was washed with 10% NH_4OH solution, dried ($MgSO_4$), filtered through a short pad of silica gel and the solvent evaporated in vaccum to afford 0.26 g of **3** (91%).

(2E,6E)-1-(4-Methyl-2,6-dimethoxyphenyl)-3,7,11-trimethyl-2,6,10-dodeca-triene (6).[2] From **5** (1.57 g, 5 mmol), trans, trans-farnesyl acetate **4** (1.32 g, 5 mmol), LiCl (0.632 g, 15 mmol) and (bis(dibenzylideneacetone)palladium (0.144 g, 0.25 mol, 5 mol%). Column chromatography (silica gel, 5% EtOAc/hexane) afforded 1.26 g of **6** (71%), $R_f = 0.53$.

STILLE Carbonyl Synthesis

Synthesis of aryl ketones or aldehydes from aryl triflates or iodides and organo stannanes in the presence of CO and a palladium catalyst.

$$\underset{\mathbf{1}}{p\text{-MeO-}C_5H_4\text{-I}} + CO + Bu_3SnH \xrightarrow[CO]{(Ph_3P)_4Pd} \underset{\mathbf{2}\ (77\%)}{p\text{-MeO-}C_6H_4\text{-CHO}}$$

MeO–C₆H₄–OTf **3** + nBu₃Sn–CH=CH–Ph **4** —CO→ MeO–C₆H₄–CO–CH=CH–Ph **5** (58%)

1	Stille, J.K.	*J. Am. Chem. Soc.*	**1983** *105* 7175
2	Stille, J.K.	*J. Am. Chem. Soc.*	**1987** *109* 5478
3	Stille, J.K.	*J. Am. Chem. Soc.*	**1988** *110* 1557

p-Methoxybenzaldehyde (2).[1] p-Methoxyiodobenzene **1** (234 mg, 1 mmol) in PhH (4.0 mL) and tetrakis(triphenylphosphine)palladium (0) (35.6 mg) were maintained under 1 atm. of CO at 50°C. A solution of tributyltin hydride (350 mg, 1.1 mmol) was added via a syringe pump over 2.5 h. Tributyltin halide was removed and purification by chromatography afforded 104 mg of **2** (77%) (CG yield 100%).

(E)-1-(p-Methoxyphenyl)-3-phenyl-2-propen-1-one (5).[3] To 4-methoxyphenyl triflate **3** (390 mg, 1.52 mmol) in DMF (7 mL) were added (E) -phenyl-tri-n-butylstyrylstannane **4** (645 mg, 1.64 mmol) LiCl (200 mg, 4.72 mmol), dichloro-1,1'-bis(diphenylphosphino) ferocene palladium (II)/$PdCl_2$(dppf)/(45 mg, 0.06 mmol), a few crystals of 2,6-di-tert-butyl-4-methylphenol and 4 Å molecular sieves (100 mg). The mixture was heated at 70°C under CO (1 atm). Work up after 23 h and chromatography (hexane: EtOAc 20:1) afforded 246 mg of **5** (58%), mp 105-106°C.

STORK Enamine alkylation

α-Alkylation and acylation of ketones via enamines or imines. Also Michael addition via enamines.

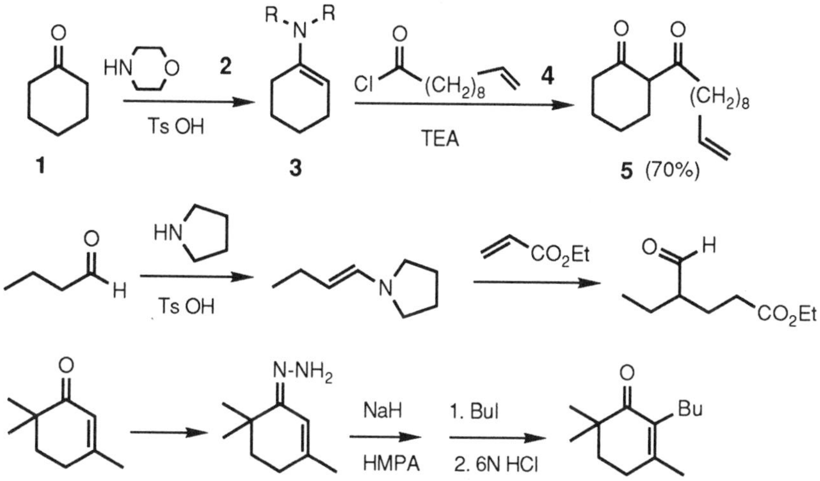

1	Stork, G.	*J. Am. Chem. Soc.*	**1954**	*76*	2029
2	Hunig, S.	*Chem. Ber.*	**1962**	*95*	2493
3	Hickmott, P.W.	*Tetrahedron*	**1982**	*38*	1975
4	Knebine, M.F.	*Synthesis*	**1970**		510
5	Stork, G.	*J. Am. Chem. Soc.*	**1971**		5939

N-(1-Cyclohexen-1-yl)morpholine (3). Cyclohexanone **1** (19.6 g, 0.2 mol), morpholine **2** (19.08 g, 0.22 mol) and TsOH (catalyst) in PhH or PhMe was refluxed with a Dean-Stark unit. After water was removed azeotrophically, distillation afforded 28 g of **3** (85%), bp 117-118°C.

2-(Δ^{10}-Undecenoyl)cyclohexanone (5).[2] To **3** (18.4 g, 0.11 mol) and TEA (15.3 mL, 0.11 mol) in $CHCl_3$ (130 mL) was added **4** (20.2 g, 0.1 mol) in $CHCl_3$ (90 mL) at 35°C during 2.5 h. After 12 h the red solution was refluxed with 32% HCl (50 mL) for 5 h. Separation of water, washing and distillation afforded 18.4 g of **5** (70%), bp 132-136°C (0.003 mm).

S T O R K Cyanohydrin alkylation

Conversion of aldehydes to ketones via cyanohydrin derivatives (ethers) by alkylation or Michael addition; also used with silyl ethers, dialkylaminonitriles (see also Stetter reaction).

KCN, 2 (OEt) → 3; 1. LDA 2. $C_6H_{13}Br$ → 4 (70%); H^+ → 5 (94%)

1 2 3 4 (70%) 5 (94%)

CHCN, O, OEt + 6 —LDA→ 7 (75%) (OR, CN)

CN, OR + 6 —LDA→ HO, CN, OR (90%)

1	Stork, G.	*J. Am. Chem. Soc.*	**1971**	*93*	5286
2	Stork, G.	*J. Am. Chem. Soc.*	**1974**	*96*	5272
3	Hunig, S.	*Chem. Ber.*	**1981**	*114*	959
4	Albright, J.D.	*Tetrahedron*	**1983**	*39*	3207
5	Watt, D.S.	*Org. React.*	**1984**	*31*	47

1-Nonen-3-one (5). Nitrile (2 g), n-Hexyl bromide (2.8 g) and HMPA (3.9 g), were added simultaneously to an LDA suspension (iPr_2 NH, nBuLi) in 5 ml THF at -78°C. After 1.5 h at -78°C and 1 h at 20°C, the mixture was extracted with ether. Distillation gave 2.12 g of **4** (70%), bp 100-101°C/0.15 mm.

4 (2.12 g) was hydrolyzed with 5% sulphuric acid (1 mL) in 5 mL of MeOH at 20°C for 15 min to afford 1.49 g (100%) of a cyanohydrin which was taken up in ether and shaken vigorously with 0.5 N NaOH (20 mL) for 5 min. Ketone **5** (0.167 g, 94%) was isolated.

S T O R K Radical cyclization

Free radical cyclization with preferential formation of cyclopentanes.

1	Stork, G.	*J. Am. Chem. Soc.*	**1982** *104* 2321
2	Stork, G.	*J. Am. Chem. Soc.*	**1983** *105* 6765
3	Stork, G.	*J. Am. Chem. Soc.*	**1987** *109* 2829
4	Stork, G.	*Tetrahedron Lett.*	**1986** *27* 4529

Diethyl 3-methyl-4-methylidenecyclopentane-1,1-dicarboxylate (2) and **diethyl cyclohexane-4,4-dicarboxylate (3).** A solution of bromide **1** (200 mg, 0.7 mmol) in PhH (34 mL), AIBN and 1.1 equivalent of tributylstannane protected by a foil of aluminium to prevent scattering of the light was irradiated with a GE 275 W sun lamp. After 3-4 h of heating, the solvent was evaporated and the residue was stirred with 5 mL of Et_2O and 5 mL of KF saturated solution. Separation and chromatography (silica gel 5% EtOAc in petroleum ether) gave a mixture of **2** and **3** of 3:1 in 87% yield and 13% recovery of **1**. While reflux with 0.02 M Bu_3SnH in PhH gave a 3:1 mixture of **2:3**, reflux with 1.7 M Bu_3SnH in PhH led to a 97:3 mixture of **2:3**.

S T O R K Reductive cyclization

Cyclization of acetylenic ketones to allyl alcohols by one electron reduction with Li/NH_3; also electrochemically (Shono) or by SmI_2 (Molander).

$K-NH_3$, $(NH_4)_2SO_4$ (60%)[3]

$C_{10}H_7Na$, THF

1 → **2** (89%)

1	Stork, G.	*J. Am. Chem. Soc.*	**1965**	*87*	1148
2	Pradhan, S.K.	*J. Org. Chem.*	**1976**	*41*	1943
3	Stork, G.	*J. Am. Chem. Soc.*	**1979**	*101*	7107
4	Shono, T.	*Chem. Lett.*	**1976**		1233
5	Molander, G.A.	*J. Am. Chem. Soc.*	**1989**	*111*	8236

4-Methylenecholestan-5β-ol (2).[2] Naphthalene sodium (from naphthalene (9.6 g, 75 mmol) and Na (2.9 g, 130 mmol) in THF (100 mL) to give a 0.6 N solution) was added to a stirred solution of **1** (198 mg, 0.5 mmol) in THF (3 mL) to a faint green end point. Approximately 2 mmol of reagent was consumed per mmol of **1**. Work up and chromatograph (alumina, pentane) gave 20 mg of **1** and 160 mg of **2** (89%), $(\alpha)_D = +21°$ (c 0.11).

STORY Macrocycle synthesis

Synthesis of large ring alkanes and lactones from smaller ring ketones via peroxides.

H_2O_2 ; $CuSO_4$

1 2 3 4 (58%)

$(CH_2)_{14}$ + $(CH_2)_{14}$

5 (14%) 6 (18%)

1	Story, P.R.	*J. Am. Chem. Soc.*	**1968**	*90*	817
2	Story, P.R.	*U.S. Patent*	**1970**		3,528,898
3	Story, P.R.	*J. Org. Chem.*	**1970**	*35*	3059
4	Story, P.R.	*Synthesis*	**1970**		181
5	Story, P.R.	*J. Org. Chem.*	**1974**	*29*	3463
6	Story, P.R.	*Synthesis*	**1975**		159

Cyclotetradecane (5) and **Pentadecanolide (6).**[4] **1** (2.94 g, 30 mmol), 90% H_2O_2 (2 mL, 60 mmol), MeCN (30 mL) and 70% $HClO_4$ (6 drops) was stirred at 25°C in an open dish for 5 h to give a solid, washed with ice water (40 mL). Crystallization (hexane) gave 3.86 g of **2** (95%), mp 83-84°C. A mixture of **2** (75.80 g, 288 mmol), **3** (140.7 g, 1.67 mol) and anh. $CuSO_4$ (95 g) was stirred at 25°C for 13 d. Water was added, the product filtered and crystallized from MeOH gave 54.78 g of **4** (58%), mp 65-67°C.

4 (49.0 g, 149 mol) was thermolyzed in refluxing n-decane (720 mL) for 3 h. The solvent was evaporated in vacuum and the product distilled on a spinning band column to afford 4.4 g of **5** (14%), bp 77-78°C (0.3 mm) and 6.62 g of **6** (18%), bp 110-114°C (0.3 mm).

STRECKER Aminoacid synthesis

Synthesis of α-amino acids from aldehydes or ketones via cyanohydrins.

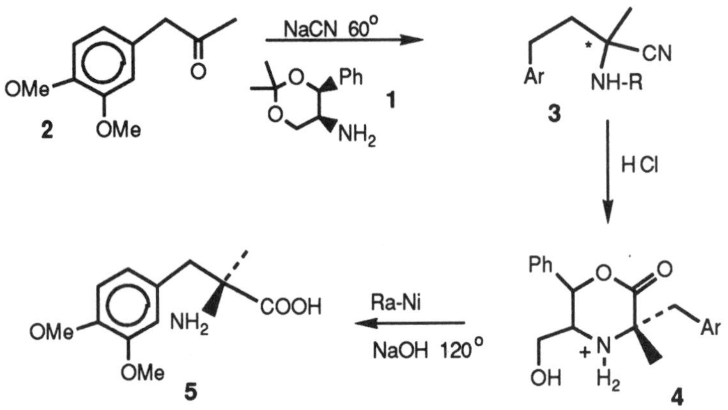

1	Strecker, A.	*Liebigs Ann.*	**1850**	*75*	27
2	Pollard, C. B.	*J. Am. Chem. Soc.*	**1955**	*77*	40
3	Weinges, K.	*Chem. Ber.*	**1971**	*104*	3594
4	Stoul, D.	*J. Org. Chem.*	**1983**	*48*	5369
5	Georgiades, M. P.	*Synthesis*	**1989**		616
6	Mowry, D. T.	*Chem. Rev.*	**1948**	*42*	236

(S)-α-Methyl-3,4-dimethoxyphenylalanine (5).[3] (4S,5S) **1** (20.7 g, 0.1 mol), ketone **2** (19.4 g, 0.1 mol) and NaCN (5.4 g, 0.11 mol) in MeOH (70 mL) was heated to 60 °C and HOAc (9mL) was added dropwise. The mixture was cooled, filtered, stirred with water (100 mL) for 1 h and filtered. Crystallization from MeOH afforded 33.6 g (82%) of **3**, mp 127-128 °C, $(\alpha)_D$ = +85.7°. **3** (14 g, 40mmol) was added to cooled conc HCl (100 mL). After stirring for 1 h at -5°C, 1 h at 20° C and 4 h at 50°C, the mixture was cooled for 2 h, filtered and recrystallized from MeOH to give 11.6 g (83%) of **4** , mp 208°C, $(\alpha)_D$ = -8.4°.

Heating of **4** with Raney-nickel and 2N NaOH at 120°C for 29 h gave **5** as **5.HCl** in 98% yield, mp 174-175°C, $(\alpha)_D$ =-4.3°.

STRYKER Regioselective reduction

Regioselective conjugate reduction and reductive silylation of α,β-unsaturated ketones, esters, and aldehydes using a stable copper (I) hydride cluster $[(Ph_3P)CuH]_6$.

1 → $[(Ph_3P)CuH]_6$, Me_3SiCl → 2 → 10% HCl → 3 (93%)

$[(Ph_3P)CuH]_6$ (85%) (94% stereoselective)

1	Stryker, J.M.	*Tetrahedron Lett.*	**1988**	*29*	3749
2	Stryker, J.M.	*J. Am. Chem. Soc.*	**1988**	*110*	291
3	Stryker, J.M.	*Tetrahedron Lett.*	**1989**	*30*	5677
4	Stryker, J.M.	*Tetrahedron Lett.*	**1990**	*31*	3237

6-Dodecenal (2).[3] Aldehyde **1** (203 mg, 1.12 mmol), chlorotrimethylsilane (366 mg, 3.37 mmol) and dry PhH (2 mL) were stirred for 5 min under N_2. $[(Ph_3P)CuH]_6$ (804 mg, 0.41 mmol) in PhH (8 mL) was added and stirred at 20°C for 3.5 h. Hexane (30 mL) was added, the precipitate filtered (Celite) and the solvents evaporated. The residue **2** (E:Z = 1:1) was treated with THF (10 mL), and 10 drops of 10% aq. HCl and stirred for 30 min. Extraction with Et_2O, washing (water, brine) and drying ($MgSO_4$) gave, after flash chromatography, 191 mg of **3** (93%).

SUZUKI Vinyl coupling

Diene synthesis via Pd catalyzed vinyl coupling of vinyl borates with vinyl halides.

$Pd(P(Ph)_3)_4$, Tl OH

Y = $SiMe_2$ tBu

3 (92%)

1 Suzuki, A. *Tetrahedron Lett.* **1979** *20* 3437

2 Suzuki, A. *Tetrahedron Lett.* **1981** *22* 127

3 Suzuki, A. *Tetrahedron* **1983** *39* 8271

4 Suzuki, A. *J. Am. Chem. Soc.* **1985** *107* 972

5 Kishi, Y. *J. Am. Chem. Soc.* **1987** *109* 4756

6 Suzuki, A. *Pur Appl. Chem.* **1991** *63* 419

Palytoxyne (3).[5] To a THF solution of **1** (92.5 mg, 0.14 mmol) (y=$SiMe_2$ t-Bu) (5 mL) was added 10% aq. TlOH (0.93 mL) under Ar. The mixture was stirred for 5 min at 20° and a THF solution of **2** (55.3 mg, 0.1 mmol) (Y=$SiMe_2$ t-Bu) (1 mL) was added followed by $Pd(PPh_3)_4$ catalyst (28.8 mg, 0.025 mmol, 0.7 mL). After 5 min stirring the mixture was diluted with Et_2O (10 mL), dried ($MgSO_4$) and filtered through Celite. The crude **3** was purified by silica gel preparative TLC to give 56.6 mg of **3** (92% based on **2**) (Y = $SiMe_2$ t-Bu).

S W A R T S Halogen exchange

Substitution of chlorine atoms with fluorine atoms by means of SbF_5

$$\underset{\mathbf{1}}{Cl_2FC\text{-}CClF\text{-}CCl_3} + SbF_5 \xrightarrow{\Delta} \underset{\mathbf{2}\ (70\%)}{Cl_2FC\text{-}CClF\text{-}CClF_2}$$

$$\underset{\mathbf{3}}{p\text{-}Cl\text{-}C_6H_4\text{-}CCl_3} + SbF_5 \xrightarrow{\Delta} \underset{\mathbf{4}\ (95\%)}{p\text{-}Cl\text{-}C_6H_4\text{-}CF_3}$$

1	Swarts, F.	*Bull. Acad. Royal Belge*	**1892**	*24*	309
2	Swarts, F.	*Rec. trav. Chim.*	**1915**	*35*	131
3	Henne, A.I.	*J. Am. Chem. Soc.*	**1941**	*63*	3478
4	Finger, G.C.	*J. Am. Chem. Soc.*	**1956**	*78*	6034
5	Finger, G.C.	*Org. React.*	**1944**	*2*	49

1,1,2,3-Tetrachloro-1,2,3,3-tetrafluoropropane (2).[3] 1,1,1,2,3-Pentachloro-2,3,3-trifluoro-propane **1** (213 g, 1 mol) was heated in a steel vessel with SbF_5 (10.8 g, 0.05 mol). From the reactor **2** (bp 112°C) was distilled and **1** was refluxed back (bp 152°C) by raising the temperature slowly and progressively from 125°C to 170°C. Finally the temperature was raised to force out the organic material with a small amount of SbF_5. The distillate was steam distilled from a 10% NaOH solution to give 117.8 g of **2** (70%) and 15% recovery of **1**.

p-Chloro-α,α,α-trifluorotoluene (4)[4] A mixture of p-chloro-α,α,α-trichlorotoluene **3** (23.0 g, 0.1 mol) and SbF_5 (29.58 g, 0.11 mol) was heated until the reaction started. After completion of the reaction, the mixture was washed with 6 N HCl and dried on BaO. Distillation afforded 17.1 g of **4** (95%), bp 136-138°C, mp -36°C, n_D^{20} = 1.4463, d = 1.353.

S W E R N Oxidation

Oxidation of alcohols to aldehydes or ketones in the presence of other groups.

Me$_3$SiO ... OMe **1** $\xrightarrow[-40^\circ,\ DMSO]{(COCl)_2}$ HO ... OMe **2** (74%)

Ph - C≡C - CH(OH) - CH_2 - CH_3 **3** $\xrightarrow[-50^\circ,\ DMSO]{TFAA}$ Ph - C≡C - C(=O) - CH_2 - CH_3 **4** (92%)

1	Swern, D.	*J. Org. Chem.*	**1976**	*41*	3329
2	Swern, D.	*J. Org. Chem.*	**1978**	*43*	2482
3	Swern, D.	*Synthesis*	**1978**		297
4	Maycock, C.D.	*J. Chem. Soc. Perkin I*	**1987**		1221
5	Tidwell, T.T.	*Org. React.*	**1990**	*39*	297

Methyl 12-Hydroxy-3-oxodeoxycholanate (2).[4] To oxalyl chloride (0.24 g, 1.93 mmol) and DMSO (0.28 mL, 3.94 mmol) in CH_2Cl_2 at -60°C, was added rapidly **1** (1.07 g, 1.93 mmol) in CH_2Cl_2 and the temperature was allowed to rise to -40°C during 15 min and maintained for 30 min at -40°C. Ttriethylamine (0.89 mL, 6.38 mmol) was added and after 5 min the temperature was allowed to rise to 20°C. The trimethylsilyl group was removed with 5% HCl in MeOH (TLC). Dilution with water, extraction with CH_2Cl_2 and chromatography gave 0.58 g of **2** (74%), mp 137-140°C.

S Z A R V A S Y - S C H Ö P F Carbomethoxylation

Carboxylation of activated CH groups with MMMC (methoxy magnesium methyl carbonate) **1** (Szarvasy) and addition of the resulting activated ketones to C=N bonds (Schöpf).

Mg —(MeOH, Δ)→ $(MeO)_2Mg$ —(CO_2, DMF)→ MeO-Mg-O-C(=O)-OMe **1** —(1-nitropropane, CO_2, 60°)→ **2** (44%)

3 —(**1**, 120°)→ —(CO_2 20°, Δ¹-pyrroline)→ **4** (63%)

1	Szarvasy, S.	*Chem. Ber.*	**1897**	*30*	1836
2	Finkbeiner, H.L.	*J. Org. Chem.*	**1963**	*28*	215
3	Schöpf, C.	*Liebigs Annl.*	**1959**	*626*	123
4	Schöpf, C.	*Angew. Chem.*	**1949**	*61*	31
5	Grisar, J.M.	*Synthesis*	**1974**		284

Methyl 2-Nitrobutanoate (2).[2] A 2N solution of MMMC **1** (100 mL) was heated to 60°C under a flow of CO_2, 1-nitropropane (9.80 g, 0.2 mol) was added and CO_2 was replaced by N_2. After 6 h at 60°C, 32% HCl (60 mL) and ice (75 g) were added, the acid was extracted with Et_2O, the solvent evaporated and the residue esterified by MeOH-HCl to afford 6.5 g of **2** (44%), bp 77°C (2.5 mm), n_D^{20} = 1.4249.

4-(2-Phenylethyl)-ω-(2-pyrrolidinyl)acetophenone (4).[5] **1** (2M, 0.4 mol) in DMF was heated under a flow of CO_2 to 120°C. **3** (22.4 g, 0.1 mol) was added and all was heated for 4 h under N_2 at 120°C to allow MeOH to escape. The mixture was cooled under CO_2 and Δ^1-pyrroline (8.3 g,0.12 mol) was added and stirred under CO_2 for 40 h. The mixture was poured into conc HCl (200 mL) and ice (800 g) to give after recrystallization (iPrOH-water) 20.7 g of **4**.HCl (63%), mp 200-201°C (dec.).

TEBBE Olefination

Synthesis of terminal olefins from ketones or esters via a Ti methylene transfer reagent.

Cp_2TiCl_2 + $AlMe_3$ → $Cp_2Ti(\mu\text{-}CH_2)(\mu\text{-}Cl)AlMe_2$ → (with 4) 5

1 **2** **3** **4** **5** (96%)

1	Tebbe, F.N.	*J. Am. Chem. Soc.*	**1978**	*100*	3611
2	Pine, S.K.	*Synthesis*	**1991**		165
3	Grubbs, R.H., Evans, D.A.	*J. Am. Chem. Soc.*	**1980**	*102*	3270
4	Tebbe, F.N.	*J. Chem. Soc. Chem. Commun.*	**1973**		227

μ-Chlorobis (η^5-2,4-cyclopentadien-1-yl)(dimethylaluminium)-μ-methylene-titanium (3).[1] Into the stirred suspension of Cp_2TiCl_2 **1** (62 g, 0.25 mol) in PhMe (100 mL), $AlMe_3$ **2** (pyrophoric!) (48 mL, 0.5 mol) is washed with PhMe (50 mL) in a dry box. The mixture is stirred 48 h at 20°C (CH_4 escapes). The product is filtered through Celite, the volatiles are evaporated at 20°C under vacuum. The crude solid is washed with hexane, filtered, dissolved in PhMe (90 mL) and filtered. The filtrate is treated with **2** (10 mL) chilled to -25°C and the reddish crystals of **3** filtered under Ar, 38 g (53%).

2-tert-Butylmethylidenecyclohexane (5)[2] To a solution of **2** and tert-butyl cyclohexanone **4** (153 mg, 1 mmol) in dry THF (3 mL) at 0°C is added a toluene solution of **3** (2 mL, 0.5 M, 1 mmol). After 15 min at 20°C, Et_2O (15 mL) is added followed by 10 drops of 0.1 M NaOH. The mixture is dried (Na_2SO_4), filtered (celite pad) and evaporated. Chromatography (alumina 2% Et_2O in pentane) afforded 144 mg of **5** (96%).

TEER MEER Dinitroalkane Synthesis

Gem dinitroderivatives from chloronitro compounds.

1. KNO_2
2. NH_2OH, HCl

1. NH_3
2. HCl

1 **2** (47%)

1 teer Meer, E. *Liebigs Ann.* **1876** *181* 4

2 Hawthorne, M.F. *J. Am. Chem. Soc.* **1956** *78* 4980

3 Kaplan, R.B. *J. Am. Chem. Soc.* **1965** *83* 3535

1,1-Dinitroethane (2).[2] A solution of 1-chloro-1-nitroethane **1** (0.313 g, 3 mmol) in EtOH (25 mL) was added to KNO_2 (10.0 g, 0.117 mol) in water (75 mL). After 1 h at 20° this mixture was added to an ice cooled mixture of hydroxylamine hydrochloride (15.0 g, 0.216 mol) in conc. HCl (25 mL) and Et_2O (75 mL). The solution was saturated with NaCl and the Et_2O layer separated and dried ($MgSO_4$). The Et_2O solution was treated with a saturated solution of NH_3 in Et_2O until separation of NH_4Cl ceased. The NH_4Cl was removed by filtration, the solvent evaporated and the concentrate (10 mL) was extracted with NH_4OH. Cautious acidification with HCl under cooling, extraction with Et_2O and solvent evaporation left a residue which was dissolved in EtOH (0.2 mL) and precipitated with saturated NH_3 in Et_2O (30 mL). The precipitate furnished 65 mg of **2** (47%), mp 127-128°C.

TEUBER Quinone synthesis

Oxidation of phenols or anilines to quinones by means of potassium nitrosodisulfonate (Fremy's salt).

OH, CH_3, 1 — $(KSO_3)_2NO$ **5**, 2 N NaOH → O, CH_3, O, **2** (94%)

OH, H_3C, CH_3, N, C_2H_5, **3** — **5**, KH_2PO_4 → O, H_3C, CH_3, N, O, C_2H_5, **4** (5%)[5]

1	Teuber, H.I.	*Chem. Ber.*	**1952**	*85*	95
2	Teuber, H.I.	*Chem. Ber.*	**1953**	*86*	1036
3	Teuber, H.I.	*Chem. Ber.*	**1955**	*88*	802
4	Teuber, H.I.	*Angew. Chem. Int. Ed.*	**1969**	*8*	218
5	Roth, R.A.	*J. Org. Chem.*	**1966**	*31*	1014
6	Zimmer, H.	*Chem. Rev.*	**1971**	*71*	229
7	Kosikowski, A.	*J. Org. Chem.*	**1981**	46	2426

Thymoquinone (2).[1] Thymol **1** (0.3 g, 2 mmol) in water (2 mL) was treated with 2N NaOH (2 mL) and potassium nitrosodisulfonate **5** (0.54 g, 1 equiv) in water (5 mL), followed by 2 more portions of **5**. The yellow oil was treated with 2M AcOH and was extracted with Et_2O. Washing the extract with NaOH and water and evaporation gave 0.31 g of **2** (94.5%), mp 43-46°C (petroleum ether).

2,6-Dimethyl-1-ethyl-4,7-indoloquinone (4).[5] **3** (1.89 g, 10 mmol) in Me_2CO.(350 mL) was added to **5** (10.24 g, 40 mmol) in 0.055 M KH_2PO_4 (60 mL). After1 h dilution with water, extraction with CH_2Cl_2 (600 mL) and chromatography on Florisil gave 95 mg of orange **4** (5%), mp 115-117°C (hexane).

THILE - WINTER Quinone acetoxylation

Synthesis of triacetoxyaryl derivatives from quinones.

Ac_2O Δ, $BF_3.Et_2O$

1 → 2 (81%)

OAc OAc OAc

1	Thile, J.	*Chem. Ber.*	**1898**	*31*	1247
2	Thile, J., Winter, E.	*Liebigs Ann.*	**1900**	*311*	341
3	Fieser, L.F.	*J. Am. Chem. Soc.*	**1948**	*70*	3165
4	Blatchly, J.M.	*J. Chem. Soc.*	**1963**		5311
5	McOmie, J.F.W.	*Org. React.*	**1972**	*19*	200

TIEMANN Rearrangement

Beckmann rearrangement of amidoximes to urea derivatives in the presence of acids (benzene sulfonyl chloride).

$$\underset{\mathbf{1}}{Ph-\overset{N-OH}{\overset{\|}{C}}-NH-C_6H_{11}} \xrightarrow[\text{pyr, dioxane}]{PhSO_2Cl} Ph-\overset{N-O-SO_2Ph}{\overset{\|}{C}}-NH-C_6H_{11} \longrightarrow \underset{\mathbf{3}\ (75\%)}{Ph-NH-\overset{O}{\overset{\|}{C}}-NH-C_6H_{11}}$$

1	Tiemann, F.	*Chem. Ber.*	**1891**	*24*	4162
2	Plapinger, R.F.	*J. Org. Chem.*	**1956**	*21*	1186
3	Boyer, J.H.	*J. Org. Chem.*	**1970**	*35*	2249
4	Smith, P.A.S.	*Org. React.*	**1946**	*3*	366

TIFFENEAU Aminoalcohol rearrangement

Cationic rearrangement (ring enlargement) of 1,2-aminoalcohols by diazotization.

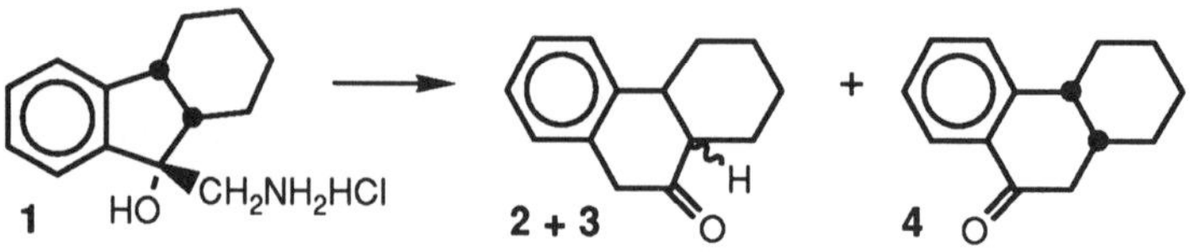

1	Tiffeneau, M.	*C.R.*	**1937**	*205*	54
2	Tiffeneau, M.	*C.R.*	**1941**	*212*	195
3	Parham, W.E.	*J. Org. Chem.*	**1972**	*37*	1975
4	Smith, P.A.S.	*Org. React.*	**1960**	*11*	157

Octahydrophenanthrones (2,3,4).[3] 9-cis-Aminomethyl-1,2,3,4,4a,-cis-9a-cis-hexahydrofluorene-9-trans-ol . HCl **1** (331.7 mg, 1.315 mmol) in water (8 mL) was treated with $NaNO_2$ (274.3 mg, 3.965 mmol) in water (3 mL) and 1 drop of HCl. The mixture was stirred for 2 h at 0° and 15 h at 25°. Extraction with Et_2O and evaporation gave ketones **2:3:4** in a ratio of 55:0.7:25, total yield: 229 mg (81%).

TISCHENKO - CLAISEN

Conversion of aldehydes to esters in the presence of metal alcoholates.

$$2\ CH_3\text{-}CH_2\text{-}CH_2\text{-}CHO \xrightarrow[CCl_4]{2\%\ Al(OiPr)_3} CH_3\text{-}CH_2\text{-}CH_2\text{-}CH_2\text{-}O\text{-}C(=O)\text{-}CH_2\text{-}CH_2\text{-}CH_3$$

1 → **2** (91%)

1	Claisen, L.	*Chem. Ber.*	**1887**	*20*	648
2	Tischenko, W.	*J. Russ. Phys. Chem. Soc.*	**1906**	*38*	355,542
3	Lin, I	*J. Am. Chem. Soc.*	**1957**	*74*	5133
4	Stapp, P.R.	*J. Org. Chem.*	**1973**	*38*	1433

TIPSON - COHEN Olefination

Conversion of sugar glycols via epoxides into sugar olefins.

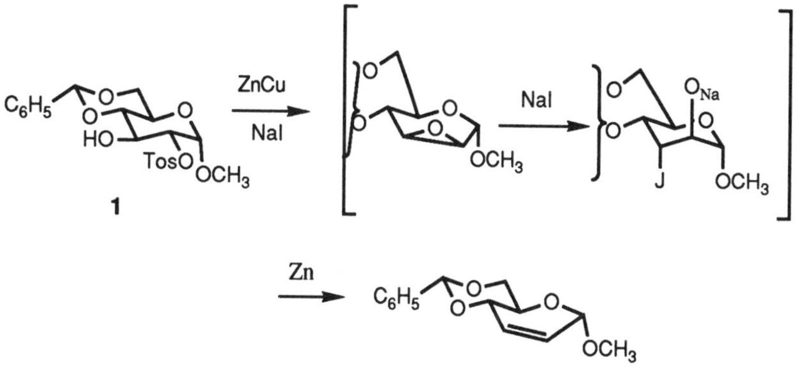

2 (78%)

1	Tipson, R.S., Cohen, A.	*Carbohydr. Res.*	**1965**	*1*	338
2	Fraser-Ried, B.	*Can. J. Chem.*	**1969**	*47*	393
3	Radatus, B.K.	*Synthesis*	**1980**		47
4	Bloch, E.	*Org. React.*	**1984**	*30*	500

Methyl 4,6-0-benzylidene-2,3-dideoxy-α-D-erythro-hex-2-enopyranoside (2).[3] A mixture of methyl 4,6-0-benzylidene-2-0-tosyl-α-D-glucapyranoside **1** (87.2 g, 0.2 mol), zinc-copper couple (65 g, 1 mol), NaI (150 g, 1 mol), DMF (500 mL) and dimethoxyethane (100 mL) was heated to reflux (130°C) for 3 h under efficient stirring. The cooled mixture was poured into water (3000 mL) stirred with PhH (1000 mL) and filtered. The organic phase and the PhH extract was evaporated to dryness to give 44.1 g of crude **2**. Recrystallization from 85% EtOH afforded 38.5 g of **2** (78%), mp 118-120°C, $[\alpha]^{30}_{D}$ = +135°C (c 1.0 $CHCl_3$).

TRAHANOVSKY Ether oxidation

Oxidation of aromatic ethers to carbonyl compounds with cerium ammonium nitrate.

$Ce(NH_4)_2(NO_3)_6$, MeCN 20°C

1 → **2** (77%)

$Ce(NH_4)_2(NO_3)_6$

1	Trahanovsky, W.S.	*J. Chem. Soc.*	**1965**		5777
2	Jacobs, P.	*J. Org. Chem.*	**1976**	*41*	3627
3	Lepage, L. & Y.	*Can. J. Chem.*	**1980**	*58*	1161
4	Lepage, L. & Y.	*Synthesis*	**1983**		1018

1,4-Diphenyl-1,4-dioxo-2-butene (2).[2] A solution of $Ce(NH_4)_2(NO_3)_6$ (1.2 g, 2.2 mol) in water (4 mL) was added dropwise to a stirred solution of 2,5-diphenylfuran **1** (0.22 g, 1 mmol) in MeCN (10 mL). The mixture was stirred at 25°C for 30 min and was extracted with $CHCl_3$ (3x40 mL). The organic extract is concentrated and a few drops of Et_2O are added to induce crystallization. There are obtained 0.1817 g of **2** (77%), mp 134-135°C.

TRAUBE Purine synthesis

Pyrimidine synthesis from guanidine and cyanoacetic ester and purine synthesis from aminopyrimidines.

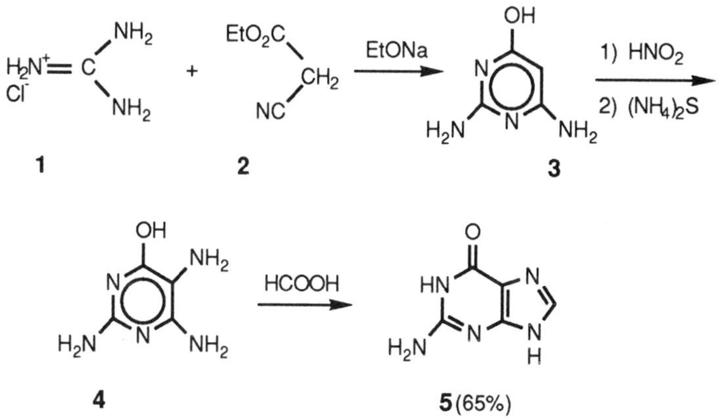

1	Traube, W.	*Chem. Ber.*	**1900**	*33*	1371,3035
2	Traube, W.	*Liebigs, Ann.*	**1904**	*331*	641
3	Katritzky, A.	*Quart. Rev. (London)*	**1956**	*10*	397

Guanine (5).[1] A suspension of guanidine.HCl **1** (40.0 g, 0.4 mol) in EtOH was treated with NaOEt (from Na 9.2 g, 0.4 at g). To this solution was added ethyl cyanoacetate **2** (48.0 g, 0.4 mol) and the mixture was heated to reflux for 6 h. The salts were removed by filtration and the filtrate was concentrated to dryness. The 2,4-diamino-6-oxypyrimidine **3** after nitrosation and reduction with $(NH_4)_2S$ gave 2,4,5-triamino-6-oxypyrimidine **4**. By refluxing **4** (10.0 g, 74 mmol) with HCOOH (190 mL) for 4-5 h there are obtained 7-8 g of **5** (60-67%).

TREIBS Allylic oxidation

Allylic oxidation of alkenes using mercuric trifluoroacetate with allylic rearrangement possible.

CH_3COO **1** $\xrightarrow{(F_3CCOO)_2Hg}$ CH_3COO OH **2**

1	Treibs, W.	*Naturwissenschaften*	**1948**	*35*	125
2	Wiberg, K.B.	*J. Org. Chem.*	**1964**	*29*	3353
3	Arzoumanian, N.	*Synthesis*	**1971**		527
4	Nassiot, G.	*Synthesis*	**1974**		722

17-Oxo-Δ⁴-androsten-3β, 6β-diol-3-acetate (2).[4] A solution of 17-oxo-Δ⁵-androsten-3β-ol-acetate **1** (1.03 g, 31 mmol) and mercury (II) trifluoroacetate (3.1 g, 72 mmol) in CH_2Cl_2 (100 mL) was stirred for 24 h at room temperature. The solvent was removed in vacuum (60 mL) and the remaining solution was filtered. The filtrate was washed with 5% $NaHCO_3$ solution, then with water and again filtered. After evaporation of the solvent, the crude residue (720 mg) was recrystallized from MeOH to give 411 mg of **2** (40%), mp 148-150°C, $[\alpha]n_D^{25}$ = +25° ($CHCl_3$ c = 1.1).

TROST Cyclopentanation

Methylenecyclopentane formation from siloxymethylallylsilane or acetoxymethylallylsilane **4** with Michael acceptor olefins and Pd catalysts (via trimethylene methane equivalent).

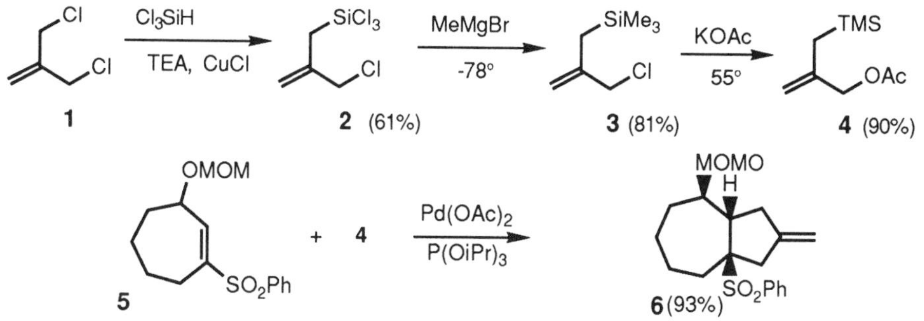

1 Trost, B.M. *J. Am. Chem. Soc.* **1979** *101* 6429

2 Trost, B.M. *J. Am. Chem. Soc.* **1983** *105* 2315

3 Trost, B.M. *Angew. Chem. Int. Ed* **1986** *25* 1

4 Trost, B.M. *J. Org. Chem.* **1988** *53* 4887

5 Trost, B.M. *J. Am. Chem. Soc.* **1989** *111* 7487

2-(Acetoxymethyl)-3-(trimethylsilyl)propene (4).[4] Cl_3SiH (27.09 g, 0.2 mol) and **1** (20 g, 0.16 mol) in Et_2O (50 mL) was added under stirring to Cu_2Cl_2 (0.158 g, 1.6 mmol) and TEA (28 mL) in Et_2O (300 mL) over 4 h. After 14 h at 20°, filtration (Ar) and distillation gave 21.83 g of **2** (61%), bp 50-54°C (2 mm). **2** (35 g, 0.156 mol) in Et_2O (700 mL) at -78° was added dropwise to 3M MeMgBr (183 mL, 0.55 mol). After 1 h at -78°C and 10 h at 20°C, work up and distillation afforded 20.41 g of **3** (81%), bp 158°C (760 mm). **3** (20.4 g, 0.125 mol) and KOAc (49 g, 0.5 mol) in DMF (200 mL) was heated at 55-60°C for 48 h. Dilution with water, extraction with Et_2O and distillation gave 20.97 g of **4** (90%), bp 95°C (7 mm).

2-Methylene-4-(methoxymethoxy)-8aβ-(phenylsulfonyl)-3aβ-decahydroazulene (6).[5] To $Pd(OAc)_2$ (15 mg, 0.06 mmol) and $P(OiPr)_3$ (101 mg, 0.487 mmol) in PhMe (2 mL) was added **5** (1.05 g, 3.54 mmol) in PhMe (2 mL) followed at 60°C by **4** (0.95 g, 5.3 mmol). After 40 h at 80°C, chromatography (3:1 hexane:EtOAc, R_f = 0.33) gave 1.15 g of **6** (93%).

TROST - CHEN Decarboxylation

Ni complex catalyzed decarboxylation of dicarboxylic acid anhydrides to form alkenes.

COO CH$_3$ — $(Ph_3P)_2Ni(CO)_2$ → COO CH$_3$

1 **2** (91%)

1	Trost, B.M., Chen, F.	*Tetrahedron Lett.*	**1971**		2603
2	Cramer, R.	*J. Org. Chem.*	**1975**	*40*	2267
3	Jennings, P.W.	*J. Org. Chem.*	**1975**	*40*	260
4	Flood, T.C.	*Tetrahedron Lett.*	**1977**		3861
5	Rose, J.D.	*J. Chem. Soc.*	**1950**		69
6	Grunewald, G.L.	*J. Org. Chem.*	**1978**	*43*	3074

1-Methoxycarbonylbicyclo[2,2,1]hept-2-ene (2).[6] To freshly distilled diglyme (75 mL) was added 1-methoxycarbonylbicyclo (2,2,1)heptane-endo-,endo-2,3-dicarboxylic anhydride **1** (9.45 g. 42 mmol) and freshly prepared dicarbonyl bis(triphenylphosphine) nickel (0)[5] (5.39 g, 8 mmol). The stirred solution was heated at reflux and the effluent gas was collected. After 6 h, ca 1880 mL of gas (1 equivalent of CO and CO_2) had been collected; the solvent, codistilling with the product (bp 40°C, 2.5 mm) was removed in vacuum. To the residue was added more diglyme (2x25 mL) and the solution was distilled to dryness. The combined distillates were poured into water (800 mL) and extracted with pentane (4x100 mL). Washing, drying and evaporation of the solvent afforded 5.82 g of **2** (91%).

TSCHUGAEF Olefin synthesis

Olefin formation (preferentially less substituted) from alcohols via xanthate pyrolysis.

1) CS_2 2) NaOH MeI → Δ

1 **2** **3** (15%)

1	Tschugaef, J.	*Ber.*	**1898**	*31*	1775
2	de Groote, A.	*J. Org. Chem.*	**1968**	*33*	2214
3	De Puy, C.H.	*Chem. Rev.*	**1960**	*60*	444
4	Nace, H.R.	*Org. React.*	**1962**	*12*	58

3-Methyl-1-butene (3).[2] A mixture of powdered NaOH (40.5 g, 1 mol), CCl_4 (50 mL), Et_2O (600 mL) and isoamyl alcohol **1** (89.0 g, 1 mol) was treated with CS_2 (76.0 g, 1 mol) for 1 h, then methyl iodide (149.0 g, 1.05 mol) was added. Distillation afforded isoamyl-S-methyl xanthate **2** was 126 g, (71%) bp 100-102°C (10 mm), n_D^{20} = 1.5234.

Boiling of **2** under partial reflux for 7-8 h was followed by distillation and washing with 30% NaOH, then with a saturated solution of $HgCl_2$. Redistillation of the product afforded 10.5 g of 3 (15%), bp 19-20°C (760 mm).

TSUJI - TROST Allylation

Direct C-allylation of ketones or of tin enol ethers with Pd(O) catalysts.

1	Tsuji, J.	*Tetrahedron Lett.*	**1965**		4387
2	Tsuji, J.	*J. Org. Chem.*	**1985**	*50*	1523
3	Tsuji, J.	*Acc. Chem. Res.*	**1969**	*2*	144
4	Trost, B.	*J. Am. Chem. Soc.*	**1973**	*95*	292
5	Trost, B.	*J. Am. Chem. Soc.*	**1980**	*102*	5699
6	Trost, B.	*Acc. Chem. Res.*	**1980**	*13*	385
7	Ukai, T.	*J. Organomet. Chem.*	**1974**	*65*	235

Geranyl acetone (4).[2] To Pd_2 $(dba)_3$ $CHCl_3$ **3**[7] (52 mg, 0.05 mmol) and dppe (78 mg, 0.2 mmol) in THF (4 mL) was added dropwise methyl geranyl carbonate **1** (427 mg, 2 mmol) followed by methyl acetoacetate **2** (925 mg, 8 mmol) in THF (2 mL). After 3 h stirring at 50°C under Ar, the solvent was removed in vacuum, the product was treated with 10% NaOH (10 mL) and MeOH (10 mL) and stirred at 20°C for 18 h. Extraction with 3N HCl and CH_2Cl_2, evaporation, refluxing the residue for 1 h with PhH (10 mL), evaporation and chromatography (SiO_2Et_2O:hexane) afforded 288 mg of **4** (74%), 92% E.

U G I Condensation

Peptide synthesis via a three or four component condensation (amino acid, imine and isocyanide).

iPr - CH(NHCbz) - CO_2H + Ph - C(Ph) = N - CH_2Ph (**2**) + C≡N - CH_2COOMe (**3**) → (9 Kbar, 14 d) → **4** (63%)

1	Ugi, I.	*Angew. Chem.*	**1977**	*89*	267
2	Yamada, T.	*J. Chem. Soc. Chem. Commun.*	**1984**		1500
3	Yamada, T.	*Chem. Lett.*	**1987**		723
4	Yamada, T.	*J. Chem. Soc. Chem. Commun.*	**1990**		1640

Peptide (4).[4] A mixture of N-carbobenzyloxy-L-valine **1** (1.104 g, 4.4 mmol), Schiff base **2** (1.139 g, 4.4 mmol), methyl isocyanidoacetate **3** (0.433 g, 4.4 mmol) in CH_2Cl_2 (4 mL) was compressed for 14 days at 9 kbar. Evaporation of the solvent and chromatography afforded 1.675 g of **4** (63%), mp 126-127°C, α_D^{25} = -16.0° (c 1.0, $CHCl_3$). The original reaction employed four components; in place of **2** the reagents were benzylamine and benzophenone.

ULLMANN - FEDVADJAN Acridine synthesis

Synthesis of polynuclear pyridines from anilines, phenols and formaldehyde.

1 + CH_2O + 2 —250°→ 3 (46%)

1	Ullmann, F., Fedvadjan,	A.*Chem. Ber.*	**1903**	*36*	1027
2	Buu Hoi, N.P.	*Bull Soc. Chim. Fr. Mem.*	**1944**	*11*	406
3	Buu Hoi, N.P.	*J. Chem. Soc. (C)*	**1967**		213

Benzo-(a)-pyrido-(4,3,h)-acridine (3).[3] To a mixture of 5-aminoisoquinoline **1** (1.5 g, 11 mmol) and β-naphtol **2** (1.5 g, 10 mmol) heated to 250°C, was added in small portions paraformaldehyde (0.45 g, 150 mmol); when steam had ceased to evolve, the mixture was heated for a few more minutes then fractionated in vacuum. The thick resin (bp 290°C-11 mm) was collected and recrystalllized from BuOH. A final purification was effected by vacuum sublimation to afford 1.3 g of **3** (46%), mp 289°C.

ULLMANN - GOLDBERG Aromatic substitution

Cu catalyzed substitution of aromatic halides in the synthesis of diaryls, diaryl ethers, diaryl amines, phenols.

1 + Br–Ph 2 → (Cu cat., KOAc) → 3 (92%)

p - I - C_6H_4 - $N(SiMe_3)_2$ → (Cu, 240°) → H_2N–C_6H_4–C_6H_4–NH_2 4 (60%)

Ph - NH - Ph 6 (86%) ← ($PhNH_2$, Cu 210°) ← Ph - I 5 → (Ph - OH, Cu) → Ph - O - Ph 7 (94%)

1	Ullmann, F.	*Ber.*	**1903**	*36*	2389
2	Goldberg, I.	*Ber.*	**1906**	*39*	1691
3	Renger, B.	*Synthesis*	**1985**		856
4	Schulenburg, J.W.	*Org. React.*	**1965**	*14*	19
5	Yamamoto, T.	*Can.J.Chem.*	**1983**	*61*	86
6	Bunnell, J.F.	*Chem. Rev.*	**1951**	*49*	392

2-Oxo-1-phenyltetrahydropyrrole (3).[3] **1** (17.00 g, 0.2 mol), KOAc (24.5 g, 0.25 mol) and copper activated catalyst RCH 60/35 (3.00 g) in **2** (200 ml) was refluxed 4 hr. After cooling and filtration, the filtrate was washed with 5% NH_4OH and water, evaporated and the residue (31 g) stirred with hexane to give 29.5 g of **3**, (92%), mp 65-67°C.

Diphenylamine (6).[5] Aniline (93 mg, 10 mmol), **5** (408 mg, 20 mmol) and copper (1.9 g, 30 mmol) were heated to 210°C for 12 h. Work up afforded 145 mg of **6** (86%), mp 53°C.

VAN BOOM Phosphorylating reagent

Phosphorylation of sugars or nucleosides by means of salicylchlorophospite **2**.

DMT-O … T … OH **1** → **2** → DMT-O … T … O-P(=O)(OH)OH **3** (88%)

BzO, BzO, OR, NH-Ac, OH → **2** → BzO, BzO, OR, NH-Ac, O-P(=O)(OH)-OH

R = 4,4'-Dimethoxytrityl

1 Anshutz, R.	*Liebigs Ann.*	**1887**	*239*	301
2 Young, R.W.	*J. Am. Chem. Soc.*	**1952**	*74*	1672
3 Van Boom, J.H.	*Rec. Trav. Chim.*	**1986**	*105*	510
4 Van Boom, J.H.	*Tetrahedron Lett.*	**1986**	*27*	2661,6271

5'-Dimethoxytrityl-2'-deoxythyimidine-3'-phosphate (3).[4] A solution of 1 mmol of **1** (T=thymine, DMT=4,4'-dimethoxytrityl) in dioxane (6 mL), pyridine (1 mL) and a slight excess of 2-chloro-5,6-benzo-1,3-dioxaphosphorin-2-one **2** (1 mL of a 1.25 M solution in dioxane) was stirred at 20°C. After 5 min TLC indicated complete conversion of **1** into **3** with zero mobility. Water was added and after work up **3** was isolated in 88% yield.

VAN LEUSEN Reagent

A one-step synthesis of nitriles from carbonyls by a reductive cyanation with tosylmethyl isocyanide (TosMIC); also synthesis of 1,3-azole or of ketones.

1 + Tos $CH_2N{\equiv}C$ 2 —(t BuOK, DME)→ 3 (90%)

Tos - CH(R) - N ≡C 4 + Ar-CHO ⟶ (Tos, H, R, Ar oxazoline) ⟶ (R, Ar oxazole)

Tos-CH_2-N ≡C —(base, R_1-Br)→ 4 ⟶ Tos-C(R)(R_1)-N≡ C ⟶ R_1-CO-R

1	van Leusen, A.M.	*Tetrahedron Lett.*	**1973**		1357
2	van Leusen, A.M.	*J. Org. Chem.*	**1977**	*42*	3114
3	van Leusen, A.M.	*Synth. Commun.*	**1980**	*10*	399
4	van Leusen, A.M.	*Lect. Heteroc. Chem.*	**1980**	*V*	S-111
5	van Leusen, A.M.	*Org. Synth.*	**1977**	*57*	102

2-Cyanoadamantane (3).[2] To **1** (15.0 g, 0.1 mol) and TosMIC **2** (25.0 g, 0.13 mol) in 1,2-dimethoxyethane (350 mL) and anh. EtOH (10 mL), under cooling was added solid t-BuOK (28.0 g, 0.24 mol) in portions. Stirring was continued for 30 min. at 20°C, then 30 min at 35-40°C. After filtration and concentration of the filtrate, the product was filtered through a 5 cm alumina (ca 200 g) layer and eluted with petroleum ether. Evaporation afforded 14-15 g of **3** (87-93%), mp 160-180°C (sealed tube).

VEDEJS Hydroxylation

Oxidation of ketones to α-hydroxyketones by means of a peroxymolybdenum reagent.

LDA

MoOPH **1**

C_6H_5 C_6H_5 → OH, C_6H_5 C_6H_5

2 **3** (53%)

1 Vedejs, E.	*J. Org. Chem.*	**1978**	*43*	194
2 Krohn, K.	*Chem.Ber.*	**1989**	*122*	2323

Oxodiperoxymolybdenum (pyridine) (hexamethylphosphoric triamide) (MoOPH).(1). [1] MoO_3 (30 g, 0.2 mol) and 30% H_2O_2 (150 mL) was heated until exothermic, kept at 35-40°C by cooling and then by heating for 3.5 h at 40°C. Filtration, cooling to 10°C, addition of HMPA (37.3 g), filtration and recrystallization from MeOH gave $MoO_5.H_2O$.HMPA 50 g, (67%). The latter was dehydrated at 0.2 mm for 24 h over P_2O_5 to give MoO_5 HMPA. When MoO_5 HMPA (18.0 g, 51.9 mmol) in THF (40 mL) was stirred, cooled and pyridine (4.11 g, 51.9 mmol) was added dropwise, MoOPH **1** was filtered, washed, dried and stored with Drierite in a refrigerator.

6-Hydroxy-4,4-diphenylcyclohex-2-en-1-one (3).[1] 4,4-Diphenylcyclohex-2-en-1-one **2** (124 mg, 0.5 mmol) in THF (61 mL) was added to LDA (0.79 mL, 0.7 N, 0.55 mmol) at -23°C. After 5 min all was added to **1** (282 mg, 0.65 mmol) in THF (10 mL) (olive green). After 5 min the mixture was quenched on sat. Na_2SO_3 (3 mL) and separated by PLC to give 70 mg of **3** (53%), and 21 mg of **2** (17%).

VILSMEIER - HAACK - VIEHE Reagent

Formylation of aromatics, alkenes, activated H compounds by $Me_2N^+{=}CHCl\ Cl^-$ (Vilsmeier-Haack) or $Me_2N^+{=}CCl_2\ Cl^-$ (Viehe) reagent.

CH_3O … OH, Ph — **1** → DMF $POCl_3$, 95° → CH_3O … CHO, Ph — **2** (81%)

3 + $Me_2N^+{=}CCl_2\ Cl^-$ **4** → $CHCl_3$, Δ → Cl, Cl, $C{=}N^+Me_2$ **5** → $NaHCO_3$ → Cl, O, NMe_2 **6** (71%)[6]

1	Vilsmeier, A., Haack, A.	*Chem. Ber.*	**1927**	*60*	119
2	Krishna-Rao, G. S.	*J. Org. Chem.*	**1981**	*46*	5371
3	Konvar, D.	*Tetrahedron Lett.*	**1987**	*28*	955
4	Ferguson, L. N.	*Chem. Rev.*	**1946**	*38*	230
5	Grundmann, A.	*Angew. Chem.*	**1966**	*78*	747
6	Viehe, H. G.	*Angew. Chem. Int. Ed.*	**1971**	*10*	575
7	Bergmann, J.	*Tetrahedron Lett.*	**1986**	*27*	1939

1-Benzyl-6-methoxy-3,4-dihydro-2-naphthaldehyde (2).[2] To **1** (1 g, 3.7 mmol) in DMF (4 mL) at 0°C was added dropwise $POCl_3$ (0.5 mL). After 10 h at 95°C, $POCl_3$ (0.5 mL) was again added at 25°C and heating was continued for 5 h. Quenching with NaOAc (7 g in 18 mL water) extraction with Et_2O and preparative TLC (PhH) gave 0.85 g of **2** (81%), mp 116-117°C (PhH-hexane).

N,N-Dimethyl-(2-chloro-1-cyclohexenyl) carboxamide (6).[6] **2** (4.9 g, 50 mmol) and **4** (17.8 g, 110 mmol) in $CHCl_3$ (60 mL) was refluxed for 3 h until **4** had dissolved and HCl evolution ceased. After half of the solvent was removed, Et_2O (40 mL) and petroleum ether (10 mL) were added to give11.8 g of **5** (98%), mp 122-125°C. A suspension of **5** (6.7 g, 27 mmol) in water (15 mL) and solid $NaHCO_3$were stirred 30 min and extracted with $CHCl_3$ to give 3.7 g of **6** (71%), bp 99-102°C (0.6 torr.).

VOIGHT α-Aminoketone synthesis

Synthesis of α-aminoketones from α-hydroxyketones.

Ph-CH-OH / Ph-C=O **1** + 2-aminopyridine **2** —(32% HCl; Ph-Me, Δ)→ 2-pyridyl-NH-CH(Ph)-C(=O)-Ph **3** (83%)

1	Voight, K.	*J. Prakt. Chem.*	**1886**	*34*	1(2)
2	Lutz, R.E.	*J. Am. Chem. Soc.*	**1948**	*70*	2015
3	Kay, J.A.	*J. Am. Chem. Soc.*	**1953**	*75*	746
4	Lutz, R.E.	*J. Org. Chem.*	**1956**	*21*	49

VOLHARDT - ERDMANN Thiophene synthesis

Thiophene synthesis from succinic acids.

$H_3C-CH(COONa)-CH(COONa)-CH_3$ **1** —(P_4S_{10}, Δ)→ 3,4-dimethylthiophene **2** (44%)

1	Volhardt, J.	*Chem. Ber.*	**1885**	*18*	454
2	Lindstead, R.	*J. Chem. Soc.*	**1937**		915
3	Wolff, E.W.	*Org. React.*	**1951**	*6*	412

3,4-Dimethylthiophene (2).[2] Disodium salt **1** (195 g, 1 mol) and phosphorus pentasulfide ~(245 g) was distilled dry under a stream of CO_2 to give 83 g of crude **2**, which after 15 h contact with NaOH and 6 h reflux over Na was distilled to afford 50 g of **2** (44.6%), bp 145-148°C.

V O R B R Ü G G E N Nucleoside synthesis

Synthesis of nucleosides by condensation of sugars with silyl heterocycles and Lewis acids such as $SnCl_4$ or trimethylsilyl triflate **3.**

$Me_3Si\,SO_3\,CF_3$ **3**, 25°; $NaHCO_3$; **4** (89%)

1	Vorbrüggen, H.	*Angew. Chem. Int. Ed.*	**1970**	*9*	461
2	Vorbrüggen, H.	*J. Org. Chem.*	**1974**	*39*	3654
3	Vorbrüggen, H.	*Chem. Ber.*	**1981**	*114*	1234
4	Schreiber, S.L.	*J. Am. Chem. Soc.*	**1990**	*112*	9657

5-Methoxyuridine-2,3,5-tri-O-benzoate (4).[3] To 5-methoxy-2,4-bis(trimethylsilyloxy)uracil **1** (11 mmol, 34 mL of 0.356 N solution in 1,2-dichloroethane) and 1-O-acetyl-2,3,5-tri-O-benzoyl-β-D-ribofuranose **2** (5.04 g, 10 mmol) in dichloroethane (75 mL) was added trimethylsilyl triflate **3** (12 mmol, 22.8 mL of 0.522 N solution in dichloroethane) and the mixture was stirred for 4 h at 25°C. The clear yellow solution was diluted with CH_2Cl_2 (50 mL) extracted with an ice cooled solution of $NaHCO_3$ (50 mL) and after washing with water (2x20 mL), the organic phase was dried (Na_2SO_4), the solvent evaporated and the residue recrystallized from EtOAc-hexane to afford 5.24 g of **4** (89%), mp 205-207°C.

WACKER -TSUJI Olefin oxidation

Oxidation of olefins to ketones by Pd (II) catalyst.

2% (Pd (OAc)$_2$ benzoquinone / 0.24 M $HClO_4$

1 → **2** (100% GC)

$PdCl_2$ - $CaCl_2$ / O_2, DMF, 50° — OMe → OMe (65%)

1	Phillips, F.C.	*Amer. Chem. J.*	**1894**	*16*	255
2	Tsuji, J.	*Tetrahedron Lett.*	**1982**	*23*	2679
3	Smidt, J. (Wacker)	*Angew. Chem.*	**1959**	*71*	176
4	Tsuji, J.	*Synthesis*	**1984**		369
5	Wayner, D.D.M.	*J. Org. Chem.*	**1990**	*55*	2924

Cyclohexanone (2).[3] A mixture of $Pd(OAc)_2$ (44.8 mg, 0.2 mmol), benzoquinone (1.06 g, 9 mmol) and an inorganic acid such as (HCl, $HClO_4$, HBF_4, H_2SO_4 or HNO_3 0.1 M) were dissolved in MeCN (43 mL) and water (7 mL). The solution was deoxygenated with Ar (minimum 30 min) and stirred until $Pd(OAc)_2$ had dissolved. Cyclohexene **1** (0.82 g, 10 mmol) was added and the mixture was stirred for 10 min. Extraction of **2** with hexane or Et_2O followed by washing with 30% NaOH and evaporation of the solvent afforded 0.98 g of **2** (100% by capillary GC).

WAGNER - MEERWEIN - NAMETKIN Rearrangement

Skeletal rearrangement via carbocations.

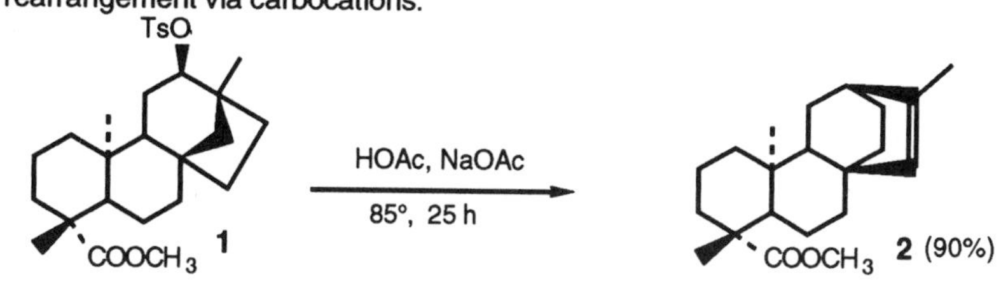

1	Wagner, G.	*J. Rus. Phys. Chem. Soc.*	**1899**	*31*	680
2	Meerwein, H.	*Liebigs Ann.*	**1914**	*405*	129
3	Nametkin, S.S.	*J. Russ. Phys. Chem. Soc.*	**1925**	*57*	80
4	Coates, R.M.	*J. Org. Chem.*	**1971**	*36*	3722
5	Zefirov, N.S.	*Tetrahedron*	**1975**	*31*	2948
6	Cristol, S.I.	*J. Org. Chem.*	**1986**	*51*	4326
7	Pliemingen, H.	*Angew. Chem. Int. Ed.*	**1976**	*15*	293

WALLACH Azoxybenzene rearrangement

Acid catalyzed rearrangement of azoxybenzenes to p-hydroxyazobenzenes.

$$\underset{\mathbf{1}}{C_6H_5\text{-}\overset{+}{N}(O^-)\text{=}N\text{-}C_6H_5} \xrightarrow[\Delta]{H_2SO_4} \underset{\mathbf{2}}{C_6H_5\text{-}N\text{=}N\text{-}C_6H_4\text{-}OH\text{-}p}$$

1	Wallach, O.	*Chem. Ber.*	**1880**	*13*	525
2	Hahr, C.S.	*J. Am. Chem. Soc.*	**1962**	*84*	946
3	Brigelow, H.E.	*Chem. Rev.*	**1931**	*9*	139

WASSERMANN - BORMANN Macrocyclic lactam synthesis

Ring expansion sequence of lactams by reaction with cyclic iminoethers followed by reductive ring opening to a macrocyclic lactam.

NaBH$_3$CN, 25° - 50°, HOAc

Na BH$_3$CN

6 (67%)

7 (93%)

R: Ph

1	Bormann, D.	*Chem. Ber.*	**1970**	*103*	1797
2	Wassermann, H.H.	*J. Am. Chem. Soc.*	**1981**	*103*	461
3	Wassermann, H.H.	*Tetrahedron Lett.*	**1983**	*24*	3669

Lactam (7).[2] 4-Phenyl-2-azetidinone **1** (2.40 mg, 1.6 mmol) was heated with iminoether **5** (448 mg, 1.56 mmol) in chlorobenzene for 21 h. Work up gave bicyclic lactam **6**, mp 195-197°C in 67% yield. Treatment of **6** with 3 equiv. of NaBH$_3$CN in acetic acid at 25°C for 3 h, then at 50° for 2 h and finally at 25°C for 12 h led to the thirteen-membered ring lactam **7** in 93% yield.

W E E R M A N Degradation

Synthesis of lower homolog aldehydes from α,β-unsaturated carboxamides (via Hofmann degradation).

1	Weerman, R.A.	*Rec. Trav. Chim.*	**1918**	*37*	1
2	Masson, C.D.	*J. Org. Chem.*	**1951**	*16*	1869
3	Lane, J.F.	*Org. React.*	**1946**	*3*	276

2-Thienylacetaldehyde (4).[2] To a suspension of **1** (44.0 g, 0.28 mol) in $CHCl_3$ (50 mL) was added $SOCl_2$ (70.0 g, 0.63 mol). After 24 h at 25° the solution was added to cooled NH_4OH (300 mL) and NaOH (35 g). Filtration and washing gave 40 g of **2** (90%), mp 152-153°C.

To a suspension of **2** (12.0 g, 80 mmol) in MeOH (100 mL) was added 0.8 N KOH (150 mL) and 0.8 M KOCl. The temperature rose to 55-60°C and after cooling crude **3** was filtered and recrystallized from EtOH to give 10 g of **3** (70%), mp 115-116°C.

3 (22.0 g, 0.12 mol) in 50% EtOH (200 mL) was treated with oxalic acid dihydrate (20.0 g, 0.4 mol) and heated on a water bath for 15 min. The solvent was evaporated and the residue steam distilled. Extraction of the distillate with Et_2O and evaporation gave an oil, which after distillation gave 2.9 g of **4** (19%), bp 69-74°C (8 mm); semicarbazone mp 131-132°C.

WEIDENHAGEN Imidazole synthesis

Imidazole synthesis from α-ketols, formaldehyde and ammonia.

$$(H_3C)_2NH\ (\mathbf{2}) + H_2C{=}CH{-}C({=}O){-}CH_2OH\ (\mathbf{1}) \longrightarrow (H_3C)_2N{-}CH_2CH_2{-}C({=}O){-}CH_2OH$$

$$\xrightarrow[Cu(OAc)_2]{NH_3,\ H_2CO} (H_3C)_2N{-}CH_2CH_2{-}\text{(imidazol-4-yl)}\quad \mathbf{3}\ (10\%)$$

$$HO{-}CH_2COCH_2CH_2OH \xrightarrow[Ph\,CHO]{NH_3} \text{2-phenyl-4-(}CH_2CH_2OH\text{)imidazole}$$

1	Weidenhagen, R.	*Chem. Ber.*	**1935**	*68*	1953
2	Weidenhagen, R.	*Chem. Ber.*	**1937**	*70*	570
3	Huebner, C.F.	*J. Am. Chem. Soc.*	**1951**	*73*	4667
4	Schobert, E.S.	*J. pr. Chem.*	**1962**	*18*	192

4-(2-Dimethylaminoethyl)imidazole (3).[3] To a solution of 6% hydroxymethyl vinyl ketone **1** (100 mL, 6.0 g, 69 mmol) was added a 32% solution of dimethylamine **2** (12 mL, 85 mmol). After 15 min at 20°, there was added a solution of $Cu(OAc)_2$ (30 g, 150 mmol) and 36% formaldehyde solution (15 mL) in 28% NH_4OH (225 mL). The mixture was kept on a steam bath for 1 h, the copper salt was decomposed by H_2O and **3** was separated as the dipicrate by addition of picric acid (10 g) to the hot solution. There was obtained 3.8 g of **3** picrate (10%), mp 230-232°C.

W E I N R E B Ketone synthesis

Synthesis of ketones from acids, acid chlorides, carbonates via N-methoxyamides.

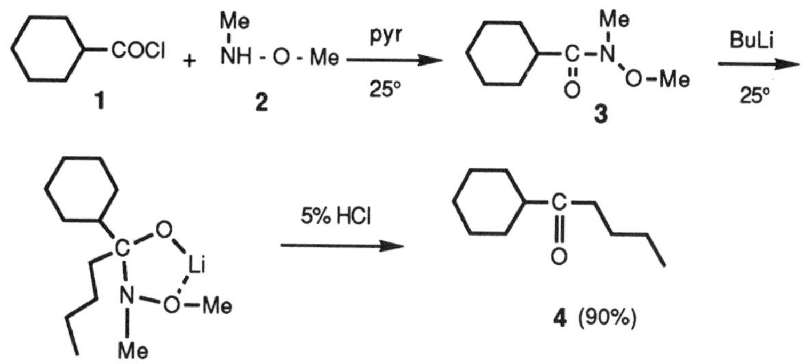

1	Weinreb, S.M.	*Tetrahedron Lett.*	**1981**	*22*	3815
2	DeBernardis, J.B.	*Tetrahedron Lett.*	**1984**	*25*	5271
3	Liebeskind, L.S.	*Tetrahedron Lett.*	**1987**	*28*	1857
4	Weinreb, S.M.	*J. Am. Chem. Soc.*	**1990**	*112*	2998
5	Weinreb, S.M.	*J. Org. Chem.*	**1991**	*56*	2911

Cyclohexyl butyl ketone (4).[1] To acid chloride **1** (146 mg, 1 mmol) and N,O-dimethylhydroxyl-amine hydrochloride **2** (67 mg, 1.1 mmol) in EtOH (10 mL) at 0°C, was added pyridine (185 mg, 2.3 mmol). After stirring for 1 h, the solvent was evaporated, brine was added followed by extraction with Et_2O:CH_2Cl_2 (1:1) to give **3** purifed by distillation or by gel chromatography. To **3** (171 mg, 1 mmol) in THF (10 mL) was added bytyllithium (1 mmol) and the mixture was poured into a 5% HCl in EtOH at 0°C and partitioned between Et_2O:CH_2Cl_2 (1:1). Evaporation of the organic layer and prep TLC gave 151 mg of **4** (90%).

WEISS Annulation

Synthesis of fused cyclopentanones (bicyclo[3.3.0] octadiones) or of propellanes from α-dicarbonyl compounds *via* the double aldol condensation with β-ketoesters **2**.

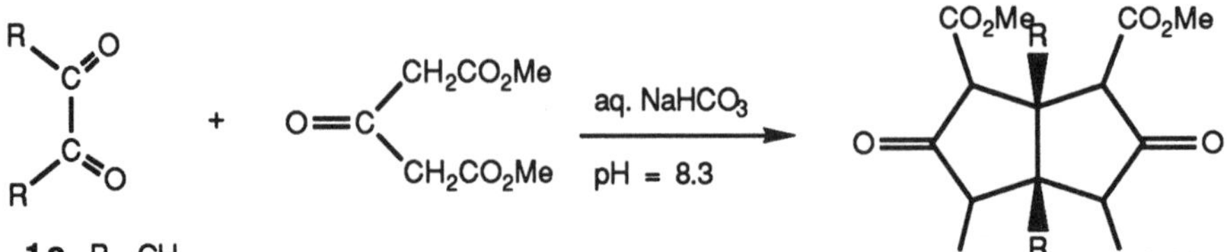

aq. $NaHCO_3$, pH = 8.3

1a. R = CH_3
1b. R = H

2

3a. R = CH_3 (75%)
3b. R = H (76%)

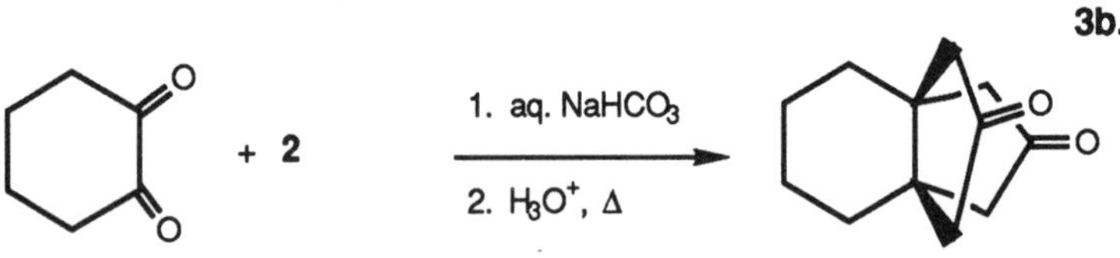

1. aq. $NaHCO_3$
2. H_3O^+, Δ

4

5 (56%)

1	Weiss, U.	*Tetrahedron Lett.*	**1968**		4885
2	Cook, J.M.	*Can. J. Chem.*	**1978**	*56*	189
3	Cook, J.M.	*Tetrahedron Lett.*	**1991**	*47*	3665

Tetramethyl 1,5-dimethyl-cis-bicyclo[3.3.0]octane-3,7-dione-2,4,6,8-tetracarboxylate (3a).[1] Biacetyl **1a** (17.2 g, 0.20 mol) was added rapidly to stirred dimethyl 3-oxoglutarate **2** (70 g, 0.40 mol) and aq. $NaHCO_3$ (5.6 g in 400 mL of H_2O) at pH = 8.3. After 24 h, **3a** (62 g), mp 155-158°C was filtered. The mother liquor with HCl (pH 5) gave additional **3a** (total yield 75%); recrystallized from hot MeOH, mp 155-157°C.

[n.3.3] Propellanes (e.g. **5**, 56%) were prepared from **2** with either 1,2-cyclohexanedione **4** or 1,2-cyclopentanedione followed by hydrolysis-decarboxylation.

W E N C K E R Aziridine synthesis

Synthesis of aziridines from 1,2-aminoalcohols.

$$H_2N-CH_2-CH_2-OH \xrightarrow{H_2SO_4} H_3\overset{+}{N}-CH_2-CH_2-OSO_3^- \xrightarrow{NaOH} \text{aziridine (NH)}\ (90\%)^2$$

$$\text{t-Bu-NH-CH}_2\text{-CH(OH)-CH}_2\text{Cl}\ (\mathbf{1}) \xrightarrow[MeSO_2Cl]{Pyr,\ MeSO_3H} \text{t-Bu-NH-CH}_2\text{-CH(OMs)-CH}_2\text{Cl}\ (\mathbf{2})$$

$$\xrightarrow[HN(CH_2CH_2NH_2)_2]{Na_2CO_3} \text{t-Bu-N(aziridine)-CH}_2\text{Cl}\quad \mathbf{3}\ (55\%)$$

1	Wencker, H.	*J. Am. Chem. Soc.*	**1935**	*57*	2328
2	Leighton, P.A.	*J. Am. Chem. Soc.*	**1947**	*69*	1540
3	Gaertner, v.R.	*J. Org. Chem.*	**1970**	*35*	3952
4	Nakagawa, Y.	*Bull. Chem. Soc. Jpn.*	**1972**	*45*	1162

1-tert-Butyl-2-chloromethylaziridine (3).[3] 1-tert-Butylamino-3-chloro-2-propanol **1** (79.1 g, 0.47 mol) in $CHCl_3$ (250 mL) at 0-10°C was treated with pyridine (71 g, 0.9 mol), $MeSO_3H$ (36 g, 0.37 mol) and then dropwise with $MeSO_2Cl$ (57.3 g, 0.46 mol). After 24 h stirring, the mixture was poured into ice-water and finally treated with Na_2CO_3 at pH=8. The $CHCl_3$ layer and the extract was dried ($MgSO_4$). Evaporation of the solvent (below 15°C) gave **2**; cold **2** was added to a stirred solution of Na_2CO_3 (53 g), diethylenetriamine (10 g) (scavenger for epoxide impurities) and water (400 mL) under cooling in an ice bath. After 24 h of stirring the mixture was extracted (Et_2O) and the solvent distilled to afford 40.6 g of **3** (55%), bp 47-48°C/10 mm, n_D^{26} = 1.4454.

WENZEL - IMAMOTO Reduction

Selective reductions of C=C with LaNi alloy.

$LaNi_5H_6$

20°, 11 h

(88%)

CHO

20°, 8 h

CH_2OH (91%)

O

0°C, 6h

(89%)

OH

20°, 33 h

(95%)

1 Wenzel, H. *Int. Met. Rev.* **1982** *27* 140

2 Imamoto, T. *J. Org. Chem.* **1987** *52* 5695

General procedure.[2] $LaNi_5$ ingot (3 g) in an autoclave was evacuated at 0.1 mm, heated to 200°C under H_2 at 30 atm for 10 min. After cooling to 20°C, the operation was repeated five times. The autoclave was cooled in dry ice-acetone, H_2 was released and N_2 was introduced. The organic compound (1 mmol) in THF:MeOH (2:1) (5 mL) was added at -78°C with stirring. The mixture was stirred under N_2 at 0°C and then at 20°C. The catalyst was filtered, the filtrate concentrated and the residue purified by preparative TLC on silica gel.

WESSELY - MOSER Rearrangement

Acid catalyzed rearrangement of dihidroxyxanthone.

1	Wessely, F., Moser, G.H.	*Monatsh.*	**1930**	*56*	97
2	Wheeler, T.S.	*J. Chem. Soc.*	**1956**		4455
3	Molho, D.	*Bull. Soc.Chim. Fr.*	**1963**		603
4	Suschitzky, H.	*J. Chem. Soc. Chem. Comm.*	**1984**		2275

WESTPHALEN - LETREE Rearrangement

Carbocation rearrangement of steroidal tert alcohols (seeWagner-Meerwein).

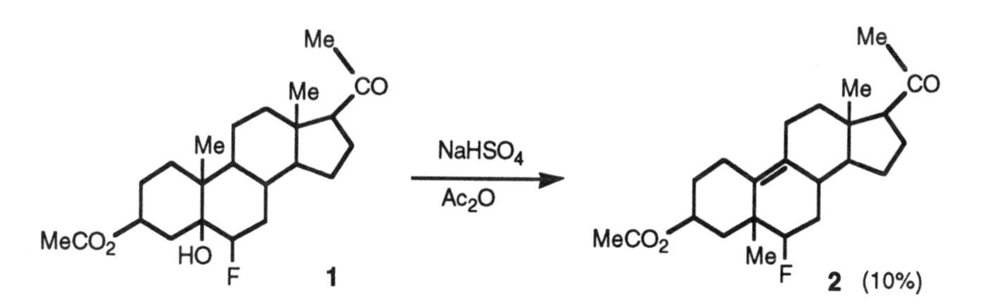

1	Westphalen Th.	*Chem. Ber.*	**1915**	*48*	1064
2	Letree, H.	*Chem. Ber.*	**1932**	*70*	1947
3	Mihina, J.S.	*J. Org. Chem.*	**1962**	*27*	2807
4	Blunth, J.W.	*Tetrahedron*	**1968**	*21*	1567
5	Szczepen, W.J.	*Acta. Chim. Hung.*	**1986**	*123*	69

W H A R T O N Olefin synthesis

Conversion of α-haloketones to olefins using hydrazine (via enediimides C=C-N=NH). Also reduction of α,β-epoxy ketones to allyl alcohols.

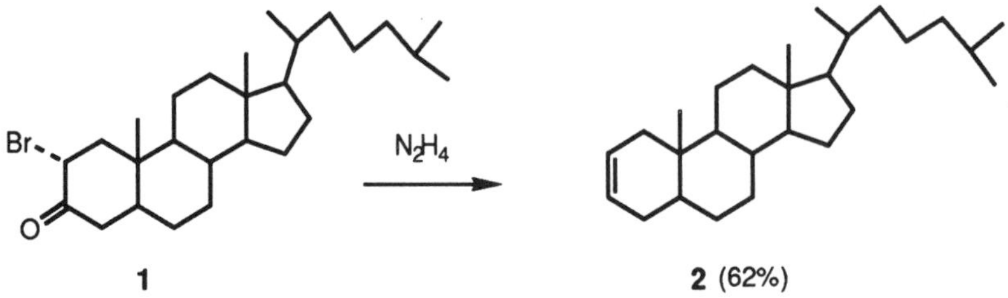

Br

O

1

N_2H_4

2 (62%)

O

O

NH_2NH_2

HOAc Δ

OH

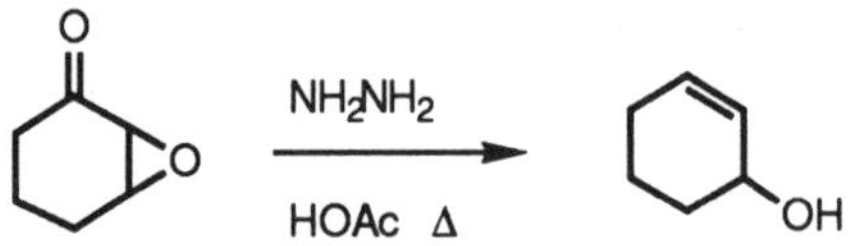

(75%)

1	Wharton, P.S.	*J. Org. Chem.*	**1961**	*26*	3615, 4781
2	Wharton, P.S.	*J. Org. Chem.*	**1964**	*29*	958
3	Stork, G.	*J. Am. Chem. Soc.*	**1977**	*99*	7067

2-Cholestene (2).[2] To hydrazine hydrate (13 mL, 0.4 mol), KOAc (2.0 g, 20 mmol) and cyclohexane (10 mL) stirred and heated to reflux was added 2α-bromocholestanone **1** (2.005 g, 4.3 mmol) in cyclohexane (30 mL) over 10 min. Heating was continued for another 30 min. After washing (H_2O) and extraction (Et_2O) the mixture was percolated through alumina (acid washed). Evaporation afforded 995 mg of **2** (62%), mp 72-74°C, $[\alpha]_D = +64°$.

WHITING Diene synthesis

Diene synthesis from 2-alkyne-1,4-diols with LAH.

1. EtMgBr
2. CHO 2

OH 1

C≡C OH OH 3 (74%)

LAH

4 (35%)

1 Whiting, M.C. *J. Chem. Soc.* **1954** 4006

2 Isler, O. *Helv. Chim. Acta.* **1956** *39* 454

2,6-Dimethylocta-1,3,5,7-tetraene (Cosmene) (4).[1] A solution of ethylmagnesium bromide (from 2.5 g Mg 0.1 at g) at 0°C was treated with a solution of 3-methylpent-1-yne-4-en-3-ol **1** (1.4 g, 14 mmol) in Et_2O (10 mL). Stirring was continued for 3 h at 0°C, a solution of α-methylacrylaldehyde **2** (3.5 g, 50 mmol) in PhH (25 mL) was added and stirring was continued for another 5 h. Saturated NH_4Cl solution was added and after usual work up, there was obtained 1.7 g of **3** (74%), bp 78-80°C/10^{-3} mm, n_D^{16} = 1.4946.

A solution of **3** (3 g, 20 mmol) in Et_2O (50 mL) was treated with LAH (2.5 g, 65 mmol) in Et_2O (200 mL). After 4 h reflux, the cooled mixture was poured on ice and tartaric acid. The Et_2O layer was dried ($MgSO_4$), the solvent evaporated under N_2 at -4°C and the residue extracted with petroleum ether and passed through alumina. After evaporation of the solvent at 0°C, there was obtained 0.8-0.9 g of **4** (33-37%), bp 50°C/ 0.2 mm, mp -7°C, n_D^{18} = 1.5852 (stable for a few months under N_2 at -60°C).

WIDEQUIST Cyclopropane synthesis

Tetracyanocyclopropane synthesis from bromomalonitrile and ketones.

$(H_3C)_2C=O$ (**2**) + $Br\text{-}HC(CN)_2$ (**1**) ⟶ $(H_3C)_2C(OH)CBr(CN)_2$ ⟶

$\xrightarrow{-HOBr}$ $(H_3C)_2C=C(CN)_2$ $\xrightarrow{1}$ $(H_3C)_2C<[C(CN)_2\text{-}C(CN)_2]$ (**3**)

1	Widequist, S.	*Arkiv. Kemi. Mineral. Geolog.*	**1937**	*12A*	8 (22)
2	Widequist, S.	*Arkiv. Kemi. Mineral. Geolog.*	**1945**	*20B*	12 (4)
3	Scribner, R.M.	*J. Org. Chem.*	**1960**	*25*	1440
4	Hart, H.	*J. Org. Chem.*	**1963**	*28*	1220
5	Hart, H.	*J. Am. Chem. Soc.*	**1963**	*85*	1161
6	Hart, H.	*J. Org. Chem.*	**1966**	*31*	2784

1,1-Dimethyl-2,2,3,3-tetracyanocyclopropane (3).[3] A mixture of bromomalononitrile **1** (1.5 g, 10 mmol) and acetone **2** (1.0 g, 17 mmol) in EtOH (10 mL) was treated with a solution of KI (3.5 g, 20 mmol) in water (10 mL) at 20°C. The product was filtered and recrystallized from aq. Me_2CO or EtOH to give 1.19 g of **3** (70%).

W I L K I N S O N Carbonylation - decarbonylation catalyst

Rh catalyst for carbonylation, decarbonylation, oxygenation, benzyl cleavage.

$RhCl_3$ **2** $\xrightarrow[\mathbf{1}]{PPh_3}$ $RhCl(PPh_3)_3$ **3**

$(CH_3)_3Si$ **4** $\xrightarrow[O_2]{\mathbf{3}}$ $(CH_3)_3Si$ ... O **5** (83%)

O, O, Ph **6** $\xrightarrow[Et_3SiH]{\mathbf{3}}$ CH_3 **7** (65%)

1	Wilkinson, G.	*J. Chem. Soc.*	**1966**		1711
2	Tsuji,J.	*J. Am. Chem. Soc.*	**1968**	*90*	99
3	Trost, B.M.	*J. Am. Chem. Soc.*	**1973**	*95*	7863
4	Reuter, J.M.	*J. Org. Chem.*	**1978**		
5	Grigg, R.	*Synthesis*	**1983**		1009
6	Liu, Hsing-Jang	*Synth. Commun.*	**1990**	*20*	557

Tris(Triphenylphosphine)chlorrhodium (3).[1] To freshly recrystallized triphenylphosphine **1** (12 g, 19 mmol) in hot EtOH (350 mL) was added $RhCl_3$. $3H_2O$ **2** (2 g, 7.6 mmol) in EtOH (70 mL) and the solution was refluxed 30 min. Hot filtration of the red crystals, washing with degassed Et_2O (50 mL) and vacuum drying gave 6.25 g of **3** (88%), mp 157-158°C.

3-(Trimethylsilyl)cyclopent-2-en-1-one (5).[4] 3-(Trimethylsilyl)cyclopent-1-ene **4** (500 mg, 3.56 mmol) and **3** (33 mg, 0.035 mmol) were heated at 97°C for 1.5 h under a flow of O_2. Removal of the catalyst and distillation gave 455 mg of **5** (83%), bp 90°C, 9.5 mm.

1-Methylnaphthalene (7).[6] Ester **6** (143 mg, 0.5 mmol), triethylsilane (475 mg, 4.1 mmol) and **3** (100 mg, 0.1 mmol) in PhH (5 mL) was refluxed for 30 h under Ar. Chromatography on silica gel afforded 46 mg of **7** (65%).

WILLGERODT - KINDLER Rearrangement

Rearrangement of ketones to amides by heating with sulfur and ammonia or amines.

$$\underset{\mathbf{1}}{Ph\text{-}CO\text{-}CH_3} + S_8 + NH_4OH \xrightarrow[\text{pyr}]{160°,\ 4.5\ h} \underset{\mathbf{2}\ (81\%)}{Ph\text{-}CH_2\text{-}CONH_2}$$

$$\underset{\mathbf{1}}{Ph\text{-}CO\text{-}CH_3} + \underset{\mathbf{3}}{HN(CH_2CH_2)_2O} + S_8 \xrightarrow[3\ h]{130°} \underset{\mathbf{4}\ (94\%)}{Ph\text{-}CH_2\text{-}C(=S)\text{-}N(CH_2CH_2)_2O}$$

1	Willgerodt, O.	*Chem. Ber.*	**1887**	*20*	2467
2	Kindler, K.	*Liebigs Ann.*	**1923**	*431*	193
3	De Tar, D.F.	*J. Am. Chem. Soc.*	**1946**	*68*	2028
4	Cavallieri, L.	*J. Am. Chem. Soc.*	**1945**	*67*	1755
5	Brown, E.W.	*Synthesis*	**1975**		358
6	Maeyer, R., Wehl, J.	*Angew. Chem.*	**1964**	*76*	861
7	Wolff,E.	*Org. React.*	**1951**	*6*	439
8	Hundt, R.N.	*Chem. Rev.*	**1961**	*61*	52

Phenylacetamide (2).[3] Acetophenone **1** (25.0 g, 0.208 mol), sulfur (37.5 g, 1.17 at. g), NH_4OH (50 mL) and pyridine (30 mL) were heated in a sealed tube at 160°C for 4.5 h. After evaporation in vacuum the residue was leached by boiling with water (500 mL). From the filtrate on cooling 20 g of **2** separated and a second crop for a total of 22.7 g of **2** (81%), mp 156-158°C. There is also obtained 1.2 g of phenylacetic acid (4.2%).

Thiophenacylmorpholinamide 4.[6] Acetophenone **1** (12.0 g, 0.1 mol), morpholine **3** (18 g, 0.2 mol) TsOH (0.5 g) and sulfur (3.2 g, 0.1 mol) were heated 3 h at 130°C. The mixture was poured into MeOH (50 mL), cooled and filtered to afford 20.7 g of **4** (94%), mp 79-79.5°C.

WILLIAMS - BEN ISHAI Amino acid synthesis

Asymmetric synthesis (Williams) of α–amino acids through C-C bond construction on an electrophylic glycine template.

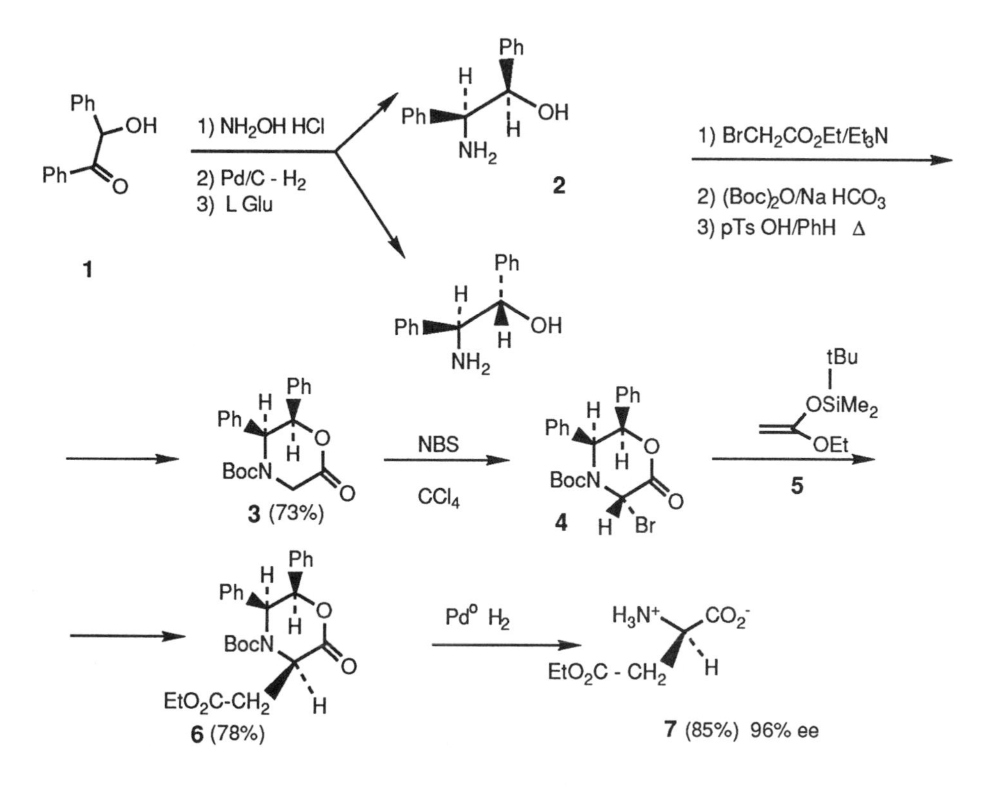

1	Ben-shai, D.	*J. Chem. Soc. Chem. Commun.*	**1975**		349, 905
2	Williams, R.M.	*J. Am. Chem. Soc.*	**1986**	*108*	1103
3	Tishler, M.	*J. Am. Chem. Soc.*	**1951**	*73*	1216
4	Williams, R.M.	*J. Org. Chem.*	**1986**	*51*	5021
5	Williams, R.M.	*J. Am. Chem. Soc.*	**1988**	*110*	1547
6	Williams, R.M.	*J. Am. Chem. Soc.*	**1988**	*110*	1553

cont. on next page

(5S,6R)-4-(tertButyloxycarbonyl)-5,6-diphenyl-2,3,5,6-tetrahydro-4H-1,4-oxazine-2-one (3).[6] A suspension of **2** (51 g, 239 mmol) in THF (1200 mL) was treated with ethyl bromoacetate (60 g, 359 mmol) and TEA (49 g, 485 mmol). After 18 h stirring, TEA.HBr was filtered and the filtrate evaporated in vacuum. The residue was washed with water and recrystallized from EtOH (250 mL) to yield 60.3 g (84%) of products. To 239 g (80 mmol) of this product in 160 mL of $CHCl_3$ was added a mixture of di-tert-butyl dicarbonate (34.9 g, 160 mmol), NaCl (32.8 g, 560 mmol) and saturated solution of $NaHCO_3$ (160 mL) and the mixture was refluxed for 20 h. The aqueous phase was extracted with $CHCl_3$, and the combined organic layers washed (water), dried (Na_2SO_4), the solvent evaporated and the residue distilled to remove the di-tert-butyl carbonate. The crude product was used for the next step. Recrystallized (hexane:EtOAc 3:1), mp 60-62°C, α_D = -20.5° (c 5.5 CH_2Cl_2). To the crude product above (32 g), in PhH (750 mL) was added pTsOH (1.5 g, 8 mmol), the mixture was refluxed (Soxhlet extractor filled with $CaCl_2$ 75 g) for 8 h. The solid after removal of solvent was dissolved in CH_2Cl_2, washed (water) and evaporated. Recrystallization from EtOH (750 mL) gave 20.7 g of **3** (73%).

(3S,5S,6R)-3-Bromo-4-(t-butoxycarbonyl)-5,6-diphenyl-2,3,5,6-tetrahydro-4H-1,4-oxazine-2-one (4). To **3** (50 mg, 0.142 mmol) in CCl_4 (15 mL) under reflux, NBS (28 mg, 0.156 mmol) was added, and the mixture was refluxed for 1 h. Filtration and concentration gave crude **4** as a white solid.

(3R,5S,6R)-4-(t-Butoxycarbonyl)-5,6-diphenyl-3-[(ethoxycarbonyl)methyl]-2,3,5,6-tetrahydro-4H-oxazin-2-one (6). To a stirred solution of crude **4** (226 mg, 0.48 mmol) in CH_2Cl_2 (11 mL) was added ketene acetal **5** (450 mg, 2.42 mmol) followed by addition of $ZnCl_2$ (575 mL, 0.44 mmol, 0.76M in THF). After 4 min it was quenched with water. Radial chromatography (silica gel EtOAc:hexane 1:4) afforded 179 mg of **6** (78%).

R-β-Ethyl aspartate (7). A solution of **6** (86.5 mg, 0.18 mmol) in THF (2 mL) and anh. EtOH (2 mL) treated with $PdCl_2$ (19 mg) was hydrogenated 24 h at 20°C and 20 psi. After usual workup there are obtained 34.2 mg of **7** (85%), 96% ee.

WILLIAMSON Ether synthesis

Synthesis of ethers from alcoholates with alkyl halides.

p - O_2N - C_6H_4 - OH + I-butyl → (KOH, Δ, 12 h) p - O_2N - C_6H_4-O-butyl

1 2 3 (55%)

cyclohexane-1,2-diol (OH, OH) → (TlOEt, PhH) (OTl, OH) → (MeI, MeCN) (OMe, OH)

4 (85%)

1	Williamson, A.W.	*J. Chem. Soc.*	**1852**	*4*	229
2	Gurney, O.	*J. Am. Chem. Soc.*	**1922**	*44*	1742
3	Dermer, O.C.	*Chem. Rev.*	**1934**	*14*	409
4	Kalinowsky, K.O.	*Angew. Chem. Int. Ed.*	**1975**	*14*	763
5	Nakatsugi, T.	*Synthesis*	**1987**		280

p-Nitrophenyl butyl ether (3).[2] To p-nitrophenol **1** (139 g, 1 mol) and KOH (56 g, 1 mol) in 50% EtOH (2000 mL) was added butyl iodide **2** (184 g, 1 mol) in BuOH (200 mL) and the mixture was refluxed for 12 h. After removal of the solvent, the residue was extracted with Et_2O and the unreacted phenol was removed by washing with NaOH. The solution was dried over KOH and the oil obtained after evaporation was recrystallized from EtOH to afford 108 g of **3** (55.3%), mp 32°C.

W I S S N E R Hydroxy ketone synthesis

Conversion of acyl chlorides to functionalized (OH, OCH_3, OPh, SCH_3) ketones by means of tris(trimethylsilyloxy)ethylene **4**.

OH–CH_2–CO_2H (**1**) —$(Me_3Si)_2NH$ (**2**)→ Me_3SiO–CH_2–CO_2SiMe_3 (**3**) —BuLi, **2**→ $(Me_3SiO)CH{=}C(OSiMe_3)_2$ **4** (80%)

5 (octanoyl chloride, COCl) —**4**, 95°→ **6** (1-hydroxy-2-nonanone)

1 Wissner, A. *Tetrahedron Lett.* **1978** 2749

2 Wissner, A. *J. Org. Chem.* **1979** *44* 4617

Tris(trimethylsilyloxy)ethylene (4).[2] To glycolic acid **1** (118 g, 1.55 mol) in pyridine was added disilazan **2** (260 g, 1.6 mol) under N_2 over 30 min while the temp. reached 75°C. Me_3SiCl (88.0 g, 0.8 mol) was added at 20°C and after 1 h work up gave 275 g of **3** (80%), bp 78--80°C/12 mm. To **2** (245 g, 1.52 mol) in THF (1200 mL) was added 2.4 M nBuLI in hexane (650 mL) at 0°C. After 30 min at 45°C, **3** (275 g, 1.25 mol) was added at -78° over 30 min. After another 30 min Me_3SiCl (205 g, 1.9 mol) was added, then petroleum ether. Filtration (Celite) and vacuum distillation gave 366 g of **4** (100%), bp 54-56°C (0.1 mm).

1-Hydroxy-2-nonanone (6)[2]. Octanoyl chloride **5** (4.0 g, 24.6 mmol) and **4** (15.5 g, 53 mmol) were heated for 3 h at 95-100°C. Dioxane (25 mL) and 0.6N HCl (10 mL) was added (exothermic, gas evolution). The mixture was heated for 30 min at 95°C, extracted with Et_2O, and distilled to afford 3.28 g of **6** (84%), bp 83-85°C/0.5 mm

WITTIG Olefin synthesis

Olefin synthesis from phosphorane ylides (e.g. **3**) with aldehydes or ketones; cis olefins predominate in aliphatic systems, trans in conjugated olefins.

$$O_2N-C_6H_4-CH_2Cl \;(\mathbf{1}) \xrightarrow{Ph_3P} Ar-CH_2\overset{+}{P}Ph_3 \;(\mathbf{2}\ (58\%)) \xrightarrow{BuLi} O_2N-C_6H_4-\overset{-}{C}H-\overset{+}{P}Ph_3 \;(\mathbf{3})$$

$$\mathbf{3} \xrightarrow{O=CH-C_6H_4OMe\ (\mathbf{4})} \left(\begin{array}{l} Ar'-CH-O \\ Ar-CH-PPh_3 \end{array}\right) \longrightarrow O_2N-C_6H_4-CH=CH-C_6H_4-OMe \;(E\text{-}\mathbf{5}\ (89\%))$$

1	Wittig, G.	*Liebigs Ann.*	**1949**	*562*	187
2	Wittig, G.	*Chem. Ber.*	**1961**	*94*	1373
3	Ketcham, R.	*J. Org. Chem.*	**1962**	*27*	4666
4	Angeletti, E.	*J. Chem. Soc. Perkin Tr. 1*	**1987**		713
5	Doudon, A.	*Tetrahedron*	**1988**	*44*	2021
6	Emmons, W.	*Angew. Chem. Int. Ed.*	**1966**	*5*	126
7	Murphy, P. B.	*Chem. Soc. Rev.*	**1988**	*17*	1
8	Maercker, A.	*Org. React.*	**1965**	*14*	333
9	Maryanoff, B. E.	*Chem. Rev.*	**1989**	*89*	863

trans 4-Nitro-4-methoxystilbene (5).[3] Triphenylphosphine (26.3 g, 0.1 mol) and p-nitrobenzyl chloride **1** (17.2 g, 0.1 mol) in PhH (50 mL) were refluxed for 2 h. After cooling the solid was collected and washed with PhH to give 25 g of **2** (58%), mp 270-276°C. Recrystallized from CCl_4-petroleum ether, mp 278-280°C.

To stirred **2** (4.3 g, 10 mmol) in PhH (50 mL) under N_2 was added butylithium (0.85 g, 13 mmol) to produce **3**. After 2 h anisaldehyde **4** (1.63 g, 12 mmol) was added, the mixture was diluted with petroleum ether and the dark solid collected. Recrystallization afforded 2.23 g of **5** (89%), mp 131-132.5°C.

W I T T I G Rearrangement

A stereoselective base catalyzed [2,3] sigmatropic rearrangement of allyl ethers to homoallylic alcohols (stereoselective).

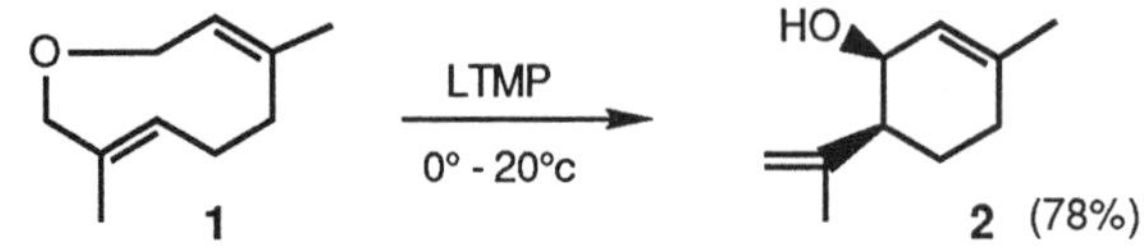

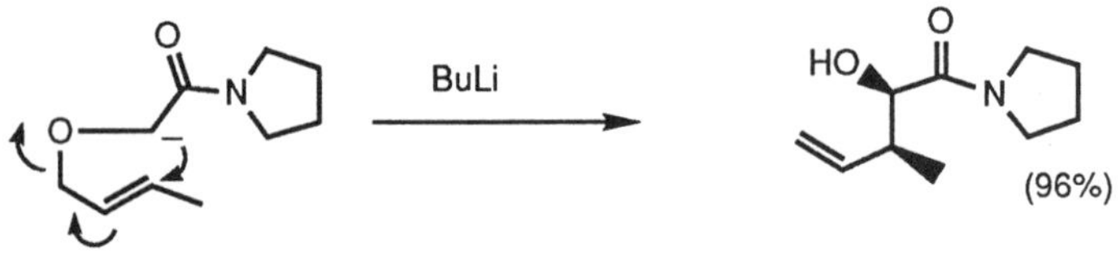

LTMP

0° - 20°c

1

2 (78%)

BuLi

(96%)

1	Wittig, G.	*Liebigs Ann.*	**1942**	*550*	260
2	Wittig, G.	*Chem. Ber.*	**1953**	*86*	151
3	Uchikawa, M.	*Tetrahedron Lett.*	**1986**	*27*	4577
4	Marshall, J.A.	*J. Org. Chem.*	**1988**	*53*	4108
5	Schöllkopf, U.	*Angew. Chem.Int.Ed.*	**1962**	*1*	126
6	Mikami, K.	*Chem. Lett.*	**1985**		1729
7	Brückner, R.	*Contakte (Darmstadt)*	**1991**	*3*	3

cis-6-Isopropenyl-3-methyl-2-cyclohexenol (2).[4] 2,2,6,6-Tetramethylpiperidine 3.2 mL and 2.5 M n-BuLi in hexane 7.9 mL was stirred in THF (20 mL) under N_2 at 0°C and for 30 min at 20°C. This lithium tetramethylpiperidide (LTMP) reagent was added to a solution of **1** (1.0 g, 658 mmol) in THF (20 mL) at 0°C. After 14 h at 20°C, the mixture was quenched with water, extracted with Et_2O and purified by chromatography (silica gel, 7-10% EtOAc in hexane), to afford 0.78 g of **2** (78%).

WOHL - AUE Phenazine synthesis

Synthesis of phenazine-N-oxide from anilines and nitrobenzene.

2 + 1 → (KOH powder; PhH, 12 h Δ) → 3 (16%)

1	Wohl, A., Aue, W.	*Chem. Ber.*	**1901**	*34*	2442
2	Maffey, S.	*Gazz. Chim. Ital.*	**1946**	*76*	239
3	Patcher, I.J.	*J. Am. Chem. Soc.*	**1951**	*73*	4958

WOHL - ZIEGLER Bromination

Allylic or benzylic bromination or chlorination with N-bromosuccinimide (NBS) or NCS.

1 → (NBS) → 2 (98%)

1	Wohl, A.	*Chem. Ber.*	**1919**	*52*	51
2	Ziegler, K.	*Liebigs Ann.*	**1942**	*551*	80
3	Dauben, H.	*J. Am. Chem. Soc.*	**1959**	*81*	4863
4	Zalusky, M.C.	*Bull. Soc. Chim. Fr.*	**1970**		1447
5	Djerassi, C.	*Chem. Rev.*	**1948**	*43*	271
6	Horner, L.	*Angew. Chem.*	**1959**	*71*	349

2-Carbomethoxy-3-bromomethylfuran (2).[3] To **1** (96 g, 0.7 mol) in PhH (600 mL) under reflux and stirring, was added a mixture of N-bromosuccinimide (137 g, 0.76 mol) and benzoyl peroxide (3 g). After completion the floating succinimide was filtered. Work up gave 151 g of **2** (98%), recrystallized from Et_2O-hexane or sublimation at 60°C (0.1 mm), mp 54°C.

WOHL - WEYGAND Aldose degradation

Degradation of sugar oximes via cyanohydrins by means of an acid chloride/pyridine (Wohl) or of fluorodinitrobenzene (Weygand).

CH = NOH, OH, OH, HO, HO, OH, OH
1
Ph COCl / Pyr
CN, OBz, OBz, BzO, BzO, OBz, OCOPh
2 (65%)
MeOH / NH_3
OH, OH, O, HO, OH, NHCOPh, NHCOPh
3 (37%)

CH = NOH, OH, HO, OH, OH, OH
4
+
F, NO_2, NO_2
5
CO_2, 60°C / Na H CO_3
CN, OH, HO, OH, OH, OH
CH = O, HO, OH, OH, OH
6 (61%)

1	Wohl, A.	*Chem. Ber.*	**1893**	*26*	730
2	Wohl, A.	*Chem. Ber.*	**1899**	*32*	3666
3	Deferrari, J.O.	*J. Org. Chem.*	**1966**	*31*	905
4	Weygand, F.	*Chem. Ber.*	**1950**	*83*	559
5	Weygand, F.	*Chem. Ber.*	**1952**	*85*	256

1,1-Bis(benzamido)-1-deoxy-D-galactitol (3).[3] **1** (3.92 g, 17 mmol) was added to 1:1 pyridine: PhCOCl (48 mL) and kept at 100°C by cooling. After 24 h at 20°C, all was poured into water (400 mL) and **2** was recrystallized from Me_2CO:EtOH (1:3), 9.45 g (65.6%), mp 190-191°C, $(\alpha)_D$ = +19.5° (c 0.7 $CHCl_3$). **2** (4 g, 4.8 mmol) was stirred with NH_3 in MeOH (100 mL) for 60 min. After 24 h evaporation afforded 770 mg of **3**, mp 194°C; from EtOH 700 mg (37%), mp 203-204°C, $(\alpha)_D$ = -5.8° (c 0.85 pyr).

Arabinose (6).[4] CO_2 was bubbled through $NaHCO_3$ (1.5 g, 17 mmol) in water (70 mL) and glucose oxime **4** (1 g, 4.5 mmol). At 55-60°C **5** (1.8 g, 10 mmol) in iPrOH (30 mL) was added. After 1.5-2 h, 420 mg of **6** (61%), mp 162°C was isolated.

WOLFF Rearrangement

Rearrangement (ring contraction) of α-diazoketones to carboxylic acids or their derivatives (esters, amides) via ketenes (see also Arndt-Eistert).

MnO_2 ; H_2O / HCl ; **1** ; **2** ; **3** (48%)

hv / MeCN, 2h, R_3NHR_3

1	Wolff, L.	*Liebigs Ann.*	**1912**	*394*	25
2	Borch, R.F.	*J. Org. Chem.*	**1969**	*34*	1481
3	Cohen, L.A.	*Angew. Chem.*	**1961**	*73*	261
4	Wynberg, H.	*J. Org. Chem.*	**1968**	*33*	4025
5	Cissy, J.	*Synthesis*	**1988**		720
6	Meier, H.	*Angew. Chem. Int. Ed.*	**1975**	*14*	32
7	Bachmann	*Org. React.*	**1942**	*1*	39

4-Carboxy-3,3,5,5-tetramethyl-1-thiacyclohexane (3).[4] 4-Hydrazo 3,3,6,6-tetramethyl-1-thiacyclohepta-5-one **1** (4.3 g, 20 mmol) in PhH (50 mL) was added to active MnO_2 (6 g) and $MgSO_4$ (10 g) in PhH (50 mL) over a period of 30 min. The solution became yellow and gas evolution started. After 24 h at 20°C, the mixture was filtered and the filtrate concentrated. 500 mg of the residue after reflux with dil HCl gave 195 mg of **2** (48%), mp 149-152°C. The rest of the residue was chromatographed (silica gel, PhMe) to give 0.55 g (16.5%) of unreacted ketene **2,** bp 103-104°C/12 mm.

WOLFF - KISHNER - HUANG MINLON Reduction

Reduction of ketones to hydrocarbons by heating with NH_2NH_2 and aqueous KOH (Wolff-Kishner) or KOH in ethylene glycol (Huang-Minlon).

NH_2NH_2, KOH; HO⁀OH, Δ

1 → **2** (95%)

NH_2NH_2 (100%); 200°

$(H_2C)_{10}$ $(CH_2)_9$ → $(H_2C)_{10}$ $(CH_2)_9$ (85%)[5]

1	Kishner, J.	*J. Russ. Phys. Chem. Soc.*	**1911**	*43*	582
2	Wolff, C.	*Liebigs Ann.*	**1912**	*394*	86
3	Huang Minlon	*J. Am. Chem. Soc.*	**1946**	*68*	2487
4	Huang Minlon	*J. Am. Chem. Soc.*	**1949**	*71*	3301
5	Nickon, A.	*J. Org chem.*	**1981**	*46*	4692
6	Todd, D.	*Org. React.*	**1948**	*4*	378

γ-(p-Phenoxyphenyl)-butyric acid (2).[3] A mixture of γ-(p-phenoxybenzoyl)propionic acid **1** (500 g, 1.85 mol), KOH (350 g, 6.25 mol), 85% hydrazine hydrate (250 mL) and di (or tri) ethylene glycol (2500 mL) was refluxed for 2 h. The condensor was removed and heating continued until the temperature reached 195°C. Refluxing was continued for 4 h, the cooled mixture was diluted with water (2500 mL) and slowly poured into 6N HCl (1500 mL). Filtration and drying gave 451 g of **2** (95%), mp 64-66°C.

WOLFRAM - SCHÖRNIG - HANSDORF Carboxymethylation

Carboxymethylation of aromatics in the presence of oxidants or photochemically.

1 + $ClCH_2$-COOH (**2**) $\xrightarrow{Fe_3O_4/KBr}$ CH_2-COOH naphthalene **3** (70%)

1 + $(CH_3CO)_2O$ (**4**) $\xrightarrow{KMnO_4}$ CH_2-COOH naphthalene **3** (66%)

1	Wolfram, A., Schörnig, L.	*Germ. Pat.*			562.391
	Hansdorf, E.	*U.S. Pat.*			1.951.686
2	Ogata, Y.	*J. Am. Chem. Soc.*	**1950**	*72*	4302
3	Ogata, Y.	*J. Org. Chem.*	**1951**	*16*	239
4	Southwick, P.L.	*Synthesis*	**1970**		628

Naphthylacetic acid (3).[4] a) Naphthalene **1** (56.6 g, 0.44 mol) monochloroacetic acid **2** (14.1 g, 0.149 mol), Fe_3O_4 (87.6 mg) and KBr (420 mg) were heated gently for 20 h so that 200°C was attained after 10 h and 218°C after 20 h. Unreacted **1** was distilled (43 g) and the residue extracted with hot NaOH solution, filtered and acidified with HCl to give 19.4 g of **3** (70%), mp 108-113°C. Recrystallized from water, mp 124-126°C.

b) To a boiling mixture of **1** (25.0 g, 0.2 mol) in Ac_2O (150 mL) was added $KMnO_4$ (3.0 g, 0.02 mol) and the mixture was boiled and stirred for 20 min. Unreacted **1** and Ac_2O were removed by distillation and the residue was steam distilled. The hot solution was filtered and acidified with H_2SO_4 to give **3** (66%) (based upon recovered naphthalene).

WOODWARD Peptide synthesis

Peptide synthesis mediated by N-ethyl-5-phenylisoxazolium-3'-sulfonate **3**.

$H_2N-CO-CH_2-CH(NH-CO_2-CH_2-Ph)-COOH$ (**1**) + $H_2N-CH_2-CO_2Et$ (**2**) $\xrightarrow[Et_3N\ 20°]{3}$ $H_2N-CO-CH_2-CH(NH-CO_2-CH_2-Ph)-CO-NH-CH_2-CO_2Et$ (**4**)

3: SO_3^-, O, N^+-Et

1	Woodward, R.B.	*Tetrahedron, Suppl. 7*	**1966**		21
2	Woodward, R.B.	*J. Am. Chem. Soc.*	**1961**	*83*	1007
3	Woodward, R.B.	*Tetrahedron, Suppl. 8*	**1966**	*22*	321

Carbobenzyloxy-L-asparaginylglycine ethyl ester (4).[3] Carbobenzyloxy-L-asparagine **1** (798 mg, 3 mmol), oxazolium salt **3** (760 mg, 3 mmol) and Et_3N (304 mg, 3 mmol) was stirred for 7 min in $MeNO_2$ at 20°C. Glycine ethyl ester hydrochloride **2** (419 mg, 3 mmol) and Et_3N (304 mg, 3 mmol) were added and the mixture was stirred for 24 h at 20°C. The solvent was evaporated and the residue was triturated with 0.5% $NaHCO_3$ solution (40 mL) under heating. Cooling and filtration gave 846 mg of **4** (80%), mp 185.5-187°C; recrystallized from Me_2CO-water **4** had mp 186-187°C.

WÜRTZ Coupling

Coupling of alkyl halides with Na, supplanted by the coupling of alkyl halides or sulfonates with Grignard reagents or RLi in the presence of Cu(I) salts.

$$2\ H_3C-CH_2-CH(CH_3)-Br\ (\mathbf{1}) \xrightarrow{Na} H_3C-CH_2-CH(H_3C)-CH(CH_3)-CH_2-CH_3\ \mathbf{2}\ (11\%)$$

$$n-C_8H_{17}-OTs\ (\mathbf{3}) + tBu-MgBr \xrightarrow{Li_2CuCl_4} n-C_8H_{17}-C\,Me_3\ \mathbf{4}\ (75\%)$$

1	Wüirtz, A.	*Liebigs Ann.*	**1855**	*96*	364
2	Bailey, W.J.	*J. Org. Chem.*	**1962**	*27*	3088
3	Horner, L.	*Angew. Chem.*	**1962**	*74*	586
4	Buntrock, R.E.	*Chem. Rev.*	**1968**	*68*	209
5	Kosolapoff, G.M.	*Org. React.*	**1951**	*6*	326
6	Whitesides, G.M.	*J. Am. Chem. Soc.*	**1969**	*91*	4871
7	Erdile, E.	*Tetrahedron*	**1984**	*40*	641
8	Schlosser, M.	*Angew. Chem. Int. Ed.*	**1974**	*13*	82

3,4-Dimethylhexane (2).[2] To sec-butyl bromide **1** (300 g, 2.18 mol) in Et_2O (700 mL) was added sodium (50 g, 2.18 at) and the mixture was refluxed for 56 h. Distillation afforded 16.5 g of **2** (11%), bp 117°C, $n_D^{25} = 1.4040$.

2,2-Dimethyldecane (4).[8] To octyl tosylate **3** (40 mmol) in 50 mL of THF at -78°C was added tBuMgBr (56 mmol in 32 mL of Et_2O) and Li_2CuCl_4 (0.2 mmol in 2 mL of THF). The mixture was warmed to 25°C during 2 h and stirred for 12 h. Acidification with 2N H_2SO_4, washing (2x50 mL water), drying and distillation gave 5.1 g (85%) of pure **4**.

YAMADA Coupling reagent

Diethylphosphoryl cyanide **3** as a reagent for amide bond formation and application to peptide synthesis free of racemization.

1 + 2 → (NC-P(O)(OEt)$_2$ **3**, 0°) → **4** (86%)

1	Yamada, S.	*Tetrahedron Lett.*	**1971**		3595
2	Yamada, S.	*J. Am. Chem. Soc.*	**1972**	*94*	6203
3	Yamada, S.	*Tetrahedron Lett.*	**1973**		1595
4	Takamizawa, A.	*Yakugaku Zasshi*	**1965**	*85I*	298

N-Benzoyl-L-leucylglycine ethyl ester (3).[3] N-Benzoy-L-leucine **1** (0.235 g, 1 mmol) and glycine ethyl ester hydrochloride **2** (0.1534 g, 1.1 mmol) in DMF (10 mL) under stirring was treated with diethylphosphoryl cyanide **3** (0.179 g, 1.1 mmol) in DMF at 0°C, followed by the addition of triethylamine (0.212 g, 2.1 mmol). The mixture was stirred for 30 min at 0°C and 4 h at 20°C. The reaction mixture was diluted with PhH-EtOAc, washed with 5% HCl, water, 5% $NaHCO_3$ solution and brine. Evaporation of the solvent gave crude **4** which after silica gel chromatography afforded 0.271 g of **4** (86%) (pure L), mp 158-160°C.

Reagent **3,** bp 93-96°C (14 mm), was prepared by condensation of triethyl phosphite and cyanogen bromide.

YAMAGUCHI Lactonization reagent

2,4,6-Trichlorobenzoyl chloride reagent for esterification of acids via a mixed anhydride, also used for large ring lactonization with DMAP (see also Steglich-Hassner).

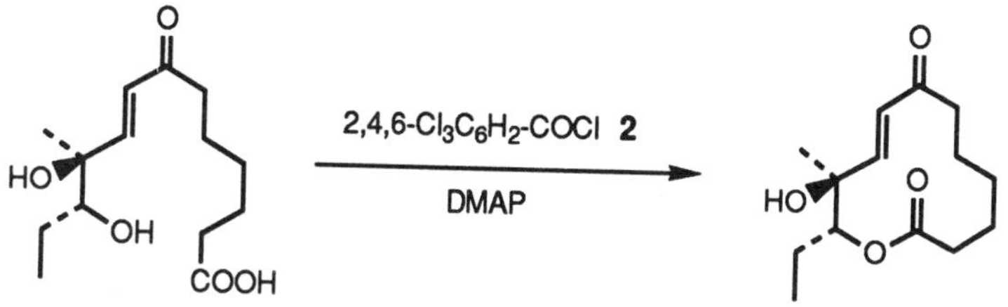

$2,4,6\text{-}Cl_3C_6H_2\text{-}COCl$ **2**, DMAP

1 → **3** (46%)

1 Yamaguchi, M.	*Bull Chem. Soc. Jpn.*	**1979**	*52*	1989
2 Yonemitsu, O.	*J. Org. Chem.*	**1990**	*55*	7

2,4,6-Tridemethyl-3-deoxymethynolide (3).[1] A mixture of seco acid **1** (272 mg, 1 mmol) and triethylamine (0.153 mL, 1.1 mmol) in THF (10 mL) was stirred for 10 min at 20°C, and then 2,4,6-trichlorobenzoyl chloride **2** (160 mL, 1 mmol) was added. After stirring for 2 h at 20°C, the resulting precipitate was filtered and washed with THF. The filtrate was diluted with PhH (100 mL) and slowly added to a refluxing solution of 4-dimethylaminopyridine (DMAP) (732 mg, 6 mmol) in PhH (100 mL). After 40 h the reaction mixture was washed with citric acid solution, water and dried (Na_2SO_4). Evaporation of the solvent afforded the crude product (247 mg). Separation by preparative TLC (silica gel Merck), Et_2O:PhH (2:1), gave 116 mg of **3** (46%), 65 mg of a dimer (26%) and 21 mg of polymer. Recrystallization of **3** (CH_2Cl_2 /diisopropyl ether) afforded colorless needles, mp 123°C.

YAMAZAKI Cyanoaniline synthesis

Synthesis of o-aminoarylnitriles (useful in pyrimidine synthesis) from nitroquinolines, nitro naphthalenes, and m-substituted (CF_3, $COCH_3$ and COC_6H_5) nitrobenzenes.

1	Yamazaki, M.	*Chem. Pharm. Bull*	**1981**	*29*	1286
2	Yamazaki, M.	*Chem. Pharm. Bull*	**1982**	*30*	851
3	Yamazaki,M.	*Chem. Pharm. Bull*	**1985**	*33*	1360

2-Amino-1-naphthalenecarbonitrile (3).[3] 2-Nitronaphthalene **1** (5.19 g, 30 mmol) was added to stirred ethyl cyanoacetate **2** (9.99 g, 90 mmol) and KOH (5.04 g, 90 mmol) in DMF (90 mL) and the mixture was stirred for 24 h. After solvent evaporation, the residue was hydrolyzed with 5% NaOH (60 mL) for 1 h at reflux. The mixture was extracted with $CHCl_3$. The $CHCl_3$ extract was dried over Na_2SO_4, and concentrated. The residue was purified by column chromatography on alumina with $CHCl_3$ as solvent to afford 3.80 g of **3** (64%), mp 131.5°C. If the reaction was carried out in the presence of KCN the yield rose to 75%.

ZEISEL - PREY Ether cleavage

Acid catalyzed cleavage of aromatic methyl or ethyl ethers. Quantitative methoxy group determination. Also ether cleavage with trimethylsilyl iodide.[6]

$$\text{Ph-OCH}_3 + \text{HI} \xrightarrow{\Delta} \text{Ph-OH} + \text{CH}_3\text{I} \xrightarrow{\text{AgNO}_3} \text{AgI}$$

$$\underset{\mathbf{1}}{\text{Ph-OMe}} \xrightarrow[\text{HOAc } 190°]{\text{pyr. HCl } \mathbf{2}} \underset{\mathbf{3}}{\text{Ph-OH}} \quad (100\%)$$

$$\mathbf{4} \xrightarrow[17\text{ h, } 100°]{\text{Ac}_2\text{O, FeCl}_3} \mathbf{5} \quad (99\%)$$

1	Zeisel, S.	*Monatsh.*	**1885**	*6*	406
2	Belcher, A.	*J. Chem. Soc.*	**1957**		4480
3	Prey, V.	*Chem. Ber.*	**1941**	*74*	350
4	Burwell, R.L.	*Chem. Rev.*	**1954**	*54*	635
5	Ganem, B.	*J. Org. Chem.*	**1974**	*39*	3728
6	Jung, M.E.	*J. Am. Chem. Soc.*	**1977**	*99*	968

Aromatic methoxy group determination [1] Anisole **1** is heated in HI (10 mL) and the methyl iodide which distills is trapped in $AgNO_3$. The AgI is determined gravimetrically.

Phenol (3).[2] Anisole **1** (107 g, 1 mol) and pyridine HCl **2** (130 g, 1.3 mol) were heated with AcOH (5-10 mL) for 5 h at 190°C. The mixture was poured into water (500 mL) and extracted with Et_2O. From the Et_2O solutions **3** was extracted with NaOH and the alkaline solution was saturated with CO_2. Extraction with ether and distllation of the extract afforded 94 g of **3** (100%), mp 35-40°C.

ZIEGLER Macrocyclic synthesis

Synthesis of macrocyclic ketones from dinitriles using high dilution.

Na, Ph NH Me

isoprene

6 N HCL

1

2

3 (85%)[3]

1	Ziegler, K.	*Chem. Ber.*	**1934**	*67*	139
2	Ziegler, K.	*Liebigs Ann.*	**1937**	*528*	143
3	Newman, M.S.	*J. Org. Chem.*	**1975**	*40*	2867

2-Cyano-7,13,16,19-pentaoxacyclotetracosanone (3).[3] A suspension of sodium sand (34.5 g, 1.5 at g) in Et_2O was treated dropwise with a solution of isoprene (102 g, 1.5 mol) and N-methylaniline (209 g, 1.95 mol) in Et_2O (300 mL). A solution of 7,10,13,16,19-pentaoxapentacosanedinitrile **1** (56.7 g, 0.15 mol) in Et_2O (400 mL) was added during 24 h. After addition, heating was continued for another 2 h. To the cooled mixture was added water, the product was distilled to remove N-methylaniline and to leave a residue (50 g) of a mixture of **2** and **3**. From a small amount of this mixture by distillation one obtains **2**, bp 235-240°C (0.5 mm). The remaining crude mixture was shaken with 6N HCl (2000 mL) for 30 min. Separation in the usual way gave 49.3 g of **3** (85%), bp 210-213°C (0.5 mm).

ZINCKE - SUHL Cyclohexadienone synthesis

Synthesis of cyclohexadienones from phenols by Friedel-Crafts alkylation.

OH; CH_3; CH_3 — $AlCl_3$, CCl_4 → O; H_3C; CH_3; CCl_3

1 **2** (72%)

1	Zincke, Th., Suhl, R.	*Chem. Ber.*	**1906**	*39*	4148
2	Newman, M.S.	*J. Org. Chem.*	**1958**	*23*	1236
3	Newman, M.S.	*J. Am. Chem. Soc.*	**1959**	*81*	6454

3,4-Dimethyl-4-trichloromethyl-2,5-cyclohexadienone (2).[3] A solution of 3,4-dimethylphenol **1** (61.0 g, 0.5 mol) in CCl_4 (300 mL) was added dropwise during 30 min to a slurry of $AlCl_3$ (133 g, 1 mol) in CCl_4 (300 mL), keeping the mixture at 5-20°C, while a stream of N_2 swept the HCl formed into a trap containing a measured amount of NaOH. After 2 h, 0.4 mol of HCl had been evolved and the reaction mixture was poured onto 1500 g of ice containing 32% HCl (100 mL). After usual work up the crude **2**, recrystallized from PhH, afforded 96.5 g of **2** (72.5%), mp 60-61°C; DNPH mp 166-167.5°C.

ZINKE - ZIEGLER Synthesis of Calixarenes

Synthesis of calixarenes (a basket-like macrocyclic compound).

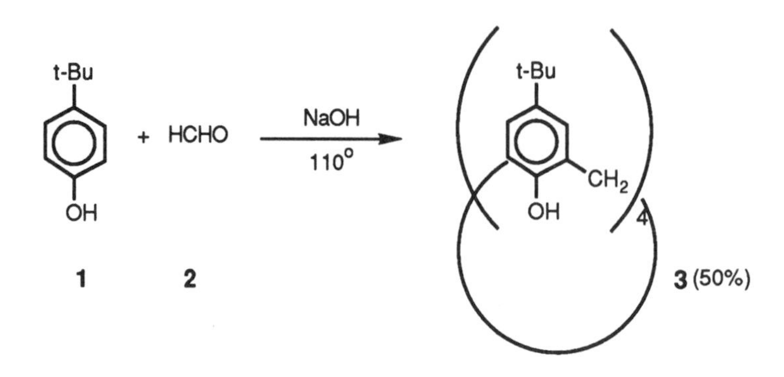

1	Zinke, A., Ziegler, E.	*Chem. Ber.*	**1941**	*74*	1729
2	Cornforth, J.W.	*Br. J. Pharmacol.*	**1955**	*10*	73
3	Gutsche, C.D.	*J. Org. Chem.*	**1986**	*51*	742
4	Gutsche, C.D.	*J. Org. Chem.*	**1990**	*55*	4487
5	Gutsche, C.D.	*Acc. Chem. Res.*	**1983**	*16*	161
6	Gutsche, C.D.	*Top. Curr. Chem.*	**1984**	*123*	1
6	Gutsche, C.D.	*Pure Appl. Chem.*	**1988**	*60*	1607

p-tert-Butylcalix(4)arenes (3).[3] A mixture of p-tert-butylphenol **1** (100 g, 0.665 mol), 37% formalin **2** (62.3 mL, 0.83 mol) and NaOH (1.2 g, 0.03 mol) in minimum quantities of water was heated for 2 h at 110-120°C. The product was dissolved in $CHCl_3$, the solution was neutralized with HCl, the organic layer was evaporated and the residue was heated in diphenyl ether (800 mL) under a flow of N_2 to remove the water. The mixture was heated to reflux for 2 h and after cooling to 20°C, EtOAc (1000 mL) was added under stirring. Filtration and washing with EtOAc (100 mL) and AcOH (200 mL) gave 66.5 g of **3** (62%). Recrystallization from PhMe gave pure **3**, 61.6 g (50%), mp 342-344°C.

Z I N I N Benzidine (semidine) Rearrangement

Acid catalyzed rearrangement of hydrazobenzenes to benzidines and semidines.

HCl

Br NH - NH Br **1** → H_2N NH_2 Br Br **2** (75%)

R NH - NH → NH R H_2N ortho semidine

+

R NH NH_2 para semidine

1	Zinin, N.	*J. Prakt. Chem.*	**1845**	*36*	93
2	Jakobson, P.	*Chem. Ber.*	**1893**	*26*	688
3	Hammond, G.S.	*J. Am. Chem. Soc.*	**1950**	*72*	220
4	Ingold, C.K.	*J. Chem. Soc.*	**1957**		1906

3,3-Dibromobenzidine (2).[3] A solution of 3,3'-dibromohydrazobenzene **1** (4.0 g, 11 mmol) in Et_2O (50 mL) was added dropwise under stirring to ice cooled conc. HCl. The white precipitate of **2** hydrochloride was filtered. Reaction with 10% NaOH, extraction with Et_2O and evaporation afforded 3 g of **2** (75%), mp 127-129°C.

NAMES INDEX

REAGENTS INDEX

REACTIONS INDEX

Acids, Acid Derivatives	Hydroxy Aldehydes or Ketones, Sugars	Amino Acids, Peptides	Heterocycles 3,4,5-Rings	Heterocycles 6,7, Large Rings	Other Heterocycles and Nucleosides	Miscellaneous
15, 137			69			
			73, 159, 285, 343	304	290	316
37, 324, 427						55
28			287	287		
12, 213			89, 187, 245, 409	14, 291, 321, 342	395	436
7			274	117, 274, 292		
), 317, 379, 433			11, 12, 104, 123, 210, 271, 291, 360	143, 228, 423	341	432
9			35, 142, 167, 169, 172, 272, 283, 284, 300, 322, 331, 365, 406, 409	17, 32, 35, 36, 71, 98, 132, 157, 175, 206, 272, 275, 299, 302, 303, 318, 322, 356, 423	396	173
						397
58, 89, 264, 9, 334, 357, 9, 384, 397	10	374	12, 88, 122, 206, 335	33, 63, 143, 157, 195, 228, 306	290, 325	233
5, 289, 334, 3, 416, 425		109, 374	48, 89, 115, 145, 168, 187, 206, 284, 310, 406	32, 132, 143, 153, 157, 195, 304, 373	185, 290	233, 408
61, 100, 162, , 254, 258, 300, , 334, 361, 383		45, 87, 120, 166	122, 197, 238, 323, 400	33, 36, 133, 140, 197, 387, 434		408
	118, 141, 203, 209, 225, 327, 385, 420, 424				168, 401	396
		166, 179, 257, 275, 337, 393, 417, 428, 430				
			38, 81, 290	9, 155, 297	237, 325	
			293	12, 40, 62, 96, 218, 256, 345, 394, 411	168, 315, 338, 387, 401	
					185	
				214, 224, 232		229, 253, 301, 329, 351, 356, 404, 411, 420, 431